Marian Mureşan
Differential Equations

Also of Interest

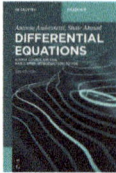

Differential Equations
A first course on ODE and a brief introduction to PDE
Antonio Ambrosetti, Shair Ahmad, 2023
ISBN 978-3-11-118524-8, e-ISBN (PDF) 978-3-11-118567-5

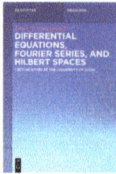

Differential Equations, Fourier Series, and Hilbert Spaces
Lecture Notes at the University of Siena
Raffaele Chiappinelli, 2023
ISBN 978-3-11-129485-8, e-ISBN (PDF) 978-3-11-130252-2

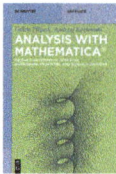

Analysis with Mathematica®
Volume 3 Differential Geometry, Differential Equations, and Special Functions
Galina Filipuk, Andrzej Kozłowski, 2022
ISBN 978-3-11-077454-2, e-ISBN (PDF) 978-3-11-077464-1

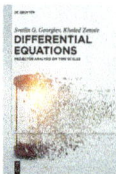

Differential Equations
Projector Analysis on Time Scales
Svetlin G. Georgiev, Khaled Zennir, 2024
ISBN 978-3-11-137509-0, e-ISBN (PDF) 978-3-11-137715-5

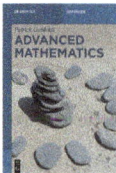

Advanced Mathematics
An Invitation in Preparation for Graduate School
Patrick Guidotti, 2022
ISBN 978-3-11-078085-7, e-ISBN (PDF) 978-3-11-078092-5

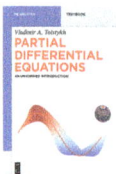

Partial Differential Equations
An Unhurried Introduction
Vladimir A. Tolstykh, 2020
ISBN 978-3-11-067724-9, e-ISBN (PDF) 978-3-11-067725-6

Marian Mureşan

Differential Equations

Solving Ordinary and Partial Differential Equations with Mathematica®

DE GRUYTER

Mathematics Subject Classification 2020
Primary: 34A05, 35A09, 33B15; Secondary: 65M12, 65Y04

Author
Marian Mureşan
Babeş-Bolyai University
Faculty of Mathematics and Computer Science
M. Kogălniceanu st. 1
400084 Cluj-Napoca
Romania
marian.muresan@ubbcluj.ro

ISBN 978-3-11-141109-5
e-ISBN (PDF) 978-3-11-141139-2
e-ISBN (EPUB) 978-3-11-141204-7

Library of Congress Control Number: 2024932806

Bibliographic information published by the Deutsche Nationalbibliothek
The Deutsche Nationalbibliothek lists this publication in the Deutsche Nationalbibliografie;
detailed bibliographic data are available on the Internet at http://dnb.dnb.de.

© 2024 Walter de Gruyter GmbH, Berlin/Boston
Cover image: Marian Mureşan
Typesetting: VTeX UAB, Lithuania
Printing and binding: CPI books GmbH, Leck

www.degruyter.com

To my lovely granddaughters, Edith and Esther, and to my children, Călin and Adela

Foreword

We live in a world in which technology continuously progresses and touches every aspect of our lives. Computer science plays a remarkable role in this process, being essential in the application of scientific knowledge for not only practical purposes. However, when it comes to fully understanding and making use of the computer science potential, the science of mathematics is indispensable. At the same time, there is an upward trend toward using computer science in mathematical problem-solving as an essential element. *Mathematica*® itself is a modern technical computing system that was built on mathematical foundations and, at the same time, engages mathematicians into the practical knowledge the use of computers.

The present book succeeds in bridging the gap between theory and practice by bringing forward a domain where mathematics and computer science are extremely significant to each other. It comes as a natural consequence of the great success of the author's preceding book, Introduction to *Mathematica*® with Applications, Springer, London, 2017, that has confirmed the interest of the scientific community in making the most out of the union of mathematics and computer science.

This book approaches the demanding subject of differential equations and highlights *Mathematica*® capabilities in addressing them. The difficulty of solving differential equations is overcome by using *Mathematica*® instruments which, by their plentiful variety of special functions and symbolic interpolating functions, can represent solutions in a manner that allows immediate manipulation and visualization. The choice of differential equations is even more important due to its close relationship with the theory and practice of dynamical systems, as there are many real-world systems that display dynamic behavior.

The first part of the book addresses ordinary differential equations and consists of four chapters describing notions and results regarding special functions, simple ordinary differential equations, first-order ordinary differential equations, higher-order ordinary differential equations, and systems of ordinary differential equations. The second part of the book deals with partial differential equations and contains six chapters. They address first-order partial differential equations, linear hyperbolic partial differential equations, sine-Gordon equations, nonlinear and higher-order hyperbolic equations, elliptic partial differential equations, parabolic partial differential Klein–Gordon equations, equations, third- and higher-order nonlinear partial differential equations. All mathematical notions introduced by the author show their value in the solutions presented in the coming chapters.

https://doi.org/10.1515/9783111411392-201

An important role in writing this book was played by the Research Centre of Modelling, Optimization and Simulation (MOS) from the Babeş-Bolyai University in Cluj-Napoca, Romania, due to its high interest in bringing close together computer science and the most theoretical chapters of mathematics.

Due to all of the above, I highly recommend Professor Marian Mureşan's book.

Prof. dr. Anca Andreica
Dean of the Faculty of Mathematics and Computer Science
Babeş-Bolyai University
Cluj-Napoca, Romania

Preface

The present book is focused on finding an answer to the question: how to realize a strong link between *Mathematica*® and a differential equation, ordinary or partial? This book tries to offer an answer.

It is largely recognized that generally it is not easy to solve a differential equation. It is also generally admitted that only a few types of differential equations are solvable in closed form. Sometimes even the numerical methods are rather difficult to handle. We only mention the stiffness phenomenon.

We chose *Mathematica*® as the main tool in solving differential equations. *Mathematica*® offers a lot of instruments and facilities to approach a solution answering to a differential equation with or without additional conditions.

This work is divided into two parts: the first deals with ordinary differential equations while the second deals with partial differential equations. Let us briefly describe the structure and the content of the book. Each chapter is structured into sections and subsections, if any. The content of each section or subsection is arranged into the elementary units that we call problems. A problem is a differential equation with or without an additional condition. Each problem is numbered and then solved. The solution of a problem shares the same number as the problem. When we propose several solutions to a problem, we distinguish them by different so-called approaches. If we have several similar problems, we group them forming a numbered list of pairs containing a statement with its solution.

The first part of the book contains four chapters.

Chapter 1 contains some notions and results on special functions because many results on differential equations are expressed by these functions. The topics of this chapter include: existence and uniqueness theorems, exponential, logarithmic, and trigonometric functions, Gaussian, error, and sign function, unit step, clip function, Dirac delta function, gradient, divergence and Laplacian, gamma and beta functions, hypergeometric equations, hyperbolic functions, Bessel functions, Legendre functions, Jacobi polynomials, Mathieu functions, and elliptic equations and functions.

Chapter 2 is a brief introduction to simple ordinary differential equations by *Mathematica*®. The sections of this chapter discuss the following: classification of certain ordinary differential equations, planar phase portrait, solid phase portrait, straight integration of first-order derivatives, differential equations with separable variables, and homogeneous equations.

Chapter 3 deals with a large number of first-order ordinary differential equations. Its sections present: first-order linear differential equations, first-order inverse linear differential equations, Bernoulli differential equations, Riccati differential equations, exact first-order ordinary differential equations, Lagrange differential equations, implicit differential equations, and other first-order differential equations.

Chapter 4 is the last chapter of the first part of the book. It deals with higher-order ordinary differential equations and with systems of ordinary differential equations. Their

https://doi.org/10.1515/9783111411392-202

list contains: second-order linear differential equations, Bessel differential equations, Legendre differential equations, Mathieu differential equations, equations with discontinuous coefficients or right-hand sides, other higher-order differential equations, and systems of differential equations.

The second part of the book contains six chapters.

Chapter 5 deals with first-order partial differential equations, more precisely: linear, quasilinear, and nonlinear first-order partial differential equations.

The next chapter focuses on linear hyperbolic partial differential equations. The equations discussed in this chapter are with constant coefficients, with variable coefficients, on curvilinear domains, in solid space, and the Klein–Gordon equation.

Chapter 7 treats the sine-Gordon equations, nonlinear, and higher dimension hyperbolic equations.

Chapter 8 focusses on the elliptic partial differential equation. The chapter starts with some considerations on harmonic function. Then the Laplace and Poisson equations are discussed on rectangles, arbitrary domains, and in higher dimensions.

Chapter 9 deals with parabolic partial differential equations. The topics include the 1D homogeneous and inhomogeneous equations, the 2D homogeneous and inhomogeneous equations, the Burgers equations, the ansatz methods, the Fisher equations, the Fitzhugh–Nagumo and the Calogero equations, the double sine-Gordon equation, and the continuous dependence on a parameter.

The last chapter of the present book concerns with third- and higher-order nonlinear partial differential equations. It is a rather large chapter discussing the following subjects: the Korteweg–de Vries equations, the Dodd–Bullough–Mikhailov equation, the Tzitzeica(Țițeica)–Dodd–Bullough equation, the modified Kawahara equation, the Benjamin equation, the Kadomtsev–Petviashvili equations, the Sawada–Kotera equation, and finally the Kaup–Kupershmidt equation.

Our teaching experience showed us that it is more profitable to ask the students to suggest or give "a solution" instead of "the solution" of a problem or exercise. Sometimes with some effort one can give a different solution highlighting different features of the problem in discussion. Therefore, we often suggest several approaches. For instance, problem 3.22 has five approaches, and problem 5.5 has three approaches.

The problems are discussed in detail helping the reader to understand the reasoning with *Mathematica*®. Sometimes the reader is left taking the benefit of the Help menu and other sources freely and generously offered by Wolfram Research at the website www.wolfram.com. A good source of ideas and discussions is offered by the Mathematica Stack Exchange site.

All the problems in this book are solved by *Mathematica*®. All the corresponding figures are also made by *Mathematica*®. A well-motivated pleading for visualization in mathematics is contained in [66].

The *Mathematica* codes are written by typewriter fonts, whereas the answers to the problems are given by gray typewriter fonts. This style is used in some cornerstone books [74], [75], [76], and [77]. Also, gray typewriter fonts are inserted with some comments that we find useful. In this way, we try to offer the inputs and outputs as close as possible to the format of notebooks in *Mathematica*.

Wolfram Research, located at Champaign, IL, USA, is the company that is the owner and the developer of *Mathematica*®.

Mathematica® is continuously enlarging and deepening. We used *Mathematica*® that is installed at the Center of Modeling, Optimization, and Simulation (MOS) for our faculty.

We have introduced notions and results on *Mathematica*® in our lectures to master students at the Faculty of Mathematics and Computer Science of the Babeş-Bolyai University in Cluj-Napoca, Romania. We did the same thing with our PhD students at three summer schools organized in the framework of the grant "Center of Excellence for Applications of Mathematics" supported by DAAD, Germany. The summer schools have been organized in Struga (Republic of North Macedonia), Sarajevo (Bosnia and Herzegovina), and Cluj-Napoca (Romania).

This book is aimed at undergraduate and graduate students as well as all those with an interest in this topic.

Here is the right place to express our gratitude to the following colleagues of us from the faculty of Mathematics and Computer Science of the Babeş-Bolyai University for their support: Anca Andreica, Valeriu Anisiu, Virginia Niculescu, and Adrian Sterca. The existence and development of the MOS (Modeling, Optimization, and Simulation) Research Center for our faculty was of much assistance during the preparation of the present book.

Cluj-Napoca,   Marian Mureşan
February 2024

Contents

Part II: *Mathematica* and partial differential equations

The references contained in the first part of this book have been very helpful; please see [2, 41, 62, 56, 63, 64, 99], and [70].

Mathematica for ordinary differential equations

The main tools of *Mathematica* for solving ordinary differential equations; please see [91, 94, 93, 95], and [92]. We will shortly review them.

The *Mathematica* command DSolve[eqn,x[t],t] solves an ordinary differential equation eqn for the function x[t] in the independent variable t.

The *Mathematica* command DSolve[eqn,x,t] solves an ordinary differential equation eqn for the function x in the independent variable t. The solution is given as a "pure function" for x. A comparison between DSolve[eqn,x[t],t] and DSolve[eqn,x,t] commands is done in [94].

It is assumed that function x[t] is at least piecewise differentiable, and thus one can check the solution.

The *Mathematica* command DSolve[{eqn$_1$,...,eqn$_n$},{x$_1$[t],...,x$_n$[t]},t] solves the list of ordinary differential equations eqn$_k$ for the functions x$_i$[t] in the independent variable t. *Mathematica* returns a list of solutions, each element of which is a solution.

The *Mathematica* command DSolve[{eqn$_1$,...,eqn$_n$},{x$_1$,...,x$_n$},t] solves a list of ordinary differential equations eqn$_k$ for the functions x$_i$ in the independent variable t. *Mathematica* returns a list of solutions as "pure functions," each element of which is a solution.

The *Mathematica* command NDSolve[{eqn$_1$,...,eqn$_n$,cond},{x$_1$[t],...,x$_n$[t]}, {t,t$_{min}$,t$_{max}$}] finds a numerical solution to a list of ordinary differential equations eqn$_k$ for the functions x$_i$[t] in the independent variable t on the given interval [t$_{min}$,t$_{max}$] satisfying condition cond. Condition cond is a list of initial conditions, boundary value conditions, mixed conditions, etc.

The *Mathematica* command NDSolve[{eqn$_1$,...,eqn$_n$,cond},{x$_1$,...,x$_n$}, {t,t$_{min}$,t$_{max}$},cond] finds a numerical solution to a list of ordinary differential equations eqn$_k$ for the functions x$_i$[t] in the independent variable t on the given interval [t$_{min}$,t$_{max}$] satisfying condition cond. *Mathematica* returns a list of solutions as "pure functions."

As a general rule, we visualize the solutions by plotting them.

Throughout the present book, we have adopted the following notational conventions; see [56, Section 1.2]:

$$\mathbb{N} = \{1, 2, \dots\}, \qquad \mathbb{N}^* = \mathbb{N} \cup \{0\},$$
$$\mathbb{Z} = \{\dots, -2, -1, 0, 1, 2, \dots\}, \quad \mathbb{Q} = \{p/q \mid p, q \in \mathbb{Z}, \quad q \neq 0\},$$
$$\mathbb{R} \text{ the set of real numbers}, \quad \mathbb{C} \text{ the set of complex numbers},$$
$$]a, b[= (a, b) \text{ the open } a, b \text{ interval}.$$

https://doi.org/10.1515/9783111411392-001

1 Certain theoretical results

The first chapter has a reminding feature and contains a short introduction to some theoretical results on the existence and uniqueness of solutions to ordinary differential equations as well as to some special functions. The functions and special functions used throughout our book include: exponential, logarithmic, trigonometric, Gaussian, error, sign, unit step, clip, Dirac delta, Laplacian, gamma, beta, hypergeometric, hyperbolic, Bessel, Legendre, Jacobi, Mathieu, and elliptic functions.

1.1 Existence and uniqueness theorems

We consider a system of ordinary differential equations of the form

$$\frac{d\,x_i}{d\,t} = f_i(t, x_1, \ldots, x_n), \quad i = 1, 2, \ldots, n, \tag{1.1}$$

denote x the column vector

$$\begin{pmatrix} x_1 \\ \vdots \\ x_n \end{pmatrix},$$

and use the Euclidean norm $\|x\| = \sqrt{x_1^2 + \cdots + x_n^2}$. System (1.1) of ordinary differential equations is often written under the more compact form

$$\frac{dx_i(t)}{dt} = f_i(t, x(t)) \tag{1.2}$$

and is supposed that $f = (f_1, \ldots, f_n)$ is *continuous* on a nonempty open, and connected domain $D \subset \mathbb{R}^{n+1}$. We recall two theorems in [23].

Theorem 1.1 ([23, p. 10]). *Suppose $a, b > 0$, $D = \{(t, x) \mid |t - t_0| \le a, \ \|x - x_0\| \le b\} \subset \mathbb{R}^{1+n}$, $M = \max_D \|f(t, x)\| > 0$, $\alpha = \min\{a, b/M\}$. Then there exists a solution of system (1.2) on the interval $|t - t_0| \le \alpha$ with $x(t_0) = x_0$.*

Theorem 1.2 ([23, p. 31]). *If the assumptions of Theorem 1.1 are satisfied and, moreover,*

$$\|f(t, x_1) - f(t, x_2)\| \le k(t)\|x_1 - x_2\|$$

in $D \subset \mathbb{R}^{n+1}$, where k is a nonnegative integrable function. Then system (1.2) has a unique solution satisfying the initial condition $x(t_0) = x_0$.

The main task of the differential equations' topic is the study of existence and of qualitative and quantitative properties of their solutions.

https://doi.org/10.1515/9783111411392-002

Commonly, we first look for analytical solutions and then for numerical solutions. If possible, we compare these solutions from a numerical point of view. A figure or several examples sustain the statements.

For more facts about ordinary differential equations, we suggest [23, 33], and [32].

1.2 Exponential, logarithmic, and trigonometric functions

1.2.1 Exponential and logarithmic functions

The exponential function has several definitions. We adhere to its series definition. The real *exponential function* exp $: \mathbb{R} \to \mathbb{R}$ is defined by the following power series [2, 4.2.1, p. 69]:

$$\exp(x) = 1 + \frac{x}{1!} + \frac{x^2}{2!} + \frac{x^3}{3!} + \frac{x^4}{4!} + \cdots = \sum_{k=0}^{\infty} \frac{x^k}{k!}, \quad x \in \mathbb{R}.$$

A definition of the exponential function by a limit is [56, p. 236]

$$\exp(x) = \lim_{h \to 0}(1 + h\,x)^{1/h}, \quad x \in \mathbb{R}.$$

The exponential function is implemented in *Mathematica* as Exp[x].

Theorem 1.3 ([56, Section 5.9.3]). *The exponential function satisfies the next properties:*
(a) *is defined for all $x \in \mathbb{R}$ and only takes strictly positive values;*
(b) *is strictly increasing on its domain and is of class C^{∞} on $\mathbb{R}$;*
(c) *satisfies the functional equation:* $\exp(x + y) = \exp(x) \times \exp(y)$, *for all $x, y \in \mathbb{R}$;*
(d) *its sided limits are:* $\lim_{x \to -\infty} \exp(x) = 0$ *and* $\lim_{x \to \infty} \exp(x) = \infty$.

Theorem 1.4. *One has that*

$$\left(\exp(x)\right)' = \exp(x), \quad x \in \mathbb{R}.$$

e in $\mathbb{R}$ is the *exponential constant* and is defined by

$$e = \exp(1) = \lim_{n \to \infty}\left(1 + \frac{1}{n}\right)^n = \sum_{k=0}^{\infty} \frac{1}{k!} \approx 2.718281828.$$

There this number exists and is transcendental [56, p. 282].

The exponential constant e is implemented in *Mathematica* as E.

More generally, for $a > 0$ and $a \neq 1$, one defines the real *exponential function to base a* as

$$a^x = 1 + \frac{\ln(a)x}{1!} + \frac{\ln^2(a)x^2}{2!} + \frac{\ln^3(a)x^3}{3!} + \cdots = \sum_{k=0}^{\infty} \frac{\ln^k(a)x^k}{k!}, \quad x \in \mathbb{R}.$$

A sketch of the exponential function with a positive nonunitary basis is given in Figure 1.1.

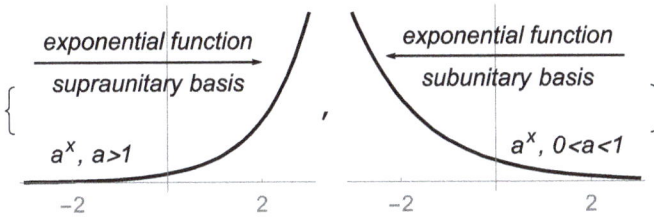

exponential function

supraunitary basis

a^x, a>1

exponential function

subunitary basis

a^x, 0<a<1

-2   2   -2   2

Figure 1.1: Exponential function.

The *exponential integral function* [6, 10.4, p. 218] is defined as

$$\mathrm{Ei}(x) = \int_{-\infty}^{x} \frac{e^t}{t} \, dt, \quad x \in \mathbb{R} \setminus \{0\},$$

and implemented in *Mathematica* as ExpIntegralEi[x].
A sketch of the exponential integral function is given in Figure 1.2.

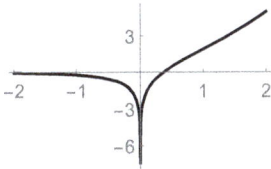

Figure 1.2: Exponential integral function.

The exponential function $\mathbb{R} \ni x \mapsto \exp(x) \in \,]0, \infty[$ being strictly monotone and onto has an inverse function, which is the *logarithmic function* (in natural basis) denoted $]0, \infty[\,\ni x \mapsto \ln(x) \in \mathbb{R}$.

The logarithmic function (to natural basis) is implemented in *Mathematica* as Log[x].

The logarithmic function has several definitions. We introduce its series definitions. The real logarithmic function (in natural basis) $\ln : \,]0, \infty[\,\to \mathbb{R}$ is defined by the following series as [2, 4.1.27, p. 68]:

$$\ln(x) = 2 \sum_{k=0}^{\infty} \frac{1}{2k+1} \left(\frac{x-1}{x+1} \right)^{2k+1}, \quad x > 0.$$

A power series related to the logarithmic function is of the form [2, 4.1.24, p. 68]

$$\ln(1+x) = \sum_{k=1}^{\infty}(-1)^{k+1}\frac{x^k}{k}, \quad -1 < x \le 1.$$

A definition of the natural logarithmic function by limit is [56, p. 231]

$$\ln(x) = \lim_{h\to 0}\frac{x^h-1}{h}, \quad x > 0.$$

Theorem 1.5 ([56, Section 5.9.2]). *The logarithmic function to natural basis satisfies the properties:*
(a) *is defined for all $x \in]0,\infty[$ and takes values in $\mathbb{R}$;*
(b) *is strictly increasing on its domain and is of class C^{∞} on $]0,\infty[$;*
(c) *satisfies the functional equation: $\ln(xy) = \ln(x) + \ln(y)$, for all $x,y > 0$;*
(d) *the sided limits are: $\lim_{x\downarrow 0}\ln(x) = -\infty$ and $\lim_{x\to\infty}\ln(x) = \infty$.*

Theorem 1.6. *One has that*
(a) $\exp(\ln(x)) = x$ *for $x > 0$ and $\ln(\exp(x)) = x$ for $x \in \mathbb{R}$;*
(b) $(\ln(x))' = \frac{1}{x}$ *for $x > 0$;*
(c) $\ln(x) < x$ *for $x > 0$ and $x < \exp(x)$ for $x \in \mathbb{R}$.*

Let a be a positive nonunitary real number and x be a positive number. The *logarithm to base a* of x is the function defined as the inverse function of exponential function to base a of x. It is denoted as $\log_a(x)$.

The logarithmic function (to basis a) is implemented in *Mathematica* as Log[a,x].
A sketch of the logarithmic function is given in Figure 1.3.

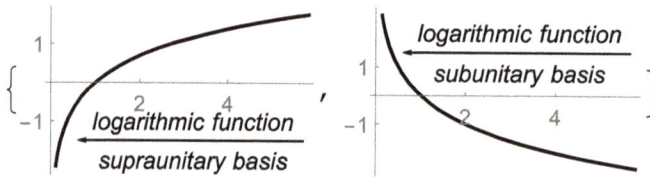

Figure 1.3: Logarithmic function.

The *logarithmic integral function* [6, 10.7, p. 219], defined as

$$\text{li}(x) = \int_0^x \frac{1}{\ln(t)}\,dt, \quad x > 0 \text{ and } x \ne 1,$$

is implemented in *Mathematica* as LogIntegral[x]. Its figure is given in Figure 1.4.

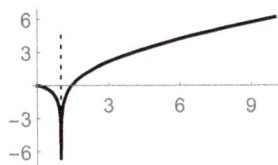

Figure 1.4: Logarithmic integral function.

1.2.2 Trigonometric functions

1.2.2.1 Sine and arcsine functions

The power series definition of the *sine function* is [2, 4.3.65, p. 74]

$$\sin(x) = x - \frac{x^3}{3!} + \frac{x^5}{5!} - \frac{x^7}{7!} + \cdots = \sum_{n=0}^{\infty} \frac{(-1)^n}{(2n+1)!} x^{2n+1}, \quad x \in \mathbb{R}.$$

The sine function is implemented in *Mathematica* as Sin[x].

Theorem 1.7. *The sine function satisfies the properties:*
(a) *is defined on $\mathbb{R}$ and is of class C^∞ on the real axis;*
(b) *is an odd function;*
(c) *is a 2π-periodic function and, therefore, bounded. Moreover, $\sin(x) \in [-1,1]$, $x \in \mathbb{R}$;*
(d) *because the restriction $\sin : [-\frac{\pi}{2}, \frac{\pi}{2}] \to [-1,1]$ is strictly increasing and onto, has an inverse;*
(e) *$\sin(x) < x$ for $x > 0$;*
(f) *$\sin(x) = \frac{e^{ix} - e^{-ix}}{2i}$, where $x \in \mathbb{R}$ and i is the imaginary unit.*

Because the restriction $\sin : [-\frac{\pi}{2}, \frac{\pi}{2}] \to [-1,1]$ is strictly increasing and onto, it admits an inverse function called the *arcsine function*, $\arcsin : [-1,1] \to [-\frac{\pi}{2}, \frac{\pi}{2}]$.
The power series definition of the arcsine function is [2, 4.4.40, p. 81]

$$\arcsin(x) = \sum_{n=0}^{\infty} \frac{(2n-1)!!}{(2n)!!} \frac{x^{2n+1}}{2n+1} = \sum_{n=0}^{\infty} \frac{\binom{2n}{n}}{4^n} \frac{x^{2n+1}}{2n+1}, \quad |x| \le 1.$$

The arcsine function is implemented in *Mathematica* as ArcSin[x].

Theorem 1.8. *The arcsine function satisfies the properties:*
(a) *is defined on $[-1,1]$ and is of class C^∞ on the open interval $]-1,1[$;*
(b) *is an odd function;*
(c) *the restriction $\arcsin : [-1,1] \to [-\frac{\pi}{2}, \frac{\pi}{2}]$ is strictly increasing;*
(d) *$\sin(\arcsin(x)) = x$ for $x \in [-1,1]$ and $\arcsin(\sin(x)) = x$ for $x \in [-\frac{\pi}{2}, \frac{\pi}{2}]$.*

A sketch of the sine and arcsine functions is given in Figure 1.5.

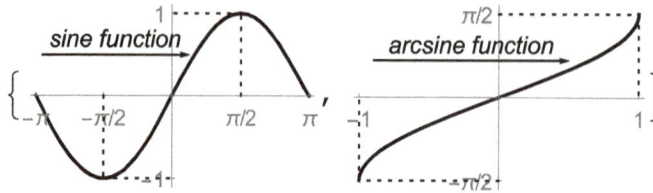

Figure 1.5: Sine and arcsine functions.

The *sine integral* function is defined by, e. g., [2, 5.2.1, p. 231] or [6, 10.10, p. 219]

$$Si(x) = \int_0^x \frac{\sin(t)}{t}\,dt, \quad x \in \mathbb{R}$$

and is implemented in *Mathematica* as $\texttt{SinIntegral[x]}$.

Theorem 1.9. *The sine integral function satisfies the properties:*
(a) *is an odd function and is of class C^∞ on $\mathbb{R}$;*
(b) $\lim_{x\to-\infty} Si(x) = -\pi/2;$
(c) $Si(0) = 0;$
(d) $\lim_{x\to\infty} Si(x) = \pi/2.$

A sketch of the sine integral function is given by the leftmost picture in Figure 1.6.

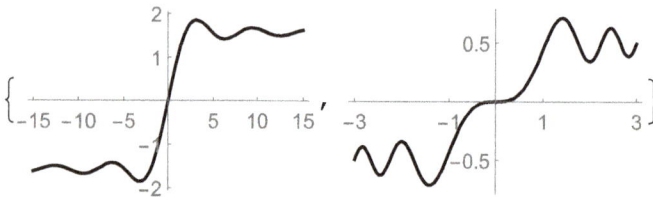

Figure 1.6: Sine integral and Fresnel sine functions.

The *Fresnel sine integral* function is defined as [2, 7.3.2, p. 300]

$$S(x) = \int_0^x \sin\left(\frac{\pi}{2} t^2\right) dt, \quad x \in \mathbb{R}$$

and is implemented in *Mathematica* as $\texttt{FresnelS[x]}$.

The power series representation of $S(x)$ below follows [2, 7.3.13, p. 301]:

$$S(x) = \sum_{k=0}^{\infty} \frac{(-1)^k}{(2k+1)!} \left(\frac{\pi}{2}\right)^{2k+1} \frac{x^{4k+3}}{4k+3}, \quad x \in \mathbb{R}.$$

Theorem 1.10. *The Fresnel sine function satisfies the properties:*
(a) *is an odd function and is of class C^∞ on $\mathbb{R}$;*
(b) $\lim_{x\to-\infty} S(x) = -1/2;$
(c) $S(0) = 0;$
(d) $\lim_{x\to\infty} S(x) = 1/2.$

A sketch of the Fresnel sine function is given by the rightmost picture in Figure 1.6.

1.2.2.2 Cosine and arccosine functions

The power series definition of the *cosine function* is [2, 4.3.66, p. 74]

$$\cos(x) = 1 - \frac{x^2}{2!} + \frac{x^4}{4!} - \frac{x^6}{6!} + \cdots = \sum_{n=0}^{\infty} \frac{(-1)^n}{(2n)!} x^{2n}, \quad x \in \mathbb{R}.$$

The cosine function is implemented in *Mathematica* as Cos[x].

Theorem 1.11. *The cosine function satisfies the properties:*
(a) *is defined on $\mathbb{R}$ and is of class C^∞ on the real axis;*
(b) *is an even function;*
(c) *is a 2π-periodic function and, therefore, bounded. Moreover, $\cos(x) \in [-1, 1]$, $x \in \mathbb{R}$;*
(d) *the restriction $\cos : [0, \pi] \to [-1, 1]$ is strictly decreasing and onto, and thus has an inverse;*
(e) $\cos(x) = \frac{e^{ix}+e^{-ix}}{2}$, *where $x \in \mathbb{R}$ and i is the imaginary unit.*

Because the restriction $\cos : [0, \pi] \to [-1, 1]$ is strictly decreasing and onto, it has an inverse function, namely the *arccosine function*, $\arccos : [-1, 1] \to [0, \pi]$.
From (d) in Theorem 1.13, the power series definition of the arccosine function is

$$\arccos(x) = \frac{\pi}{2} - \arcsin(x) = \frac{\pi}{2} - \sum_{n=0}^{\infty} \frac{(2n-1)!!}{(2n)!!} \frac{x^{2n+1}}{2n+1}, \quad |x| \le 1.$$

The arccosine function is implemented in *Mathematica* as ArcCos[x].

Theorem 1.12. *The arccosine function satisfies the next properties:*
(a) *is defined on $[-1, 1]$ and is of class C^∞ on the open interval $]-1, 1[$;*
(b) *the restriction $\arccos : [-1, 1] \to [0, \pi]$ is strictly decreasing and onto, and thus has an inverse;*
(c) $\cos(\arccos(x)) = x$ *if $x \in [-1, 1]$ and $\arccos(\cos(x)) = x$ if $x \in [0, \pi]$.*

A sketch of the cosine and arccosine functions is given in Figure 1.7.

Theorem 1.13. *The following relations between the sine and cosine functions hold:*
(a) $\sin^2(x) + \cos^2(x) = 1$, *for all $x \in \mathbb{R}$;*

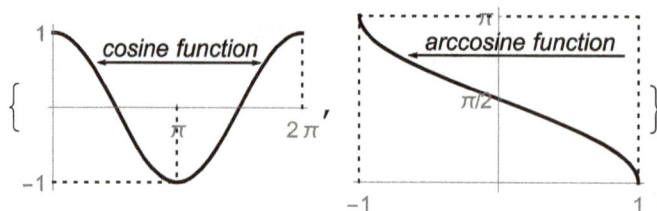

Figure 1.7: Cosine and arccosine functions.

(b) $\sin(x + y) = \sin(x)\cos(y) + \cos(x)\sin(y)$, *for all* $x, y \in \mathbb{R}$;

(c) $\cos(x + y) = \cos(x)\cos(y) - \sin(x)\sin(y)$, *for all* $x, y \in \mathbb{R}$;

(d) $\arcsin(x) + \arccos(x) = \pi/2$, *for all* $x \in [0, \pi/2]$.

The *cosine integral function* is defined by, e. g., [2, 5.2.27, p. 231] or [44, p. 33]

$$\mathrm{Ci}(x) = \int_{\infty}^{x} \frac{\cos(t)}{t}\, dt, \quad x > 0$$

and is implemented in *Mathematica* as CosIntegral[x].

Theorem 1.14. *The cosine integral function satisfies the properties:*

(a) *is of class* C^{∞} *on* $]0, \infty[$;

(b) $\lim_{x \downarrow 0} \mathrm{Ci}(x) = -\infty$;

(c) $\lim_{x \to \infty} \mathrm{Ci}(x) = 0$.

A sketch of the cosine integral function is given by the leftmost picture in Figure 1.8.

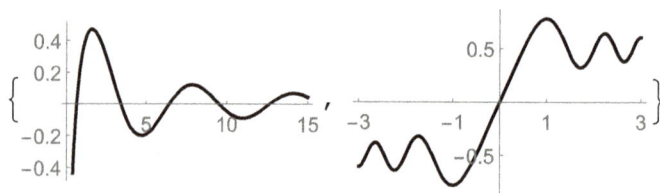

Figure 1.8: Cosine integral and Fresnel cosine functions.

The *Fresnel cosine integral* function is defined by [2, 7.3.1, p. 300]

$$C(x) = \int_{0}^{x} \cos\left(\frac{\pi}{2} t^2\right) dt, \quad x \in \mathbb{R}$$

and is implemented in *Mathematica* as FresnelC[x].

The power series representation of $C(x)$ below follows [2, 7.3.11, p. 301]:

$$C(x) = \sum_{k=0}^{\infty} \frac{(-1)^k}{(2k)!} \left(\frac{\pi}{2}\right)^{2k} \frac{x^{4k+1}}{4k+1}, \quad x \in \mathbb{R}.$$

Theorem 1.15. *The Fresnel cosine integral function satisfies the properties:*
(a) *is an odd function and is of class C^∞ on $\mathbb{R}$;*
(b) $\lim_{x \to -\infty} C(x) = -1/2$;
(c) $C(0) = 0$;
(d) $\lim_{x \to \infty} C(x) = 1/2$.

A sketch of the Fresnel cosine function is given by the rightmost picture in Figure 1.8.

1.2.2.3 Tangent and arctangent functions
The power series definition of the *tangent function* is [2, 4.3.67, p. 75]

$$\tan(x) = \sum_{n=1}^{\infty} (-1)^{n+1} 2^{2n} (2^{2n} - 1) b_{2n} \frac{x^{2n-1}}{(2n)!}, \quad |x| < \frac{\pi}{2},$$

where b_n is the nth Bernoulli number, e. g., [56, Section 5.8], [21, Section 6.5], or [89].
We are reminded that the classical definition of the tangent function is

$$\tan(x) = \frac{\sin(x)}{\cos(x)}.$$

The tangent function is implemented in *Mathematica* as Tan[x].

Theorem 1.16. *The tangent function satisfies the properties:*
(a) *is a π-periodic function and unbounded. Moreover, $\tan(x) \in \mathbb{R}$, $|x| < \frac{\pi}{2}$;*
(b) *is an odd function;*
(c) *is extended by periodicity on $\mathbb{R} \setminus \{(2k+1)\frac{\pi}{2} \mid k \in \mathbb{Z}\}$ and is of class C^∞ on this set;*
(d) *the restriction $\tan : \left]-\frac{\pi}{2}, \frac{\pi}{2}\right[\to \mathbb{R}$ is strictly increasing and onto, and thus has an inverse.*

Because the restriction $\tan : \left]-\frac{\pi}{2}, \frac{\pi}{2}\right[\to \mathbb{R}$ is strictly increasing and onto, it has an inverse function, namely the *arctangent function*, $\arctan : \mathbb{R} \to \left]-\frac{\pi}{2}, \frac{\pi}{2}\right[$.
The power series definition of the arctangent function is [2, 4.4.42, p. 81]

$$\arctan(x) = \sum_{n=0}^{\infty} \frac{(-1)^n x^{2n+1}}{2n+1}, \quad -1 < x \le 1.$$

The arctangent function is implemented in *Mathematica* as ArcTan[x].

Theorem 1.17. *The arctangent function satisfies the properties:*

(a) *is defined on the real axis and bounded. Moreover,* $\arctan(x) \in \left]-\frac{\pi}{2}, \frac{\pi}{2}\right[, x \in \mathbb{R}$;
(b) *is an odd function;*
(c) *is of class* C^∞ *on* $\mathbb{R}$;
(d) *is strictly increasing on* $\mathbb{R}$.

A sketch of the tangent and arctangent functions is given in Figure 1.9.

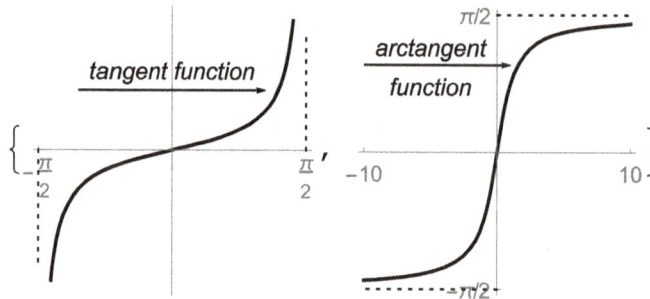

Figure 1.9: Tangent and arctangent functions.

1.2.2.4 Cotangent and arccotangent functions

The power series definition of the *cotangent function* is [2, 4.3.70, p. 75]

$$\cot(x) = \sum_{n=0}^{\infty} \frac{(-1)^n 2^{2n} b_{2n} x^{2n-1}}{(2n)!}, \quad 0 < |x| < \pi,$$

where b_n is the nth Bernoulli number.

The classical definition of the cotangent function is

$$\cot(x) = \frac{1}{\tan(x)} = \frac{\cos(x)}{\sin(x)}.$$

The cotangent function is implemented in *Mathematica* as Cot[x].

Theorem 1.18. *The cotangent function satisfies the properties:*
(a) *is a π-periodic function and unbounded. Moreover,* $\cot(x) \in \mathbb{R}, 0 < x < \pi$;
(b) *is an odd function;*
(c) *is of class* C^∞ *on* $\mathbb{R} \setminus \{k\pi \mid k \in \mathbb{Z}\}$;
(d) *the restriction on* $\cot: \left]0, \pi\right[\rightarrow \mathbb{R}$ *is strictly decreasing and onto.*

Because the restriction $\cot: \left]0, \pi\right[\rightarrow \mathbb{R}$ is strictly decreasing and onto, it has an inverse function, namely the *arccotangent function*, $\operatorname{arccot}: \mathbb{R} \rightarrow \left]0, \pi\right[$.

The arccotangent function is implemented in *Mathematica* as ArcCot[x].

Theorem 1.19. *The arccotangent function satisfies the property* [2, 4.4.8, p. 79]

$$\mathrm{arccot}(x) = \arctan\left(\frac{1}{x}\right), \quad x \neq 0.$$

A sketch of the cotangent and arccotangent functions is given in Figure 1.10.

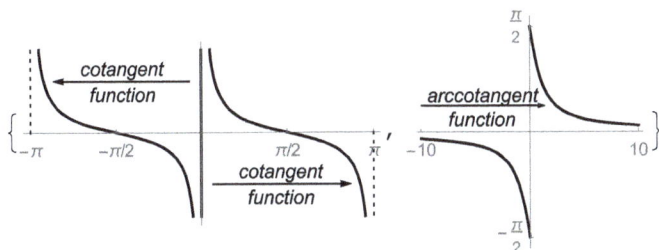

Figure 1.10: Cotangent and arccotangent functions.

1.2.3 Gaussian, error, and sign functions

A *Gaussian function* is a function of the form:

$$f(x) = a\, e^{-\frac{(x-b)^2}{2c^2}}, \quad x \in \mathbb{R}$$

for arbitrary real constants a, b, and c, with $a, c \neq 0$. The simplest form with $a = 1$, $b = 0$, and $c = 1/\sqrt{2}$ is represented by the left-hand side picture in Figure 1.11. For $a = 1$, $b = 0$, and $c = 1/\sqrt{2}$, it is implemented in *Mathematica* as GaussianWindow[x].

The simplest two-dimensional Gaussian function is of the form:

$$f(x, y) = e^{-x^2 - y^2}, \quad x, y \in \mathbb{R}.$$

This function is represented by the right-hand side picture in Figure 1.11.

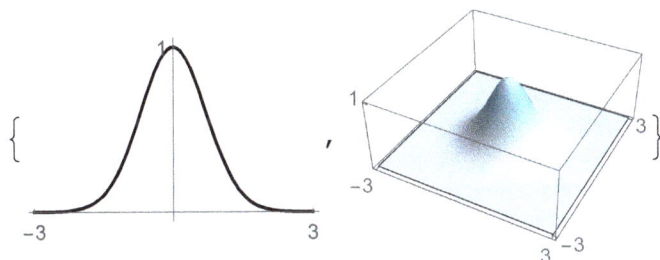

Figure 1.11: Gaussian functions.

The *error function* [44, p. 17] is defined by

$$\mathrm{erf}(x) = \frac{2}{\sqrt{\pi}} \int_0^x \exp(-t^2)\, dt, \quad x \in \mathbb{R},$$

and implemented in *Mathematica* as `Erf[x]`.

The *complementary error function* is defined by

$$\mathrm{erfc}(x) = 1 - \mathrm{erf}(x)$$

and implemented in *Mathematica* as `Erfc[x]`.

Theorem 1.20. *Some particular values of the error and the complementary error functions follow:*

(a) $\lim_{x \to -\infty} \mathrm{erf}(x) = -1$, $\mathrm{erf}(0) = 0$, $\lim_{x \to \infty} \mathrm{erf}(x) = 1$;

(b) $\lim_{x \to -\infty} \mathrm{erfc}(x) = 2$, $\mathrm{erfc}(0) = 1$, $\lim_{x \to \infty} \mathrm{erfc}(x) = 0$;

(c) *The error function is strictly increasing, whereas the complementary error function is strictly decreasing.*

The "*imaginary error function*" is a function defined by

$$\mathrm{erfi}(x) = -i\,\mathrm{erf}(ix) = \frac{2}{\sqrt{\pi}} \int_0^x \exp(t^2)\, dt, \quad x \in \mathbb{R},$$

where $\mathrm{erf}(x)$ is the error function. It is implemented in *Mathematica* as `Erfi[x]`.

The graphs of the error function (black dotted), the complementary error function (black dashed), and the imaginary error function (black) are given in Figure 1.12.

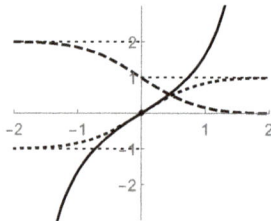

Figure 1.12: Error, complementary error, and imaginary error functions.

1.2.4 Sign, unit step, clip functions

The *sign function* or *signum function* extracts the sign of a real number, i. e.,

$$\mathrm{sign}(x) = \begin{cases} -1, & x < 0, \\ 0, & x = 0, \\ +1, & x > 0. \end{cases}$$

In some mathematical expressions, the sign function is often represented as sgn. It is implemented in *Mathematica* as Sign[x].

The graph of the sign function appears at the leftmost picture in Figure 1.13.

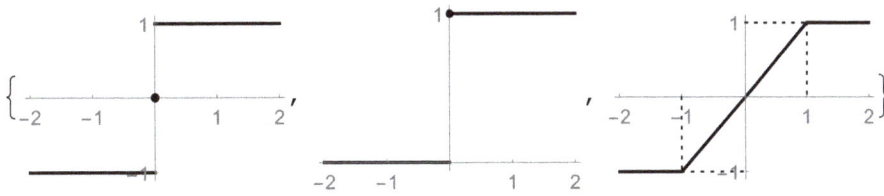

Figure 1.13: The sign, unit step, and clip functions.

The *unit step function* or the *Heaviside function* is defined by

$$H(x) = \begin{cases} 0, & x < 0, \\ +1, & x \geq 0, \end{cases}$$

and implemented in *Mathematica* as UnitStep[x]. Details on the unit step function may be found in, e. g., [65, p. 80].

The graph of the unit step function appears at the mid picture in Figure 1.13.

The clip(t) gives t clipped to be between −1 and +1 and is implemented in *Mathematica* as Clip[x].

The clip function is represented at the rightmost picture in Figure 1.13.

1.2.5 Dirac delta functions

The *Dirac delta function* is implemented in *Mathematica* as DiracDelta[x]. It is defined in the sense of distributions as

$$\delta(x) = \begin{cases} 0, & x \neq 0, \\ \infty, & x = 0, \end{cases} \quad \text{and} \quad \int_{-\infty}^{\infty} \delta(x) = 1.$$

The Dirac delta function can be defined as the pointwise limit of a sequence of functions. The general term of such a sequence of functions is

$$f_n(x) = \frac{\sqrt{n}}{\sqrt{\pi}} e^{-nx^2}, \quad n \in \mathbb{N}, \quad x \in \mathbb{R}.$$

We show some properties of this sequence of functions by means of a short *Mathematica* code:

```
Clear[n,x]
f[n_,x_]:=√n̄/√π̄ e^-nx²
{Assuming[n∈Integers&&x∈Reals,Limit[f[n,x],n→ ∞]],
Limit[f[n,0],n→ ∞],Assuming[n>0, ∫_-∞^∞ f[n,x]dx]}
{0,∞,1}
```

The process of convergence to the Dirac delta function of the above given sequence of functions is suggested by the following code and in Figure 1.14:

```
Animate[
Plot[f[n,x],{x,-5,5},ImageSize→175,PlotStyle→Black,
Ticks→{2Range[-2,2],Automatic},PlotRange→All],
{n,1,101,5},SaveDefinitions→True,DefaultDuration→20,
AnimationDirection→ForwardBackward,
AnimationRunning→False]
```

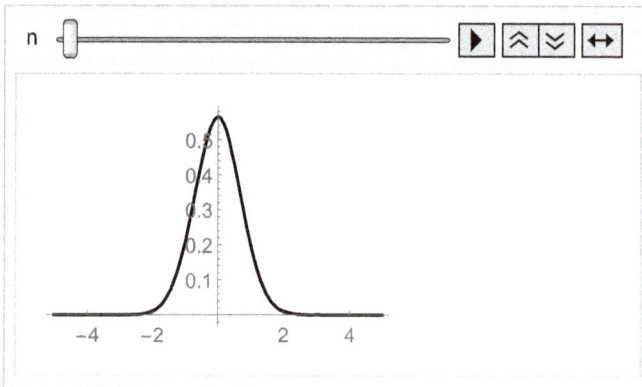

Figure 1.14: Convergence to the Dirac delta function.

Details on the Dirac delta function may be found in, e. g., [65, Section 6.1].

1.2.6 Gradient, divergence, and Laplacian

Let f be a real function of class $C^1(D)$, where D is a nonempty, open, and connected domain in $\mathbb{R}^3$. The *gradient* of f on D is the operator

$$\operatorname{grad} f(x,y,z) = \frac{\partial f(x,y,z)}{\partial x}\mathbf{i} + \frac{\partial f(x,y,z)}{\partial y}\mathbf{j} + \frac{\partial f(x,y,z)}{\partial z}\mathbf{k},$$

where $\{\mathbf{i}, \mathbf{j}, \mathbf{k}\}$ are the versors of the axes of a Cartesian coordinate system.

Another largely used notation is the ∇, *nabla* notation,

$$\nabla f(x,y,z) = \mathrm{grad}\, f(x,y,z).$$

The result is a vector in $\mathbb{R}^3$.

The gradient operator is implemented in *Mathematica* as $\mathrm{Grad}[f,\{x_1,\dots,x_n\},$ chart] and gives the gradient in the coordinates chart.

Let $f = (f_1, f_2, f_3)$ be a real function of class $C^1(D)$, where D is a nonempty open domain in $\mathbb{R}^3$. The *divergence* of f on D is the operator

$$\mathrm{div}\, f(x,y,z) = \frac{\partial f_1(x,y,z)}{\partial x} + \frac{\partial f_2(x,y,z)}{\partial y} + \frac{\partial f_3(x,y,z)}{\partial z}.$$

The result is a scalar.

The divergence operator is implemented in *Mathematica* as $\mathrm{Div}[\{f_1,\dots,f_n\},\{x_1,$ $\dots,x_n\},$ chart] and gives the divergence in the coordinates chart.

The *Laplacian* of a twice differentiable real valued function f of two real variables is the operator

$$\nabla^2_{x,y} f(x,y) = \frac{\partial^2 f(x,y)}{\partial x^2} + \frac{\partial^2 f(x,y)}{\partial y^2},$$

where (x,y) belongs to a nonempty open connected domain of f in $\mathbb{R}^2$. The result is a scalar.

If f is a real valued function of n variables that is twice differentiable on a nonempty open connected domain D in $\mathbb{R}^n$, $n \in \mathbb{N}$, $n \geq 2$, then its Laplacian is the operator

$$\nabla^2_{x_1,x_2,\dots,x_n} f(x_1, x_2, \dots, x_n) = \sum_{k=1}^{n} \frac{\partial^2 f(x_1,x_2,\dots,x_n)}{\partial^2_{x_k}} \quad \text{on } D.$$

The operator $\nabla^2_{x_1,x_2,\dots,x_n} f(x_1, x_2, \dots, x_n)$ is implemented in *Mathematica* as $\mathrm{Laplacian}[f$ $[x_1,x_2,\dots,x_n],\{x_1,x_2,\dots,x_n\}]$, and $\nabla^2_{\{x_1,x_2,\dots,x_n\}} f[x_1,x_2,\dots,x_n]$. If the result is expected in another coordinate chart, then there exists $\mathrm{Laplacian}[f[x_1,x_2,\dots,x_n],\{x_1,x_2,\dots,$ $x_n\},$ chart].

For example, using a short code in *Mathematica*, one has that

$\{\nabla^2_{\{t\}}(t^2+x^2+y^2+z^2), \nabla^2_{\{t,x,y,z\}}(t^2+x^2+y^2+z^2)\}$
$\{2,8\}$

A real valued function that is twice differentiable on a nonempty open connected domain D in $\mathbb{R}^n$, $n \in \mathbb{N}$, $n \geq 2$, is said to be *harmonic* on D if its Laplacian vanishes on D.

1.3 Gamma and beta functions

1.3.1 Gamma function

L. Euler introduced a C^∞ function that interpolates the factorial whenever the argument of the function is a positive integer. This function is called the *Euler integral of the second kind* or the *gamma function* and is defined as [2, 6.1.1, p. 255]

$$\Gamma(x) = \int_0^\infty t^{x-1}e^{-t}dt, \quad x > 0.$$

The Euler gamma function $\Gamma(x)$ is implemented in *Mathematica* as Gamma[x].

Theorem 1.21 ([56, p. 271]). *The gamma function is defined for $x > 0$ and is of class C^∞ on its domain of definition.*

Theorem 1.22 ([2, 6.1.6, p. 255 and 6.1.15, p. 256]). *One now has the next properties of the gamma function,*

$$\Gamma(1) = 1,$$
$$\Gamma(x+1) = x\,\Gamma(x), \quad x > 0, \quad \text{(Reduction formula)} \tag{1.3}$$
$$\Gamma(n+1) = n!, \quad n \in \mathbb{N},$$
$$\Gamma\left(\frac{1}{2}\right) = \sqrt{\pi}.$$

The reduction formula (1.3) is said to be the *functional equation* of the gamma function Γ.

The gamma function can be extended on the whole real axis except on the non-positive integers $\{\ldots, -2, -1, 0\}$ by considering its functional equation under the form

$$\Gamma(x) = \frac{\Gamma(x+1)}{x}.$$

From the previous recurrence relation, it follows that

$$\Gamma(x) = \frac{\Gamma(x+n)}{x(x+1)\cdots(x+n-1)}, \quad \text{if } x + n > 0.$$

The previous equality allows calculation of the gamma function for negative and non-integer values of the argument.

A sketch of the gamma function is given in Figure 1.15.

We now introduce the *complement formula* of the Γ function.

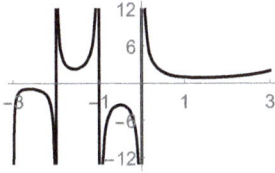

Figure 1.15: Gamma function.

Theorem 1.23 ([44, Section 1.2]). *Suppose $x \notin \mathbb{Z}$, then*

$$\Gamma(x)\Gamma(1-x) = \frac{\pi}{\sin(\pi x)}.$$

Then one has

$$\Gamma\left(\frac{1}{3}\right)\Gamma\left(\frac{2}{3}\right) = \frac{2\pi\sqrt{3}}{3} \quad \text{and} \quad \Gamma\left(\frac{1}{4}\right)\Gamma\left(\frac{3}{4}\right) = \pi\sqrt{2}.$$

The gamma function admits an asymptotic expansion [44, (1.4.23), p. 12].

Theorem 1.24. *For $x > 0$, one has*

$$\Gamma(x) = \sqrt{2\pi}\, x^{x-1/2}\, e^{-x}\left(1 + \frac{1}{12\,x} + \frac{1}{288\,x^2} - \frac{139}{51840\,x^3} + \frac{O(1)}{x^4}\right).$$

It follows at once the existence of the *Stirling formula* (1.4).

Theorem 1.25 ([56, p. 275]). *For $n \in \mathbb{N} \setminus \{0\}$, one has*

$$n! = \sqrt{2\pi n}\, n^n\, e^{-n}(1 + a_n), \tag{1.4}$$

where $a_n \to 0$ as $n \to \infty$.

For details on gamma function, we suggest [6, Chapter 2], [44, Chapter 1], [56, Section 6.4.1], and [87].

1.3.2 Beta function

The *Euler integral of the first kind* or the *beta function* is the integral defined as [2, 6.2.1, p. 258]

$$B(x,y) = \int_0^1 t^{x-1}(1-t)^{y-1}\mathrm{d}\,t, \quad x,y > 0.$$

Theorem 1.26 ([56, Section 6.4]). *The beta function is defined for all $x,y > 0$ and is continuous on $]0,\infty[\times]0,\infty[$.*

The Euler beta function $B(x,y)$ is implemented in *Mathematica* as Beta[x,y].

Theorem 1.27. *For $x, y > 0$, one has that:*

(a) $B(x, y) = B(y, x)$;

(b) $B(x, 1) = B(1, x) = \dfrac{1}{x}$;

(c) $B(x + 1, y) + B(x, 1 + y) = B(x, y)$;

(d) $B(x, y) = \dfrac{x + y}{y} B(x, y + 1)$.

The equation in (d) is said to be the *functional equation* of the beta function. The two Euler functions are tied up by the next result.

Theorem 1.28. *For $x, y > 0$, one has that*

$$B(x, y) = \frac{\Gamma(x)\,\Gamma(y)}{\Gamma(x + y)}.$$

From here, by $m, n, p \in \mathbb{N}$ one has that [62, p. 163]

$$\frac{1}{\binom{m\,n}{p\,n}} = (m\,n + 1) \int_0^1 x^{p\,n}(1 - x)^{(m-p)n}\,dx.$$

This formula is useful for the study of BBP formulas of evaluation of π, [5] and [56, Section 9.4].

Theorem 1.29. *For $x, y > 0$ and $m, n \in \mathbb{N}$, one has:*

(a) $B(1/2, 1/2) = \pi$;

(b) $B(1/3, 2/3) = 2\sqrt{3}\pi/3$;

(c) $B(1/4, 3/4) = \pi\sqrt{2}$;

(d) $B(x, 1 - x) = \dfrac{\pi}{\sin(\pi x)}$;

(e) $B(x, n) = \dfrac{(n - 1)!}{x(x + 1)\ldots(x + n - 1)}$;

(f) $B(m, n) = \dfrac{(m - 1)!(n - 1)!}{(m + n - 1)!}$.

A sketch of the beta function is given in Figure 1.16.

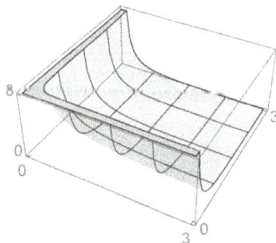

Figure 1.16: Beta function.

For details on beta function, we suggest [2, Section 6.2], [6, Chapter 2], [56, Section 6.4.2], and [86].

1.4 Hypergeometric equation

1.4.1 Hypergeometric functions

Let λ be a real number. We introduce the notation (*Pochhammer symbol*)

$$(\lambda)_0 = 1 \text{ and } (\lambda)_k = \lambda(\lambda + 1)\ldots(\lambda + k - 1) = \frac{\Gamma(\lambda + k)}{\Gamma(\lambda)}, \quad k \in \mathbb{N}^*$$

and define the general *hypergeometric function*

$$\begin{aligned}
{}_mF_n(a_1, a_2, \ldots, a_m; \beta_1, \beta_2, \ldots, \beta_n; x) &= {}_mF_n\left[\begin{matrix} a_1, a_2, \ldots, a_m; \\ \beta_1, \beta_2, \ldots, \beta_n; \end{matrix} x\right] \\
&= \sum_{k=0}^{\infty} \frac{(a_1)_k (a_2)_k \cdots (a_m)_k}{(\beta_1)_k (\beta_2)_k \cdots (\beta_n)_k} \frac{x^k}{k!}.
\end{aligned}$$

The rightmost series is said to be the *hypergeometric series* [6, Section 9.1].

The hypergeometric function ${}_2F_1(a, b; c; x)$ is implemented in *Mathematica* as Hypergeometric2F1[a,b,c,x], whereas ${}_1F_1(a; b; x)$ is implemented in *Mathematica* as Hypergeometric1F1[a,b,x].

Theorem 1.30. [6, p. 204] *The hypergeometric series* ${}_mF_n(a_1, \ldots, a_m; \beta_1, \ldots, \beta_n; x)$
(a) *converges*
    (a_1) *for all real x, if there exists* $j \in \{1, \ldots, m\}$ *such that* a_j *is a nonpositive integer; in this case, it reduces to a polynomial;*
    (a_2) *for all real x, if* $m < n + 1$;
    (a_3) *for* $|x| < 1$, *if* $m = n + 1$;
(b) *diverges*
    (b_1) *for all nonzero real x, if* $m > n + 1$;
    (b_2) *for* $|x| > 1$, *if* $m = n + 1$.

Theorem 1.31. *Let* $\{i_1, i_2, \ldots, i_m\}$ *be a permutation of* $\{1, 2, \ldots, m\}$ *and let* $\{j_1, j_2, \ldots, j_n\}$ *be a permutation of* $\{1, 2, \ldots, n\}$. *Then*

$$ {}_mF_n\left[\begin{matrix} a_1, a_2, \ldots, a_m; \\ \beta_1, \beta_2, \ldots, \beta_n; \end{matrix} x\right] = {}_mF_n\left[\begin{matrix} a_{i_1}, a_{i_2}, \ldots, a_{i_m}; \\ \beta_{j_1}, \beta_{j_2}, \ldots, \beta_{j_n}; \end{matrix} x\right]. $$

Below we introduce a list of certain elementary functions that are special cases of the hypergeometric function ${}_2F_1(\alpha, \beta; \gamma; x)$.

Theorem 1.32 ([44, § 9.8]). *One has that*
(a) ${}_2F_1(\alpha, 0; \gamma; x) = 1$;
(b) ${}_2F_1(\alpha, -2; \gamma; x) = 1 - 2\frac{\alpha}{\gamma}x + \frac{\alpha(\alpha+1)}{\gamma(\gamma+1)}x^2$;
(c) ${}_2F_1(\alpha, \beta; \beta; x) = (1 - x)^{-\alpha}$;

(d) $x \cdot {}_2F_1(1,1;2;x) = -\ln(1-x)$;

(e) $x \cdot {}_2F_1(\frac{1}{2},1;\frac{3}{2};-x^2) = \arctan(x)$.

1.4.2 Hypergeometric equation

One considers the second-order linear and homogeneous ordinary differential equation:

$$x(1-x)u''(x) + (\gamma - (\alpha + \beta + 1))u'(x) - \alpha\beta\, u(x) = 0, \tag{1.5}$$

where x is a real variable and α, β, and γ are real parameters. Equation (1.5) is said to be the *hypergeometric equation*.

We suppose that $\gamma \notin \{\ldots, -3, -2, -1, 0\}$. Then a particular solution of equation (1.5) is [44, Section 7.2]

$$u(x) = {}_2F_1(\alpha, \beta; \gamma; x), \quad |x| < 1. \tag{1.6}$$

Another particular solution of (1.5) is [44, Section 7.2]

$$u(x) = x^{1-\gamma}{}_2F_1(1 - \gamma + \alpha, 1 - \gamma + \beta; 2 - \gamma; x), \quad |x| < 1. \tag{1.7}$$

If γ is not an integer, these two particular solutions exist simultaneously and are linearly independent.

1.5 Hyperbolic functions

In Table 1.1, we introduce the hyperbolic sine, cosine, tangent, cotangent, secant, and cosecant functions [2, 4.5.1–4.5.6, p. 83].

Table 1.1: Hyperbolic circular functions.

no.	function	definition	implemented as
1	hyperbolic sine	$\sinh(x) = \frac{\exp(x)-\exp(-x)}{2}$	$\mathrm{Sinh}[x]$
2	hyperbolic cosine	$\cosh(x) = \frac{\exp(x)+\exp(-x)}{2}$	$\mathrm{Cosh}[x]$
3	hyperbolic tangent	$\tanh(x) = \frac{\sinh(x)}{\cosh(x)}$	$\mathrm{Tanh}[x]$
4	hyperbolic cotangent	$\coth(x) = \frac{\cosh(x)}{\sinh(x)}$	$\mathrm{Coth}[x]$
5	hyperbolic secant	$\mathrm{sech}(x) = \frac{1}{\cosh(x)}$	$\mathrm{Sech}[x]$
6	hyperbolic cosecant	$\mathrm{csch}(x) = \frac{1}{\sinh(x)}$	$\mathrm{Csch}[x]$

Thus, one has that sinh, cosh, tanh, and sech : $\mathbb{R} \to \mathbb{R}$ while coth and csch : $\mathbb{R} \setminus \{0\} \to \mathbb{R}$.

A sketch of the hyperbolic sine function, hyperbolic cosine function, and hyperbolic tangent function are given by the left-hand side picture in Figure 1.17. A sketch of the hyperbolic cotangent function, the hyperbolic secant function, and hyperbolic cosecant function is given by the right-hand side picture in Figure 1.17.

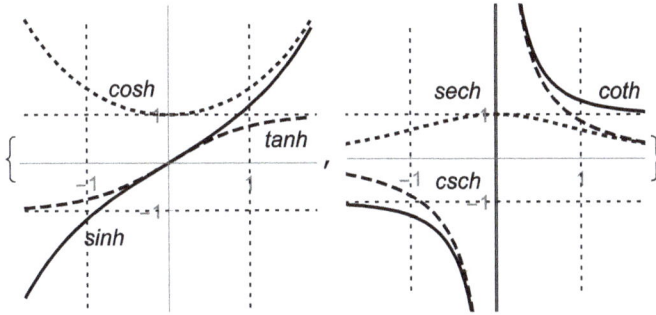

Figure 1.17: Sinh, cosh, tanh, coth, sech, and csch functions.

Remark. We note the following inequalities: $\cosh(x) \geq 1$, $|\tanh(x)| < 1$ and $0 < \operatorname{sech}(x) \leq 1$ are true for all $x \in \mathbb{R}$. □

Theorem 1.33 ([2, 4.5.16- -4.5.25, p. 83]). *One has the next obvious identities:*
(a) $\cosh(x) + \sinh(x) = \exp(x)$;
(b) $\cosh(x) - \sinh(x) = \exp(-x)$;
(c) $\cosh^2(x) - \sinh^2(x) = 1$;
(d) $\sinh(-x) = -\sinh(x)$;
(e) $\cosh(-x) = \cosh(x)$;
(f) $\sinh(x + y) = \sinh(x)\cosh(y) + \cosh(x)\sinh(y)$;
(g) $\cosh(x + y) = \cosh(x)\cosh(y) + \sinh(x)\sinh(y)$.

Theorem 1.34 ([2, 4.5.71- -4.5.76, pp. 85–86]). *One has the next obvious identities:*
(a) $\sinh'(x) = \cosh(x)$;
(b) $\cosh'(x) = \sinh(x)$;
(c) $\tanh'(x) = \operatorname{sech}^2(x)$;
(d) $\coth'(x) = -\operatorname{csch}^2(x)$;
(e) $\operatorname{sech}'(x) = -\tanh(x)\operatorname{sech}(x)$;
(f) $\operatorname{csch}'(x) = -\coth(x)\operatorname{csch}(x)$.

Remark. We note the following limits:

$$\lim_{x \to 0,\, x<0} \coth(x) = -\infty \quad \text{and} \quad \lim_{x \to 0,\, x>0} \coth(x) = +\infty.$$

Indeed, by *Mathematica* one has

```
Limit[Coth[x],x→0,Direction→#]&/@{0,1,-1}
{ComplexInfinity,-∞,∞}.
```

□

In Table 1.2, the hyperbolic arcsine, arccosine, arctangent, arccotangent, arcsecant, and arccosecant functions are introduced.

Table 1.2: Hyperbolic circular arc functions.

no.	function	definition	implemented as
1	hyperbolic arcsine	$\text{arcsinh}(x) = \ln(x + \sqrt{x^2 + 1}), x \in \mathbb{R}$	ArcSinh[x]
2	hyperbolic arccosine	$\text{arccosh}(x) = \ln(x + \sqrt{x^2 - 1}), x \geq 1$	ArcCosh[x]
3	hyperbolic arctangent	$\text{arctanh}(x) = \frac{1}{2} \ln \frac{1+x}{1-x}, |x| < 1$	ArcTanh[x]
4	hyperbolic arccotangent	$\text{arccoth}(x) = \frac{1}{2} \ln \frac{x+1}{x-1}, |x| > 1$	ArcCoth[x]
5	hyperbolic arcsecant	$\text{arcsech}(x) = \ln \frac{1+\sqrt{1-x^2}}{x}, 0 < x \leq 1$	ArcSech[x]
6	hyperbolic arccosecant	$\text{arccsch}(x) = \ln \frac{1+\sqrt{1+x^2}}{x}, x \neq 0$	ArcCsch[x]

A sketch of the hyperbolic arcsine function, hyperbolic arccosine function, and hyperbolic arctangent function is given by the left-hand side picture of Figure 1.18, whereas a sketch of the hyperbolic arccotangent function, hyperbolic arcsecant function, and hyperbolic arccosecant function are given by the right-hand side picture of the same figure.

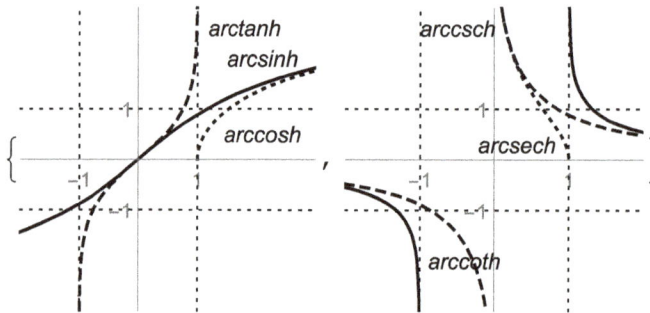

Figure 1.18: Arcsinh, arccosh, arctanh, arccoth, arcsech, and arccsch functions.

1.6 Bessel functions

The second-order linear differential equation [2, 9.1.1, p. 358]

$$x^2 y''(x) + x y'(x) + (x^2 - v^2) y(x) = 0, \quad x \in \mathbb{R}, \tag{1.8}$$

where v is a real parameter, is said to be the *Bessel equation* and its solutions the *Bessel functions* or the *cylinder functions*. Because v appears at the power 2, we may admit that $v \geq 0$.

Theorem 1.35. *Let J_v be the function defined by*

$$J_v(x) = \sum_{k=0}^{\infty} \frac{(-1)^k}{k!\Gamma(v+k+1)} \left(\frac{x}{2} \right)^{v+2k}, \quad x \in \mathbb{R}.$$

(a) *The series in the right-hand side converges for all $x \in \mathbb{R}$.*
(b) *The series is differentiable term by term converging to the differential of J_v.*
(c) *J_v is a solution of ordinary differential equation (1.8).*
(d) *The function*

$$J_{-v}(x) = \sum_{k=0}^{\infty} \frac{(-1)^k}{k!\Gamma(-v+k+1)} \left(\frac{x}{2} \right)^{-v+2k}, \quad x \in \mathbb{R}$$

is a solution of (1.8).

Functions J_v and J_{-v} are said to be *Bessel functions of the first kind*. The Bessel function of the first kind is implemented in *Mathematica* as BesselJ[v,x].
A sketch of some Bessel functions of the first kind is given by the leftmost picture in Figure 1.19.

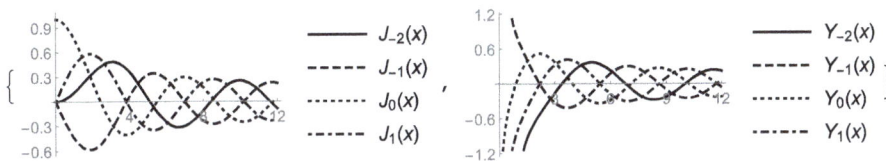

Figure 1.19: Bessel functions of the first and the second kind.

Theorem 1.36 ([2, 9.1.15, p. 360]). *The Wronskian of J_v and J_{-v} functions is of the form,*

$$W(x) = \begin{vmatrix} J_v(x) & J_{-v}(x) \\ J_v'(x) & J_{-v}'(x) \end{vmatrix} = -\frac{2\sin(v\pi)}{\pi x}.$$

Thus, J_v and J_{-v} are linearly dependent whenever $v \in \mathbb{N}$.

Theorem 1.37 ([2, 9.1.2, p. 358]). *Let Y_v be the function defined as*

$$Y_v(x) = \frac{1}{\sin(v\pi)} (J_v(x)\cos(v\pi) - J_{-v}(x)), \quad v \notin \mathbb{N}. \tag{1.9}$$

Then:

(a) *The Wronskian* $W(J_v(x), Y_v(x)) = \frac{2}{\pi x}$, $v \notin \mathbb{N}$;

(b) *If* $v \notin \mathbb{N}$, *there exists* $Y_n(x) = \lim_{v \to n} Y_v(x)$ *for all* $n \in \mathbb{N}$, *namely*

$$Y_n(x) = \frac{1}{\pi}\left(\frac{\partial J_v(x)}{\partial v} - (-1)^n \frac{\partial J_{-v}(x)}{\partial v} \right)\bigg|_{v=n} ; \qquad (1.10)$$

(c) Y_n *is a solution of* (1.8).

The function Y_v defined by (1.9) for $v \notin \mathbb{N}$ and by (1.10) for $v = n \in \mathbb{N}$ is said to be *Bessel functions of the second kind*.

The Bessel function of the second kind is implemented in *Mathematica* as `BesselY[v,x]`.

A sketch of some Bessel functions of the second kind is given by the rightmost picture in Figure 1.19.

Theorem 1.38 ([44, Chapter 5]). *The following relations hold:*

(a) $J_{v-1}(x) + J_{v+1}(x) = \frac{2v}{x}J_v(x)$;

(b) $J_{v-1}(x) - J_{v+1}(x) = 2J_v'(x)$;

(c) $J_{-n}(x) = (-1)^n J_n(x)$, $n \in \mathbb{N}^*$;

(d) $Y_{-n}(x) = (-1)^n Y_n(x)$, $n \in \mathbb{N}^*$;

(e) *For* $n \in \mathbb{N}$, *one has the generating function*

$$\exp\left(\frac{x}{2}\left(t - \frac{1}{t} \right) \right) = \sum_{n=-\infty}^{+\infty} J_n(x)t^n, \quad t \in \mathbb{R}\setminus\{0\};$$

(f) *For* $v > -1$, *all the roots of function* J_v *are real;*

(g) *For* $v > -1$, *the equation* $J_v(x) = 0$ *has countable roots;*

(h) *For* $v > -1$, *one has*

$$(p^2 - q^2)\int_0^1 xJ_v(qx)J_v(px)dx = qJ_v'(q)J_v(p) - pJ_v'(p)J_v(q).$$

1.6.1 Modified Bessel functions

Similar to (1.8), one may consider the second-order linear differential equation of the form [2, 9.6.1, p. 374]

$$x^2 y''(x) + x y'(x) + (x^2 + v^2)y(x) = 0, \quad x \in \mathbb{R}, \qquad (1.11)$$

where v is a real parameter. It is said to be the *modified Bessel equation* and its solutions are called the *modified Bessel functions*. Because v appears at the power 2, we may admit that $v \geq 0$.

Theorem 1.39. *Let I_ν be the function defined as*

$$I_\nu(x) = \sum_{k=0}^{\infty} \frac{1}{k!\Gamma(\nu + k + 1)}\left(\frac{x}{2}\right)^{\nu+2k}, \quad \nu \in \mathbb{R}.$$

(a) *The series in the right-hand side converges for all $x \in \mathbb{R}$.*
(b) *The series is differentiable term by term converging to the differential of I_ν.*
(c) *I_ν is a solution of ordinary differential equation (1.11).*

Function I_ν defined for $\nu \in \mathbb{R}$ is said to be *modified Bessel functions of the first kind*. The modified Bessel function of the first kind is implemented in *Mathematica* as BesselI[ν,x].

Equation (1.11) admits another particular solution, namely

$$K_\nu(x) = \begin{cases} \frac{\pi\csc(\pi\nu)}{2}(I_{-\nu}(x) - I_\nu(x)), & \nu \notin \mathbb{Z}, \\ \lim_{\mu \to \nu} K_\mu(x), & \nu \in \mathbb{Z}. \end{cases}$$

Function K_ν defined for $\nu \in \mathbb{R}$ is said to be the *modified Bessel function of the second kind*. The modified Bessel function of the second kind is implemented in *Mathematica* as BesselK[ν,x].

More details on Bessel functions may be found in [2, Chapter 9], [84], and [85].
Some sketches of modified Bessel functions are presented in Figure 1.20.

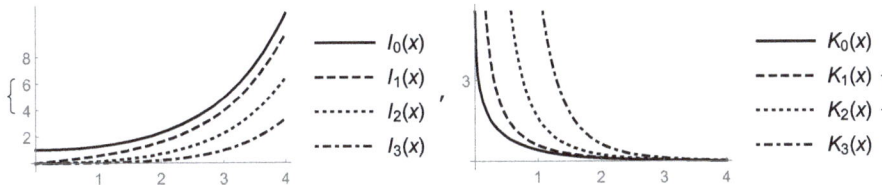

Figure 1.20: Modified Bessel functions.

1.7 Legendre functions

The differential equation [44, Section 7.3.]

$$(1 - x^2)u''(x) - 2x\,u'(x) + \nu(\nu + 1)u(x) = 0, \quad x \in {]-1, 1[},$$

where ν is a real constant, and is said to be the *Legendre equation*. This equation can be reduced to a hypergeometric equation by the substitution $t = (1 - x)/2$ getting

$$t(1 - t)u''(t) + (1 - 2t)u(t) + \nu(\nu + 1)u(t) = 0. \tag{1.12}$$

Equation (1.12) is a particular case of equation (1.5) with $\alpha = -\nu$, $\beta = \nu + 1$, and $\gamma = 1$.

Based on (1.6) and (1.7), equation (1.12) has two linearly independent solutions, [44, 7.3.6–7.3.7],

$$u(x) = P_\nu(x) = {}_2F_1\left(-\nu, \nu + 1; 1; \frac{1-x}{2}\right), \quad \left|\frac{1-x}{2}\right| < 1, \text{ and}$$

$$u(x) = Q_\nu(x) = \frac{\sqrt{\pi}\,\Gamma(\nu + 1)}{\Gamma(\nu + \frac{3}{2})(2x)^{\nu+1}}\,{}_2F_1\left(\frac{\nu}{2} + 1, \frac{\nu}{2} + \frac{1}{2}; \nu + \frac{3}{2}; \frac{1}{x^2}\right),$$

$$|x| > 1, \quad -\nu \notin \{1, 2, 3, \dots\}.$$

$P_\nu(x)$ is said to be the *Legendre function of degree ν of the first kind*, whereas $Q_\nu(x)$ is said to be the *Legendre function of degree ν of the second kind*.

The following series expansion of the Legendre function of the first kind is valid for $|x| < 1$ and arbitrary ν, [44, p. 178],

$$P_\nu(x) = \frac{\Gamma(\frac{\nu}{2} + \frac{1}{2})}{\sqrt{\pi}\,\Gamma(\frac{\nu}{2} + 1)} \cos\frac{\nu\pi}{2}\,{}_2F_1\left(\frac{\nu}{2} + \frac{1}{2}, -\frac{\nu}{2}; \frac{1}{2}; x^2\right)$$

$$+ \frac{2\Gamma(\frac{\nu}{2} + 1)}{\sqrt{\pi}\,\Gamma(\frac{\nu}{2} + \frac{1}{2})} \sin\frac{\nu\pi}{2}\,x\,{}_2F_1\left(\frac{1}{2} - \frac{\nu}{2}, \frac{\nu}{2} + 1; \frac{3}{2}; x^2\right).$$

Similarly, for $|x| < 1$ and $-\nu \notin \mathbb{N}^*$ the following series expansion of the Legendre function of the second kind is valid [44, p. 179]:

$$Q_\nu(x) = \frac{\Gamma(\frac{\nu}{2} + 1)\sqrt{\pi}\cos\frac{\nu\pi}{2}}{\Gamma(\frac{\nu}{2} + \frac{1}{2})}\,x\,{}_2F_1\left(\frac{1}{2} - \frac{\nu}{2}, \frac{\nu}{2} + 1; \frac{3}{2}; x^2\right)$$

$$- \frac{\Gamma(\frac{\nu}{2} + \frac{1}{2})\sqrt{\pi}\sin\frac{\nu\pi}{2}}{2\Gamma(\frac{\nu}{2} + 1)}\,{}_2F_1\left(\frac{\nu}{2} + \frac{1}{2}, -\frac{\nu}{2}; \frac{1}{2}; x^2\right)$$

The *Legendre polynomial of order $n \in \mathbb{N}^*$ and first kind* is defined by

$$P_n(x) = \sum_{k=0}^{\lfloor n/2 \rfloor} \frac{(-1)^k(2n - 2k)!}{2^n k!(n - k)!(n - 2k)!}\,x^{n-2k}, \quad x \in \mathbb{R}$$

and is implemented in *Mathematica* as LegendreP[n,x].

We list the first four Legendre polynomials of the first kind

$$P_0(x) = 1, \quad P_1(x) = x, \quad P_2(x) = \frac{1}{2}(3x^2 - 1), \quad P_3(x) = \frac{1}{2}(5x^3 - 3x).$$

A sketch of some Legendre polynomials of the first kind P_n is given by the leftmost picture in Figure 1.21.

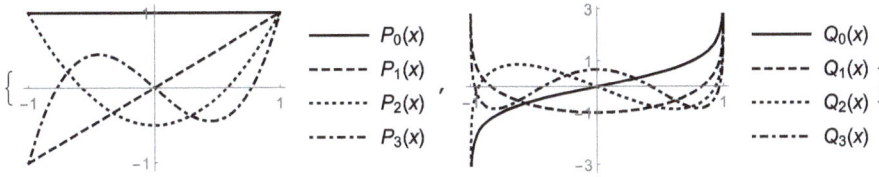

The *Legendre polynomial of order* $n \in \mathbb{N}^*$ *and second kind* [44, Section 3.10] is defined for $|x| < 1$ as

$$\begin{cases} Q_0(x) = \operatorname{arctanh}(x), \\ Q_n(x) = P_n(x) \operatorname{arctanh}(x) - \sum_{k=0}^{\lfloor \frac{n-1}{2} \rfloor} \frac{2n-4k-1}{(2k+1)(n-k)} P_{n-2k-1}(x), \quad n \in \mathbb{N} \end{cases}$$

and is implemented in *Mathematica* as LegendreQ[n,x].
We list the first four Legendre polynomials of the second kind:

$$\begin{cases} Q_0(x) = \operatorname{arctanh}(x), \quad Q_1(x) = x \operatorname{arctanh}(x) - 1, \\ Q_2(x) = \frac{1}{4}(6x^2 - 6x - 2) \operatorname{arctanh}(x), \\ Q_3(x) = \frac{1}{6}(15x^4 - 15x^2 - 9x + 4) \operatorname{arctanh}(x). \end{cases}$$

A sketch of some Legendre polynomials of the second kind Q_n is given by the rightmost picture in Figure 1.21 in [83] and the references therein.

1.8 Jacobi polynomials

The Legendre polynomial of order $n \in \mathbb{N}^*$ and first kind is further generalized to the *Jacobi polynomial* $P_n^{(\alpha,\beta)}$ [6, Section 8.2].
A definition of the Jacobi polynomial is given by a generating function, namely

$$\frac{2^{\alpha+\beta}}{\sqrt{1 - 2xt + t^2}(1 - t + \sqrt{1 - 2xt + t^2})^{\alpha}(1 + t + \sqrt{1 - 2xt + t^2})^{\beta}} = \sum_{n=0}^{\infty} P_n^{(\alpha,\beta)}(x)t^n.$$

The Legendre polynomial $P_n(x)$ is a particular case of the Jacobi polynomial $P_n^{(\alpha,\beta)}(x)$, i. e., $P_n(x) = P_n^{(0,0)}(x)$.
The Jacobi polynomial $P_n^{(\alpha,\beta)}(x)$ is implemented in *Mathematica* as JacobiP[n,α, β,x].
Some graphs of Jacobi polynomials are given in Figure 1.22.

Theorem 1.40. *The Jacobi polynomials* $P_n^{(\alpha,\beta)}$ *fulfill the properties:*
(a) $P_n^{(\alpha,\beta)}(-x) = (-1)^n P_n^{(\alpha,\beta)}(x);$

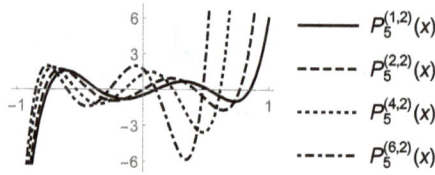

Figure 1.22: The Jacobi polynomials $P_5^{(1,2)}$, $P_5^{(2,2)}$, $P_5^{(4,2)}$, and $P_5^{(6,2)}$.

(b) *Is a solution of the differential equation*

$$(1-x^2)u''(x) + (\beta - \alpha - (\alpha + \beta + 2)x)u'(x) + n(n + \alpha + \beta + 1)u(x) = 0;$$

(c) *The Rodrigues-like formula*

$$P_n^{(\alpha,\beta)}(x) = \frac{(-1)^n}{2^n n!}(1-x)^{-\alpha}(1+x)^{-\beta}\frac{d^n}{dx^n}\left((1-x)^{\alpha+n}(1+x)^{\beta+n}\right)$$

is true;

(d) *The next recurrence is valid:*

$$2n(\alpha + \beta + n)(\alpha + \beta + 2n - 2)P_n^{(\alpha,\beta)}(x)$$
$$- (\alpha + \beta + 2n - 1)(\alpha^2 - \beta^2 + x(\alpha + \beta + 2n)(\alpha + \beta + 2n - 2))P_{n-1}^{(\alpha,\beta)}(x)$$
$$+ 2(\alpha + n - 1)(\beta + n - 1)(\alpha + \beta + 2n)P_{n-2}^{(\alpha,\beta)}(x) = 0.$$

1.9 Mathieu functions

E. L. Mathieu in [53] has introduced certain functions that are now called the *Mathieu function* in connection to the vibrating movement of an elliptical-shaped membrane. Some comprehensive introductions to the Mathieu functions may be found in [54], [33, Section 7.4], and [4, Chapter 32]. See also [2, Chapter 20] and [77, Section 3.11].

1.9.1 Mathieu equation

Suppose that an elliptical-shaped membrane with vertical movement $u = u(t,x,y)$ oscillates according to the wave equation,

$$\frac{\partial^2 u}{\partial x^2} + \frac{\partial^2 u}{\partial y^2} = \frac{1}{v^2}\frac{\partial^2 u}{\partial t^2},\tag{1.13}$$

where v is the wave velocity. If we are looking for solutions satisfying $u(t,x,y) = w(t)z(x,y)$ and substituting in (1.13), then we get

$$\begin{cases} w(t)\nabla_{x,y}^2 z(x,y) = \frac{1}{v^2}z(x,y)\frac{\partial^2 w}{\partial t^2} \implies \\ \frac{1}{z(x,y)}\nabla_{x,y}^2 z(x,y) = \frac{1}{v^2}\frac{1}{w(t)}\frac{\partial^2 w(t)}{\partial t^2} = -k^2. \end{cases}$$

Thus, we derived the *Helmholtz equation* in Cartesian system of coordinates,

$$\nabla^2_{x,y} z(x,y) + k^2 \, z(x,y) = 0. \tag{1.14}$$

Since we mentioned that the domain of the unknown function is an elliptical-shaped membrane, we pass from the rectangular coordinates (x,y) to the elliptical coordinates (ξ, η) by the transformations

$$x = c \cosh(\xi) \cos(\eta), \quad y = c \sinh(\xi) \sin(\eta), \quad 0 \le \xi, \quad 0 \le \eta < 2\pi,$$

with $c > 0$.

Because

$$\frac{x^2}{c^2 \cosh^2(\xi)} + \frac{y^2}{c^2 \sinh^2(\xi)} = 1 \quad \text{and} \quad \frac{x^2}{c^2 \cos^2(\eta)} - \frac{y^2}{c^2 \sin^2(\eta)} = 1,$$

it is obvious that the ξ constant supplies an ellipsis and the η constant supplies a hyperbola. A sketch of the elliptic coordinates is given in Figure 1.23.

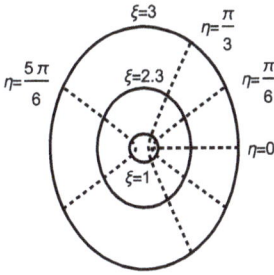

Figure 1.23: Elliptical coordinates.

Substituting the elliptical coordinates into (1.14), one gets

$$\frac{1}{z}\nabla^2_{\xi,\eta} z(\xi, \eta) + c^2 k^2 \cosh^2(\xi) - c^2 k^2 \cos^2(\eta) = 0. \tag{1.15}$$

Supposing that function z has separable variables, i. e., $z(\xi, \eta) = R(\xi)\Phi(\eta)$, from (1.15) it further follows that

$$\begin{cases} \dfrac{1}{R(\xi)}\dfrac{d^2 R}{d\xi^2} + c^2 k^2 \cosh^2(\xi) = a + \dfrac{1}{2} c^2 k^2, \\[2mm] -\dfrac{1}{\Phi(\eta)}\dfrac{d^2\Phi}{d\eta^2} + c^2 k^2 \cos^2(\eta) = a + \dfrac{1}{2} c^2 k^2, \end{cases}$$

where a is a constant. Denote $q = c^2 k^2/4$ to get the second-order linear differential equations

$$\frac{d^2\Phi}{d\eta^2} + (a - 2q\cos(2\eta))\Phi(\eta) = 0, \tag{1.16}$$

$$\frac{d^2R}{d\xi^2} - (a - 2q\cosh(2\xi))R(\xi) = 0. \tag{1.17}$$

Equation (1.16) is said to be the *Mathieu equation*, whereas equation (1.17) is said to be the *modified Mathieu equation*. The Mathieu equation is also called the *angular Mathieu equation*, whereas the modified Mathieu equation is also called the *radial Mathieu equation*. The *Mathematica* offers us the next results:

```
Clear[a,q,t,x] (* Clean the values and definitions *)
{DSolve[x''[t]+(a-2q Cos[2t])x[t]==0,x,t],
DSolve[x''[t]-(a-2q Cos[2t])x[t]==0,x,t]}//Flatten
{x→Function[{t},C[1] MathieuC[a,q,t]
+C[2] MathieuS[a,q,t]],
x→Function[{t},C[1] MathieuC[-a,-q,t]
+C[2] MathieuS[-a,-q,t]]}
```

The next two theorems indicate the behavior of the solutions of the (1.16) and (1.17) equations.

Theorem 1.41 ([33, p. 175]). *Let*

$$x''(t) + (a + b\cos(2t))x(t) = 0, \quad a, b, t, x \in \mathbb{R}, \tag{1.18}$$

be a differential equation. Equation (1.18) has no periodic solution unless $a = 0$, *and although the equation*

$$x''(t) + n^2x(t) = 0$$

always has a periodic general solution, the period is not π *unless* n *is an even integer.*

Theorem 1.42 ([33, p. 175]). *Let*

$$x''(t) + (a + 2q\cos(2t))x(t) = 0, \quad a, q, t, x \in \mathbb{R}, \tag{1.19}$$

be a differential equation. If $f(x)$ *is a solution, which is neither even nor odd, then* $f(-x)$ *is a distinct solution and* $(f(x) + f(-x))/2$ *is an even solution, not identically zero, and* $(f(x) - f(-x))/2$ *is an odd solution, not identically zero.*

Two distinct even solutions, and likewise two distinct odd solutions, cannot exist. Thus, one fundamental solution is even and the other is odd.

The even fundamental solution is denoted $Ce_a(q, t)$ and implemented in *Mathematica* as MathieuC[a,q,t] and the odd fundamental solution is denoted $Se_a(q, t)$ and implemented in *Mathematica* as MathieuS[a,q,t], [77, p. 1086].

Because Mathieu functions are solutions to second-order differential equation (1.19), the derivatives with respect to t are available. The derivative of the $Ce_a(q, t)$ function is implemented in *Mathematica* as MathieuCPrime[a,q,t] and the derivative of the $Se_a(q, t)$ function is implemented in *Mathematica* as MathieuSPrime[a,q,t].

1.9.2 Mathieu functions

Remarks.
1. The solutions of both equations (1.16) and (1.17) have complicated dynamics. We ex-
    hibit a simple case.

```
Block[{a=7.5,q=2},
Plot[{MathieuC[a,q,t],MathieuS[a,q,t]},{t,0,4Pi},
Ticks→{5Range[3],{-1,1}},PlotLegends→"Expressions",
PlotStyle→{Black,{Black,Dashed}},ImageSize→175]]
```

See Figure 1.24.

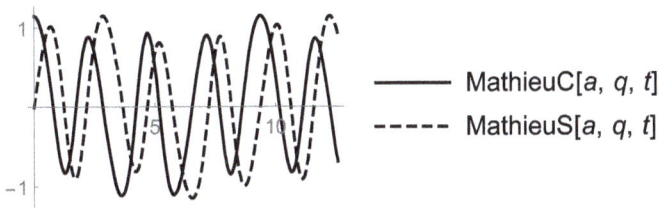

Figure 1.24: Mathieu functions.

2. The algebraic form of the Mathieu ordinary differential equation follows by the
    transformation $\zeta = \cos^2(\eta)$:

$$4\zeta(1-\zeta)\frac{d^2\Phi}{d\zeta^2} + 2(1-2\zeta)\frac{d\Phi}{d\zeta} + (\lambda + 2q(1-2\zeta))\Phi = 0.$$

It is clear that this equation is not of hypergeometric type; see [4, Section 32.1].
3. There are the following particular cases:

```
{MathieuC[m,0,t],MathieuS[m,0,t]}
{Cos[Sqrt[m] t],Sin[Sqrt[m] t]}
```

Remark. Some other Mathieu differential equations are introduced in Section 4.4. □

1.10 Elliptic integrals, elliptic functions

Details on elliptic integrals and elliptic functions may be found in many references; we
just mention [43] and [2, Chapter 17].

1.10.1 Elliptic integrals

An *elliptic integral* is an integral of the form

$$\int_0^x \frac{A(t) + B(t)}{C(t) + D(t)\sqrt{S(t)}}\, dt,$$

where $A(t)$, $B(t)$, $C(t)$, and $D(t)$ are polynomials in t and $S(t)$ is a polynomial of degree 3 or 4. Commonly, an elliptic integral cannot be evaluated by elementary methods.

We introduce the *elliptic integral of the first kind*.

Suppose that *modulus k* satisfies $0 \le k^2 < 1$ (sometimes written in terms of the parameter $m = k^2$) or *modular angle* $\phi = \arcsin(k)$. The *incomplete elliptic integral of the first kind* is written as

$$F(\phi, k) = \int_0^{\sin(\phi)} \frac{dt}{\sqrt{(1 - t^2)(1 - k^2 t^2)}}, \qquad 0 \le k^2 \le 1 \text{ and } 0 \le \sin(\phi) \le 1$$

or

$$F(\phi, m) = \int_0^{\sin(\phi)} \frac{dt}{\sqrt{(1 - t^2)(1 - m\, t^2)}}, \qquad 0 \le m \le 1 \text{ and } 0 \le \sin(\phi) \le 1.$$

If we set $t = \sin(\theta)$ and $dt = \cos(\theta)d\theta = \sqrt{1 - t^2}d\theta$, then

$$F(\phi, k) = \int_0^{\phi} \frac{d\theta}{\sqrt{1 - k^2 \sin^2(\theta)}}, \qquad 0 \le k^2 \le 1, \text{ and } 0 \le \phi \le \frac{\pi}{2}$$

or

$$F(\phi, m) = \int_0^{\phi} \frac{d\theta}{\sqrt{1 - m\, \sin^2(\theta)}}, \qquad 0 \le m \le 1, \text{ and } 0 \le \phi \le \frac{\pi}{2}. \tag{1.20}$$

The incomplete elliptic integral of the first kind is implemented in *Mathematica* as EllipticF[ϕ,m].

The *complete elliptic integral of the first kind* is obtained by setting the upper bound of the integral to its maximum range, i. e.,

$$K(k) = \int_0^1 \frac{dt}{\sqrt{(1 - t^2)(1 - k^2 t^2)}} = \int_0^{\pi/2} \frac{d\theta}{\sqrt{1 - k^2 \sin^2(\theta)}}, \qquad 0 \le k^2 \le 1$$

or

$$K(m) = \int_0^1 \frac{dt}{\sqrt{(1-t^2)(1-mt^2)}} = \int_0^{\pi/2} \frac{d\theta}{\sqrt{1-m\sin^2(\theta)}}, \quad 0 \leq k^2 \leq 1.$$

The complete elliptic integral of the first kind is implemented in *Mathematica* as EllipticK[m].

The *incomplete elliptic integral of the second kind* is written as

$$E(\phi, m) = \int_0^{\sin(\phi)} \frac{\sqrt{1-mt^2}}{\sqrt{1-t^2}}\, dt, \quad 0 \leq m \leq 1 \text{ and } 0 \leq \sin(\phi) \leq 1.$$

If we set $t = \sin(\theta)$ and $dt = \cos(\theta)d\theta = \sqrt{1-t^2}d\theta$, then

$$E(\phi, m) = \int_0^\phi \sqrt{1-m\sin^2(\theta)}d\theta, \quad 0 \leq m \leq 1 \text{ and } 0 \leq \phi \leq \frac{\pi}{2}.$$

The incomplete elliptic integral of the second kind is implemented in *Mathematica* as EllipticE[ϕ,m].

The *complete elliptic integral of the second kind* is obtained by setting the upper bound of the integral to its maximum range, i. e.,

$$E(m) = \int_0^1 \frac{\sqrt{1-mt^2}}{\sqrt{1-t^2}}\, dt = \int_0^{\pi/2} \sqrt{1-m\sin^2(\theta)}d\theta, \quad 0 \leq m \leq 1.$$

The complete elliptic integral of the second kind is implemented in *Mathematica* as EllipticE[m].

1.10.2 Elliptic functions

We discuss the most common elliptic functions, namely the *Jacobian elliptic functions*.

For $\phi \in [0, \pi/2]$, function F in (1.20) is invertible with respect to ϕ being a strictly increasing function. If we denote $u = F(\phi, m)$, then the inversion of the incomplete elliptic integral gives us

$$\phi = F^{-1}(u, m) = \text{amp}(u, m).$$

Thus, we may define the functions

$$\text{sn}(u, m) = \sin(\phi) = \sin(\text{amp}(u, m)),$$
$$\text{cn}(u, m) = \cos(\phi) = \cos(\text{amp}(u, m)),$$

$$dn(u, m) = \sqrt{1 - m \, \sin^2(\phi)}.$$

$sn(u, m)$ is said to be the *sine amplitude elliptic function*, $cn(u, m)$ is said to be the *cosine amplitude elliptic function*, $dn(u, m)$ is said to be the *amplitude elliptic function*, and $amp(u, m)$ is said to be the *Jacobi amplitude*.

These elliptic functions are implemented in *Mathematica* as `JacobiSN[u,m]`, `JacobiCN[u,m]`, `JacobiDN[u,m]`, and `JacobiAmplitude[u,m]`, respectively.

The elliptic functions satisfy:

1. $sn(u, 0) = \sin(\phi)$;
2. $cn(u, 0) = \cos(\phi)$;
3. $dn(u, 0) = 1$.

2 Elementary ordinary differential equations

The second chapter deals with the next topics: classification of certain ordinary differential equations, planar phase portrait, solid phase portrait, straight integration of first-order derivatives, differential equations with separable variables, and homogeneous differential equations.

2.1 Classification of certain ordinary differential equations

We introduce a table on the classification of certain ordinary differential equations; see Table 2.1.

Table 2.1: List of certain first- and second-order ordinary differential equations.

Type of equation	General form
separable variables	$x'(t) = f(t)g(x)$
homogeneous	$x'(t) = f(x/t)$
first-order linear	$x'(t) + f(t)x(t) = g(t)$
Bernoulli	$x'(t) + f(t)x(t) = g(t)x(t)^a, a \neq 0, a \neq 1$
Riccati	$x'(t) = f(t)x^2(t) + g(t)x(t) + h(t), f(t)h(t) \neq 0$
exact first-order	$f(t,x)dx + g(t,x)dt = 0, \partial_t f = \partial_x g$
Lagrange	$x(t) = tf(x'(t)) + g(x'(t))$
Bessel	$t^2 x''(t) + tx'(t) + (t^2 - v^2)x(t) = 0$
Legendre	$(1 - t^2)x''(t) - 2tx'(t) + v(v + 1)x(t) = 0$
Mathieu	$x''(t) + (\lambda - 2q\cos(2t))x(t) = 0$

The general solutions of the above ordinary differential equations, if any, are given in Table 2.2, [33, 41, 70], and [99].

Table 2.2: List of solutions to certain first- and second-order differential equations.

Type of equation	General solution
separable variables	$\int (1/g(x))dx = \int f(t)dt + C$
homogeneous	$F(y) = \int dy/(f(y) - y), F(y) - F(y_0) - \ln(t/t_0) = 0, x = t * y$
first-order linear	$x(t) = e^{-\int f(t)dt}(C + \int g(t)e^{\int f(t)dt}dt)$
Bernoulli	substitution $y(t) = (x(t))^{1-a}$
exact first-order	$u(t,x) = \int_{(t_0,x_0)}^{(t,x)} f(t,x)dx + \int_{(t_0,x_0)}^{(t,x)} g(t,x)dt$
Lagrange	$t = h(p, C)$ and $x = f(p)h(p, C) + g(p), g(p) = 0$
Bessel first kind	$x(t) = c_1 J_v(t) + c_2 J_{-v}(t), J_v(t) = \sum_{k=0}^{\infty} \frac{(-1)^k (t/2)^{2k+v}}{k!\Gamma(v+k+1)}, v \in \mathbb{R}\backslash\mathbb{Z}$
Mathieu	$x(t) = c_1 C(a, q, t) + c_2 S(a, q, t),$

https://doi.org/10.1515/9783111411392-003

We introduce several examples of simple ordinary differential equations. Then a systematic study of them starts with Section 2.2.

2.1. In the first problem of this chapter, we assume that the derivative of function $x(t)$ exists and is equal to function $x(t)$. The ordinary differential equation below is said to be of *first-order linear with constant coefficient and homogeneous*,

$$x'(t) = x(t), \quad t \in \mathbb{R}.$$

2.1 We introduce the *Mathematica* code.

```
Clear[t,x] (* Clean the values and definitions *)
x'[t]==x[t];
sol=DSolve[%,x,t]//Flatten (* Solve the equation *)
%%/.%//First (* Check the solution *)
{x→Function[{t},e^tC[1]]}
True
```

The general solution depends on a constant denoted C[1], that is an arbitrary real or complex number. The general solution of this ordinary differential equation is given under one of the next forms.

```
{sol,x[t]/.sol,x[t]/.sol[[1]],x[t]/.First[sol],
First[x[t]/.sol]}
{{x→Function[{t},e^tC[1]]},e^tC[1],e^tC[1],e^tC[1],e^tC[1]}
```

A particular solution is a general solution satisfying some extra condition. If the assumptions of Theorem 1.2 are satisfied, the uniqueness of the solution is guaranteed. There are several ways to get a particular solution. We show one of them.

```
{DSolve[{x'[t]==x[t],x[0]==1},x[t],t][[1,1,2]];
Plot[%,{t,-3,2},PlotStyle→Black,ImageSize→150,
Ticks→{{-3,2},{2,6}}]}
```

See Figure 2.1.

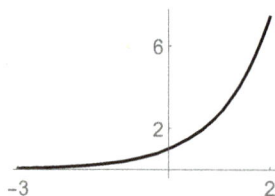

Figure 2.1: Representation of the solution for problem 2.1.

2.2. The next problem (a first-order linear inhomogeneous ordinary differential equation with constant coefficients) exhibits an application of Theorem 1.2. The assumptions of the theorem are satisfied, and thus there exists a unique particular solution passing through the initial point $(t_0, x_0) = (1, 2)$,

$$x'(t) = x(t) + \sin(2t), \quad x(1) = 2, \quad t \in \mathbb{R}.$$

2.2

```
Clear[t,x] (* Clean the values and definitions *)
partsol=DSolve[{x'[t]==x[t]+Sin[2t],x[1]==2},x[t],t]//
Flatten
```
$$\{x[t] \to -\frac{-10e^t - 2e^t \cos[2] + 2e\cos[2t] - e^t \sin[2] + e\sin[2t]}{5e}\}$$

Now one plots the particular solution below:

```
Plot[x[t]/.partsol,{t,0,2},ImageSize→175,
Ticks→{{1,2},{2,5}},PlotStyle→Black,
Epilog→{{Arrowheads[.03],Arrow[{{1,3.3},{1,2.2}}],
Arrow[{{{.7,4.6},{1.65,4.6}},{{.7,4.63},{.5,.95}}}]},
{Dashed,Line[{{0,2},{1,2},{1,0}}]},
Text[Style["initial point",Italic,Black,12],
{1.05,3.8}],Text[Style["particular solution",Italic,
Black,12],{1.18,5.05}],PointSize[.03],Point[{1,2}]}]
```

See Figure 2.2.

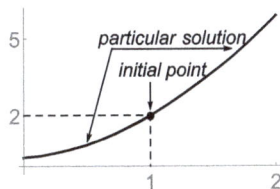

Figure 2.2: Solution for problem 2.2.

2.3. Below there is a first-order second-degree ordinary differential equation with an initial value condition. It exhibits a nonuniqueness behavior at point $(-.5, 1.5)$,

$$x(t)x'^2(t) - (t - x(t))x'(t) - t = 0, \quad x(-.5) = 1.5, \quad t \in \mathbb{R}.$$

This initial value problem appears in [41, p. 66].

2.3

```
Clear[t,x] (* Clean the values and definitions *)
{x[t]x'[t]²-(t-x[t])x'[t]-t==0,x[-.5]==1.5};
sol=DSolve[%,x,t] (* Two solutions are returned, both
as pure functions *)
{{x→Function[t,1.-t]},{x→Function[{t}, √2.+t²]}}
```

We check the solutions and plot them.

```
%% /.% Simplify
{{True,True},{True,True}}
```

```
Plot[x[t]/.sol,{t,-.65,.2},ImageSize→200,Ticks→
{{-.5,0},{1.5}},PlotStyle→Black,
PlotRange→{{-.65,.2},{0,1.65}},
Epilog→{{Dashed,Line[{{-.5,0},{-.5,1.5},{0,1.5}}]},
{PointSize[.03],Black,Point[{-.5,1.5}]},
{Arrowheads[.03],Arrow[{{-.45,.9},{-.49,1.44}}],
Arrow[{{-.13,1.25},{-.22,1.4}}],
Arrow[{{-.13,1.25},{-.19,1.22}}]},
Text[Style["point (-.5,1.5)",Italic,Black,12],
{-.28,.8}],Text[Style["two solutions",Italic,Black,
12],{.04,1.25}]}]
```

See Figure 2.3.

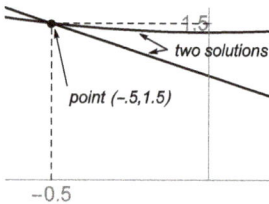

Figure 2.3: Solutions for problem 2.3.

Next, we rotate the figure around the vertical axis.

```
{RevolutionPlot3D[x[t]/.Last[sol],##,Ticks→{{-1,0,1},
{-1,0,1},{1.5,1.7}}],
RevolutionPlot3D[x[t]/.First[sol],##,Ticks→{{-1,0,1},
{-1,0,1},{.4,1.7}}]}&[{t,-1,1},Mesh→2,MeshStyle→
Black,ImageSize→175,PlotStyle→{Gray,Opacity[.2]}]
```

See Figure 2.4.

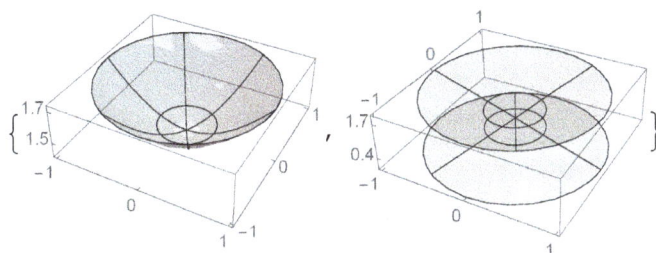

Figure 2.4: Rotations of some solutions for Problem 2.3.

Remark. A similar differential equation is discussed in [94, p. 10]. ☐

2.4. We consider a first-order nonlinear homogeneous ordinary differential equation with an initial condition and write its solution under five forms,

$$x'(t) = x^2(t), \quad x(0) = 1, \quad t \in \mathbb{R}.$$

2.4

```
Clear[t,x] (* Clean the values and definitions *)
eqn={x'[t]==x[t]²,x[0]==1};
listofsols={sol=DSolve[eqn,x,t],Flatten[sol],
x[t]/.sol,x[t]/.First[sol],sol[[1,1,2]][t]}
(* Solve the problem and show the solution
under different forms *)
{{{x→Function[{t},1/(1-t)]}},{x→Function[{t},1/(1-t)]},{1/(1-t)},
1/(1-t),1/(1-t)}
```

The first two pictures are exhibited here.

```
Plot[x[t]/.listofsols[[#]],{t,-.3,3/2},ImageSize→
170,Ticks→{{-.3,1},{-8,8}},TicksStyle→Black,
PlotLabel→Style[listofsols[[#]],10],PlotStyle→Black]
&/@Range[2]
```

See Figure 2.5.

2.5. Let us study the case of a system of linear ordinary differential equations with constant coefficients. Its characteristic equation has distinct real roots,

$$\begin{cases} x'(t) = 3x(t) - y(t) + z(t), \\ y'(t) = -x(t) + 5y(t) - z(t), \qquad t \in \mathbb{R}. \\ z'(t) = x(t) - y(t) + 3z(t), \end{cases}$$

$$\left\{\left\{x \to \left(\{t\} \mapsto \frac{1}{1-t}\right)\right\}\right\} \qquad \left\{x \to \left(\{t\} \mapsto \frac{1}{1-t}\right)\right\}$$

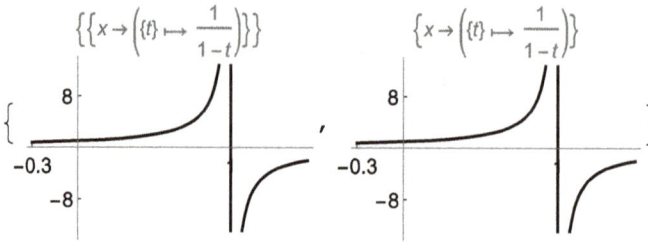

Figure 2.5: Two figures for problem 2.4.

2.5 First approach. We use the power of *Mathematica*.

```
Clear[t,x,y,z]
{x'[t]==3x[t]-y[t]+z[t], (* First equation *)
y'[t]==-x[t]+5y[t]-z[t], (* Second equation *)
z'[t]==x[t]-y[t]+3z[t]}; (* Third equation *)
DSolve[%,{x,y,z},t]//Flatten;
%%/.%//Simplify (* Check the solutions *)
{True,True,True}
```

We set some values of the constants of integration and plot the corresponding particular solutions.

```
{funcn={x[t],y[t],z[t]}/.%%/.{C[1]→1,C[2]→-1,
C[3]→-1}//Simplify;
Plot[funcn,{t,-.22,.42},PlotLegends→"Expressions",
PlotStyle→{Black,{Black,Dashed},{Black,Dotted}},
Ticks→{{{-.2,"-.2"},{.2,".2"},{.4,".4"}},{-5,5}},
TicksStyle→Directive[Black,13],ImageSize→175]}
```

See Figure 2.6.

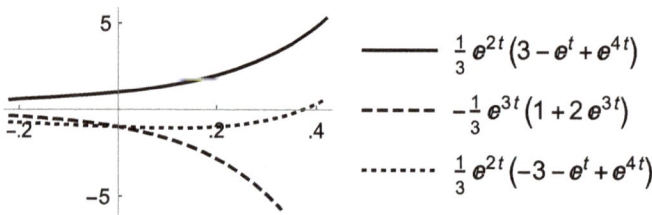

Figure 2.6: Figure for problem 2.5.

$$\frac{1}{3}e^{2t}\left(3 - e^{t} + e^{4t}\right)$$

$$-\frac{1}{3}e^{3t}\left(1 + 2\,e^{3t}\right)$$

$$\frac{1}{3}e^{2t}\left(-3 - e^{t} + e^{4t}\right)$$

Now we find the limits of the vector solution at $-\infty$ and $+\infty$.

```
Limit[funcn,t→#]&/@{-∞, ∞}
{{0,0,0},{∞, -∞, ∞}}
```

2.6. Let us examine the next system of first-order nonlinear ordinary differential equations,

$$\begin{cases} x'(t) = x^2(t) + y^2(t), \\ y'(t) = 2\,x(t)\,y(t), \end{cases} \quad t \in \mathbb{R}.$$

2.6

```
Clear[t,x,y]
{x'[t]==x[t]²+y[t]²,y'[t]==2x[t]y[t]}; (* List of
equations *)
DSolve[%,{x,y},t]; (* Solve the system of equations *)
%%/.%/.{t→0,C[1]→0,C[2]→0} (* Set a particular
solution of the system and check it *)
{{False,True},{True,True}}
```

Remark. We found a point t and some integration constants $C[1]$ and $C[2]$ so that the first equation is not satisfied. We conclude the entire system of equations is not satisfied by the returned solutions. Thus, the solutions yielded above are not correct. □

The correct solutions are given below:

```
Clear[t,a,b]
x[t_]:=½(1/(a-t) + 1/(b-t)); (* The first solution *)
y[t_]:=½(1/(a-t) + 1/(t-b)); (* The second solution *)
{x'[t]==x[t]²+y[t]²,y'[t]==2x[t]y[t]}//Simplify
(* Check the solutions *)
{True,True} (* The equations are satisfied *)
```

We select a particular solution,

```
func={x[t],y[t]}/.{a→1,b→2}; (* Set a particular
solution *)
```

and plot it below:

```
Plot[func,{t,0,2.75},ImageSize→200,PlotStyle→{Black,
{Dashed,Black}},Ticks→{Range[0,2.5,.5],{-7,-3,3,7}},
PlotLegends→"Expressions"]
```

See Figure 2.7.

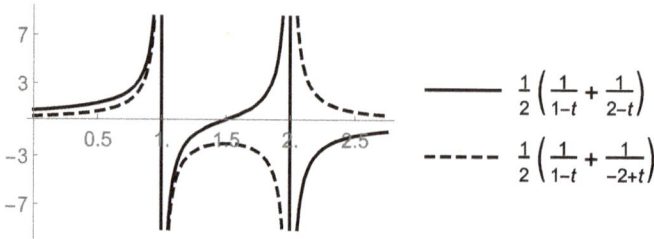

$$\frac{1}{2}\left(\frac{1}{1-t} + \frac{1}{2-t}\right)$$

$$\frac{1}{2}\left(\frac{1}{1-t} + \frac{1}{-2+t}\right)$$

Figure 2.7: Figure for problem 2.6.

2.7. The next nonlinear ordinary differential equation of the first order shows that in certain cases its solution is not continuable,

$$x'(t) = 1 + x^2(t), \quad x(0) = 0, \quad t \in \mathbb{R}.$$

2.7 First approach.

```
Clear[t,x]
sol=DSolve[{x'[t]==1+x[t]²,x[0]==0},x,t]//First
{x→Function[{t},Tan[t]]} (* This function is defined
on the real axis except the points (2k+1)Pi/2 with k
an integer. Thus it is not continuable on the entire
real axis *)
```

One introduces the graph of the tangent function on interval $[0, 2\pi]$ and its rotation.

```
{Plot[x[t]/.sol,{t,0,2Pi},Ticks→{Pi/2Range[4],None},
PlotStyle→Black,ImageSize→175],
RevolutionPlot3D[x[t]/.sol,{t,0,2Pi},Mesh→{7,2},
ImageSize→140,MeshStyle→{Dashed,Dotted},
PlotStyle→Directive[Black,Opacity[.03]],
PlotPoints→35,RevolutionAxis→"X"]}
```

See Figure 2.8.

Second approach. One separates the variables and supposes that all operations are valid. Then

$$\frac{dx(t)}{1 + x^2(t)} = dt \implies \int \frac{dx(t)}{1 + x^2(t)} = \int dt \implies \arctan(x(t)) = t + C$$

$$x(t) = \tan(t + C) \implies x(t) = \tan(t) \text{ because } x(0) = 0.$$

We obtained the same solution as before.

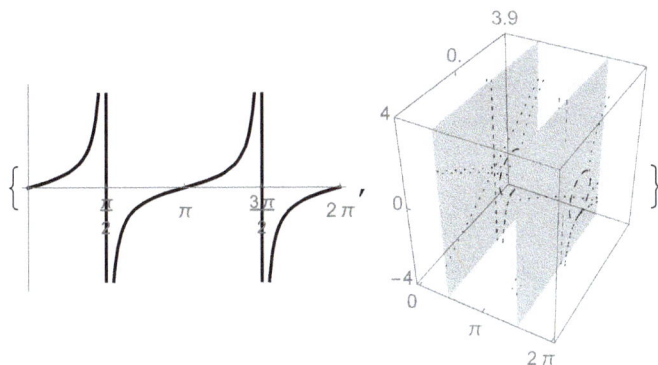

2.8. We bring into discussion a simple nonlinear differential equation, its solution depending on a parameter introduced by the first-order derivative,

$$x'(t) = 1 + x^2(t), \quad x'(0) = p, \quad t, p \in \mathbb{R}.$$

2.8

```
Clear[t,x,p,q]
sol=DSolve[{x'[t]==1+x[t]²,x'[0]==p},x,t]//Simplify;
(* Analytical solution *)
Flatten[ReIm[x[t]/.sol]]/.p→q
{Re[Tan[t-ArcSec[-√q]]],Im[Tan[t-ArcSec[-√q]]],
Re[Tan[t+ArcSec[-√q]]],Im[Tan[t+ArcSec[-√q]]]}
Manipulate[Plot[Tooltip[%],{t,-3π/2,3π/2},
(* Plot the solutions *)
Ticks→{π/2Range[-3,3],Range[-3,3,2]},
ImageSize→180,FillingStyle→Opacity[.1],Filling→Axis,
PlotStyle→{Black,{Black,Dashed},{Black,Dotted}},
{q,-10,5}],SaveDefinitions→True]
```

See Figure 2.9.

2.9. This is a nonlinear first-order ordinary differential equation,

$$t^2(1 - t^2)x'(t) = (t - 3t^3 - x(t))x(t), \quad t \in \mathbb{R}.$$

2.9

```
Clear[t,x]
sol=DSolve[t²(1-t²)x'[t]==(t-3t³-x[t])x[t],x[t],t]
//Flatten (* Solve the equation *)
{x[t]→ (t-t³)/(C[1]+Log[t])}
```

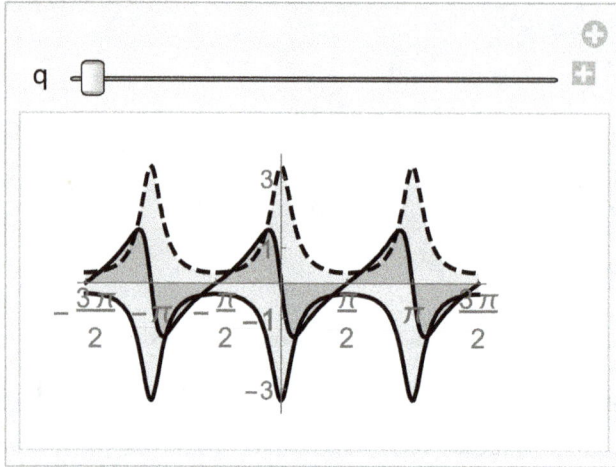

Figure 2.9: Dynamical plot of the solution for problem 2.8.

We set a particular solution and plot it.

```
{Plot[x[t]/.sol/.C[1]→-1,{t,1,5},ImageSize→175,
PlotStyle→Black,Ticks→{Range[2,5],200Range[-2,1]}],
(* Rotating this picture around the Oy axis,
we get the other figure. Here we consider C[1]=-1.*)
RevolutionPlot3D[x[t]/.sol/.C[1]→-1,{t,0,5},Mesh→3,
PlotStyle→{Gray,Opacity[.15]},ImageSize→200,
Ticks→{5Range[-1,1],5Range[-1,1],100Range[-4,2,3]}]}
```

See Figure 2.10.

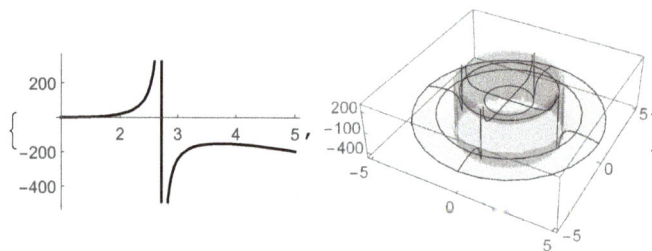

Figure 2.10: Figures for problem 2.9.

2.10. The next nonlinear first-order ordinary differential equation has only an implicit solution,

$$(x(t) + t - 1)x'(t) - x(t) + 2t + 3 = 0, \quad t \in \mathbb{R}.$$

2.10

```
Clear[t,x]
DSolve[(x[t]+t-1)x'[t]-x[t]+2t+3==0,x[t],t];
functx=C[1]/.Solve[First[%],C[1]]/.x[t]→x
{2/3(Sqrt[2]ArcTan[(3+2t-x)/(Sqrt[2](-1+t+x))]
-2Log[2+3t]-Log[(11+8t+6t²-10x+3x²)/(2+3t)²])}
```

Here, we plot the solution.

```
Plot3D[functx,{t,0,10},{x,1,5},ImageSize→175,
Mesh→{{.4,.8,1.3,1.8, 5},{1.3,1.8,2.3,3}},
Ticks→{5{0,1,2},{1,5},{-3.68,0}},
PlotStyle→{Gray,Opacity[.15]}]
```

See Figure 2.11.

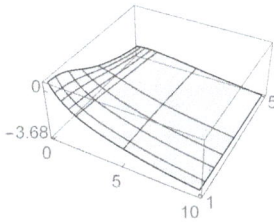

Figure 2.11: Figure for problem 2.10.

2.2 Planar phase portrait

A common method to study the solutions of a planar ordinary differential equation, if any, relies on the inspection of the corresponding phase portrait [17, p. 20], [15], and [32].

2.11. We introduce some problems so that their right-hand sides are polynomials of degree at most 2:

$$x'(t) = 1, \quad x'(t) = t, \quad x'(t) = -x(t), \quad x'(t) = t^2, \quad x'(t) = -x^2(t).$$

2.11

```
Clear[t,x]
With[{l=4},
StreamPlot[{1,#},{t,-1,1},{x,-1,1},ImageSize→130,
StreamStyle→Black]&/@{1,t,-x,t²,-x²}]
```

See Figure 2.12.

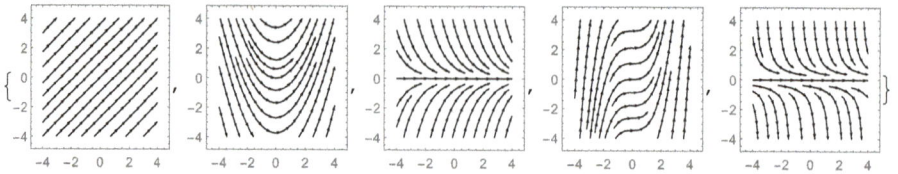

Figure 2.12: Figures for problem 2.11.

2.12. We introduce some problems so that their right-hand sides are simple rational functions:

$$x'(t) = 2t - x(t), \quad x'(t) = \frac{x(t)}{t^2 - 1}, \quad x'(t) = \frac{4 - t^2 - x^2(t)}{t(x - 1)}.$$

2.12

```
Clear[t,x]
With[{l=4},
StreamPlot[{1,#},{t,-1,1},{x,-1,1},ImageSize→130,
StreamStyle→Black]&/@{2t-x, x/(t²-1), (4-t²-x²)/(t(x(t)-1))}]
```

See Figure 2.13.

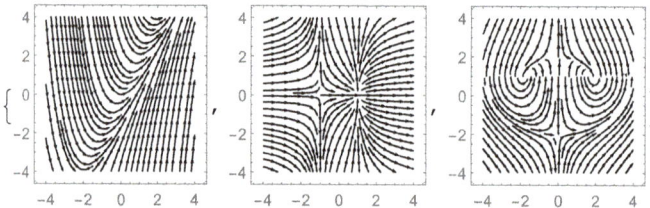

Figure 2.13: Figures for problem 2.12.

2.13. We introduce some problems so that their right-hand sides are simple trigonometric functions:

$$x'(t) = \sin(t\,x(t)), \quad x'(t) = t\cos(x(t)) - 1, \quad x'(t) = \sin(x^3(t) - t^3).$$

2.13

```
Clear[t,x]
With[{l=4},
StreamPlot[{1,#},{t,-1,1},{x,-1,1},ImageSize→130,
StreamPoints→Medium,StreamStyle→Green]&/@
{Sin[t x],t Cos[x]-1,Sin[x³-t³]}]
```

See Figure 2.14.

Figure 2.14: Figures for problem 2.13.

Below we introduce several problems and use both `StreamPlot` and `VectorPlot` built-in functions.

2.14. The solution of this ordinary differential equation with separable variables and with an initial condition is a hyperbola:

$$x'(t) = \frac{-x(t)}{t}, \quad x(.5) = .5, \quad t \in \mathbb{R} \setminus \{0\}.$$

2.14

```
Clear[t,x]
vt=t;vx=-x;
sol=DSolve[{x'[t]==-x[t]/vt,x[.5]==.5},x,t]//Flatten
(* Solution as a pure function *)
{x→Function[{t}, 0.25/t ]}

{Show[##,StreamPlot[{vt,vx},{t,-1,1},{x,-1,1},
(* Graphs *)
StreamPoints→Coarse,StreamScale→.25,
StreamColorFunction→Black,RegionFunction→
Function[{t,x},t x>0&&t x<.5&&t x>.08]]],
Show[##,VectorPlot[{vt,vx},{t,-1,1},{x,-1,1},
RegionFunction→Function[{t,x},t x>0]]]}&[
Plot[x[t]/.sol,{t,-1,1},PlotStyle→Black,
AspectRatio→1,PlotRange→Outer[Times,{1,1},{-1,1}],
ImageSize→175]]}}
```

See Figure 2.15.

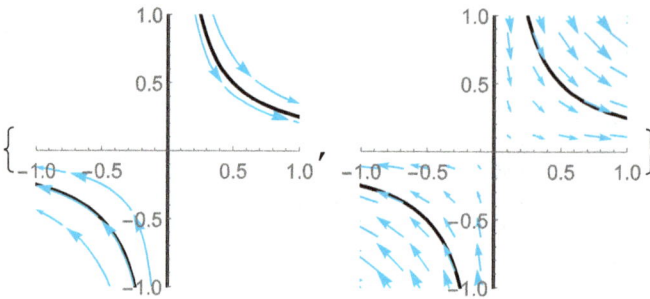

Figure 2.15: Figures for problem 2.14.

2.15. This problem studies a second-order nonlinear ordinary differential equation with initial conditions:

$$x'' + \sin(x) = 0, \quad x(1) = x'(1) = 1, \quad t \in \mathbb{R}.$$

2.15

```
Clear[t,x,y]
vx=y;vy=Sin[x];
sol=DSolve[{x''[t]+Sin[x[t]]==0,x[1]==x'[1]==1},x,t]
//Flatten; (* Solution as a pure function *)
{x→Function[{t},2JacobiAmplitude[½(t√3-2Cos[1]
+ (√3-2Cos[1](3-2Cos[1]-2√3-2Cos[1]EllipticF[½,4/(3-2Cos[1])]))/(-3+2Cos[1])), 4/(3-2Cos[1])]]}
```

The Jacobi amplitude function is defined in Section 1.10.
    Two figures are now introduced.

```
{Show[#,StreamPlot[{vx,-vy},{x,-2,2},{y,-2,2},
(* Graphs *)
StreamPoints→Coarse,StreamScale→.2,
StreamColorFunction→Function[RGBColor[0.,0.,0.]]]],
Show[#,VectorPlot[{vx,-vy},{x,-2,2},{y,-2,2},
VectorPoints→10,VectorScale→.1,
VectorColorFunction→Function[RGBColor[0.,0.,0.]]]}
&/@{ParametricPlot[{x[t],x'[t]}/.sol,{t,-4,4},
PlotStyle→{Black,Thick},AspectRatio→1,
PlotRange→Outer[Times,{1,1},{-2,2}],ImageSize→175]}
//Flatten
```

See Figure 2.16.

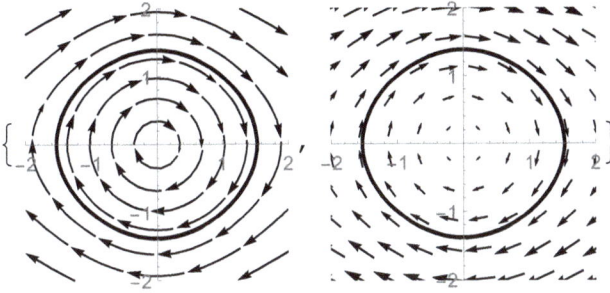

Figure 2.16: Figures for problem 2.15.

2.16. Here, we collect other problems of the same kind.

2.16

```
Clear[t,x]
streams={{1,x²},{1,1+t x},{1,√t²+x²},{x,-Sin[t]},
{t(x-1),4-t²-x²}};  (* Right-hand sides *)
arrows={"Arrow","ArrowArrow","Dart","Toothpick",
"PinDart","DoubleDart","Line"};
icons=Table[str→Image[StreamPlot[str,{t,-3,3},
{x,-3,3},Axes→False,Frame→False,StreamPoints→Coarse,
StreamStyle→"Line"],ImageSize→42],{str,streams}];
Manipulate[StreamPlot[stream,{t,-3,3},{x,-3,3},
(* Graphs *)
StreamPoints→Coarse,ImageSize→140,
StreamStyle→arrow,StreamScale→scale,
StreamColorFunction→color],{arrow,arrows},
{color,ColorData["Gradients"]},{{scale,.1},.1,.2,.02},
{{stream,icons[[1,1]]},icons,Setter}]
```

See Figure 2.17.

2.3 Solid phase portrait

One can study the solutions of some systems of spatial ordinary differential equations by inspecting the corresponding phase portrait. Below we introduce several cases.

2.17. We present a system of first-order linear ordinary differential equations in $\mathbb{R}^3$ with initial conditions,

$$\begin{cases} x'(t) = x(t), \\ y'(t) = y(t), \\ z'(t) = z(t), \end{cases} \quad x(1) = 0, \quad y(1) = z(1) = 1, \quad t \in \mathbb{R}.$$

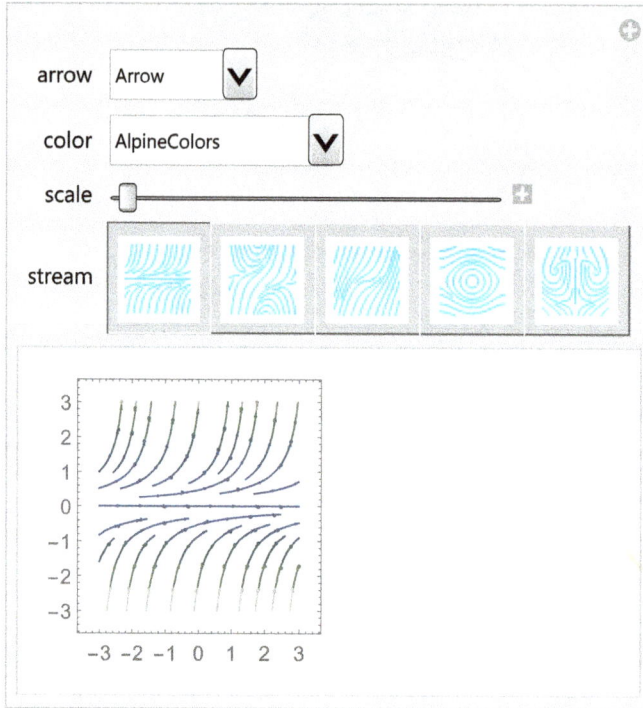

Figure 2.17: Figures for problem 2.16.

2.17

```
Clear[t,x,y,z]
vx=x;vy=y;vz=z;
sol=DSolve[{x'[t]==vx[t],y'[t]==vy[t],z'[t]==vz[t],
x[1]==0,y[1]==z[1]==1},{x,y,z},t]//Flatten
(* Solutions as pure functions *)
{x→Function[{t},0], y→Function[{t},e⁻¹⁺ᵗ],
z→Function[{t},e⁻¹⁺ᵗ]}
```

The graph is introduced below.

```
Show[VectorPlot3D[{vx,vy,vz},{x,-1,1},{y,-1,1},
{z,-1,1},VectorColorFunction→Function[RGB[1.,1.,1.]],
VectorPoints→Coarse,VectorScale→.15,
ImageSize→150,Ticks→Outer[Times,{1,1,1},{-1,0,1}]],
ParametricPlot3D[{x[t],y[t],z[t]}/.sol,{t,-1,1},
PlotStyle→{BlackThickness[.018]},AspectRatio→1],
Graphics3D[{Arrowheads[.01],Arrow[{{x[#],y[#],
z[#]}/.sol,{x[1],y[1],z[1]}/.sol}]}]&[1-.0001]]
```

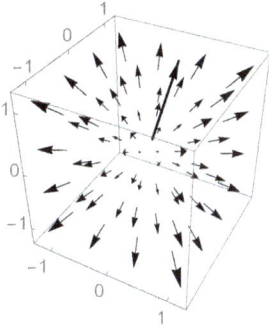

Figure 2.18: Figure for problem 2.17.

See Figure 2.18.

2.18. The first and the third components of the solution to the initial value problem that follow tend to $-\infty$ whenever $t \to -\infty$ and to 0 as $t \to \infty$,

$$\begin{cases} x'(t) = 3x(t) - y(t) + z(t), & x(1) = 1, \\ y'(t) = -x(t) + 5y(t) - z(t), & y(1) = 1, \\ z'(t) = x(t) - y(t) + 3z(t), & z(1) = -1, \end{cases} \quad t \in \mathbb{R}.$$

2.18

```
Clear[t,x,y,z]
v={3x-y+z,-x+5y-z,x-y+3z};
sol=DSolve[{x'[t]==3x[t]-y[t]+z[t],
y'[t]==-x[t]+5y[t]-z[t],z'[t]==x[t]-y[t]+3z[t],
x[1]==y[1]==-z[1]==1},{x,y,z},t]//Flatten
(* Solutions as pure functions *)
{{x→Function[t,⅓e⁻⁶⁺²ᵗ(3e⁴-e⁴ᵗ+e³⁺ᵗ)],
y→Function[t,⅓e⁻⁶⁺³ᵗ(e³+2e³ᵗ)],
z→Function[t,-⅓e⁻⁶⁺²ᵗ(3e⁴+e⁴ᵗ-e³⁺ᵗ)]}}
```

The graph is introduced below.

```
Show[VectorPlot3D[v,{x,-1,1},{y,-1,1},{z,-1,1},
VectorColorFunction→RGB[0.,0.,0.],ImageSize→150,
Ticks→Outer[Times,{1,1,1},{-1,0,1}],
VectorPoints→Coarse],
ParametricPlot3D[{x[t],y[t],z[t]}/.sol,{t,-1,1},
PlotStyle→{Black,Thick},AspectRatio→1],
Graphics3D[{Arrowheads[.07],Arrow[{{x[#],y[#],z[#]}
/.sol//First,{x[1],y[1],z[1]}/.sol//First}]}]&[1-.001]]
```

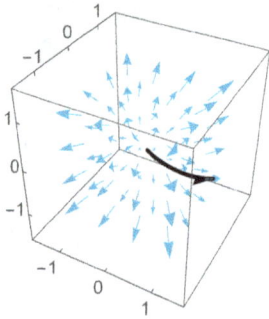

Figure 2.19: Figure 2.19: Figure for problem 2.18.

See Figure 2.19.

```
Limit[{x[t],y[t],z[t]}/.sol,t→#]&/@{-∞, ∞}
(* Limits of the solutions at -∞ and +∞ *)
{{{0,0,0}},{{-∞, ∞, -∞}}}
```

2.19. Here, we collect certain problems of the same kind.

2.19

```
Clear[x,y,z]
streams={{x,y,z},{-x,-y,-z},{3x-y+z,-x+5y-z,x-y+3z},
{8y,-2z,2x+8y-2z}};
arrows={"Arrow3D","Segment"};
icons=Table[stream→Image[VectorPlot3D[stream,{x,-1,1},
{y,-1,1},{z,-1,1},Axes→False,VectorPoints→Coarse],
VectorScale→.15,ImageSize→55],{stream,streams}];
```

The graph follows in the sequel.

```
Manipulate[VectorPlot3D[stream,{x,-1,1},{y,-1,1},
{z,-1,1},ImageSize→140,VectorPoints→Coarse,
VectorScale→.15,Ticks→Outer[Times,{1,1,1},{-1,0,1}],
VectorStyle→arrow,VectorColorFunction→color],
{arrow,arrows},{color,Gray(*ColorData["Aquamarine"]*)},
{{stream,icons[[1,1]]},icons,Setter}]
```

See Figure 2.20.

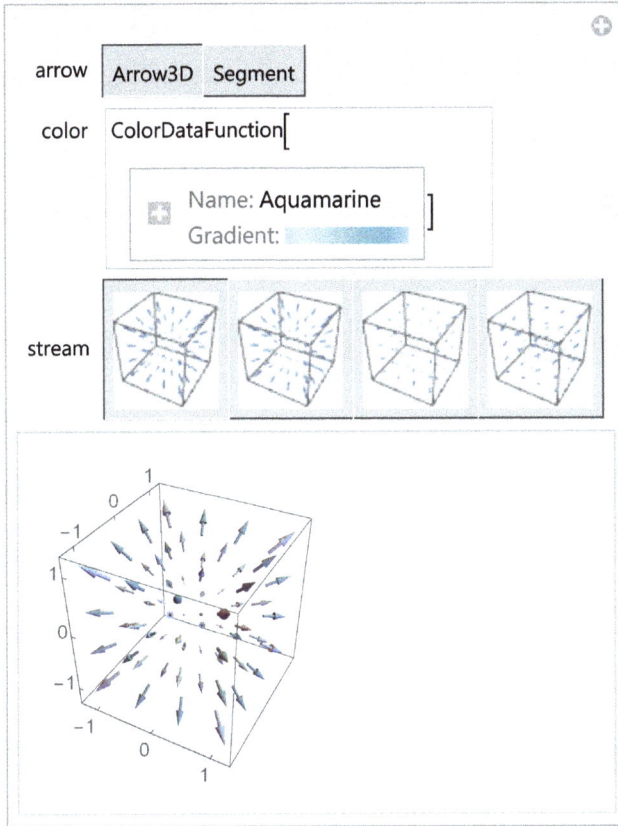

Figure 2.20: Figure for problem 2.19.

2.4 Straight integration of first-order derivatives

We encounter straight integration whenever the ordinary differential equation is a derivative.

2.20. Our first problem of straight integration follows right now,

$$x'(t) = t^2 \exp(t^2) + \frac{1}{\ln(t)}, \quad t > 0, \quad t \neq 1.$$

2.20

```
Clear[t,x]
sol=DSolve[x'[t]==t^2 E^{t^2}+1/Log[t],x[t],t]//Flatten
(* Solve the equation *)
{x[t]→ e^{t^2}/t +C[1]-1/4 √π Erfi[t]+LogIntegral[t]}
```

Several solutions corresponding to different values of the constant of integration are plotted together.

```
tab=Table[x[t]/.sol/.C[1]→k,{k,3Range[-1,1]}];
Plot[Evaluate[tab],{t,0,1.5},ImageSize→160,
Ticks→{.5Range[3],3Range[-3,2]},
PlotLegends→"Expressions",LabelStyle→10,
PlotStyle→{Black,{Black,Dashed},{Black,Dotted}}]
```

See Figure 2.21.

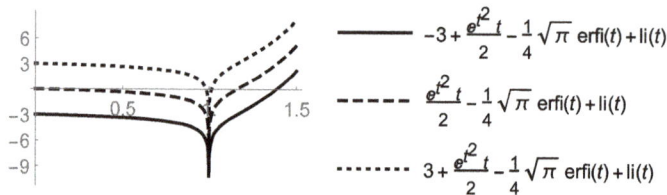

Figure 2.21: Figure for problem 2.20.

2.21. Another problem of the same kind of differential equation is introduced,

$$x'(t) = t^2 \sin(t) + \sqrt{1 + t^2}, \quad t \in \mathbb{R}.$$

2.21

```
Clear[t,x,p]
DSolve[x'[t]==t²Sin[t]+√1+t²,x[t],t]//Flatten
(* Solve the equation *)
Table[x[t]/.%/.C[1]→k,{k,-100,200,150}];
{x[t]→ ½t√1+t²+ ArcSinh[t]/2 +C[1]-(-2+t²)Cos[t]+2tSin[t]}
```

The code and the graph follow at once.

```
Plot[Evaluate[%],{t,0,18},ImageSize→150,
AxesOrigin→{0,0},Ticks→{2{5,9},100{-1,2,5.8}},
PlotLegends→Placed["Expressions",Above],
PlotStyle→{Black,{Black,Dashed},{Black,Dotted}},
LabelStyle→10]
```

See Figure 2.22.

$$-100 + \frac{1}{2}t\sqrt{1+t^2} + \frac{1}{2}\sinh^{-1}(t) - \left(-2+t^2\right)\cos(t) + 2t\sin(t)$$

$$50 + \frac{1}{2}t\sqrt{1+t^2} + \frac{1}{2}\sinh^{-1}(t) - \left(-2+t^2\right)\cos(t) + 2t\sin(t)$$

$$200 + \frac{1}{2}t\sqrt{1+t^2} + \frac{1}{2}\sinh^{-1}(t) - \left(-2+t^2\right)\cos(t) + 2t\sin(t)$$

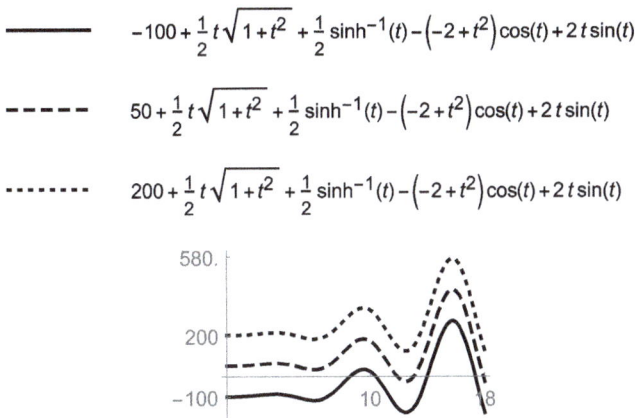

Figure 2.22: Figure for problem 2.21.

2.22. We solve the upcoming initial value problem. The right-hand side of the differential equation is a derivative,

$$\begin{cases} x'(t) = \sqrt{1 + \tan(t) + \tan^2(t)}, \\ x(0) = -\frac{1}{\sqrt{2}}\arctan(2\sqrt{2}), \end{cases} \quad t \in \mathbb{R} \setminus \left\{ (2k+1)\frac{\pi}{2} \mid k \in \mathbb{Z} \right\}.$$

2.22

```
Clear[t,x]
sol=DSolve[{x'[t]==Sqrt[1+Tan[t]+Tan[t]^2],
x[0]==-1/√2 ArcTan[2√2]},x,t]//Simplify;
(* Solution of the problem *)
```

The graph of the solution together with the rotated figures come next.

```
{Plot[x[t]/.sol,{t,0,1.4},ImageSize→150,
PlotStyle→Black,Ticks→{.4Range[1,3],.5Range[-1,4]}],
{RevolutionPlot3D[x[t]/.First[sol],##,ImageSize→160,
Ticks→Outer[Times,{1,1,.7},{-1,0,1}]],
RevolutionPlot3D[x[t]/.First[sol],##,ImageSize→140,
Ticks→Outer[Times,{1,.8,.8},{-1,0,1}],
RevolutionAxis→"X"]}&[{t,0,1.4},Mesh→2,MeshStyle→
Black,PlotStyle→Opacity[.2]]}
```

See Figure 2.23.

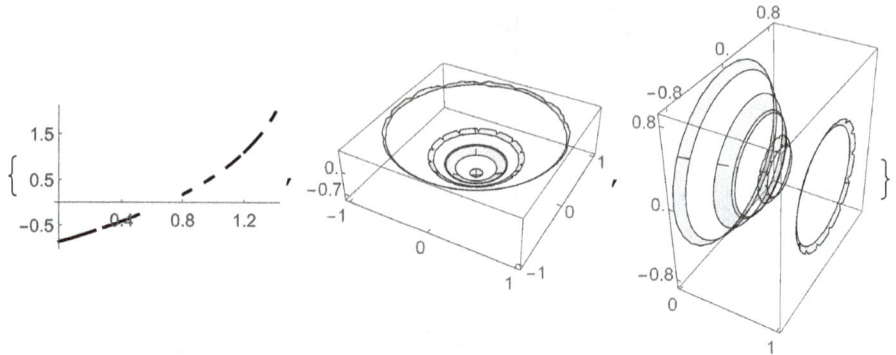

Figure 2.23: Figure of problem 2.22.

2.5 Differential equations with separable variables

2.23. Another problem of an ordinary differential equation with separable variables is exposed below:

$$x'(t) = \frac{t^2 \exp(x(t))}{\sqrt{3 - t^2}}, \quad -\sqrt{3} < t < \sqrt{3}.$$

2.23

```
Clear[t,x]
sol=DSolve[x'[t]==t²Exp[x[t]]/√3-t²,x,t]//Flatten (* Solve the
equation *)
{x→Function[t,-Log[½t √3-t²-3/2ArcSin[t/√3]-C[1]]]}
```

The graph is given below for two values of the constant of integration.

```
Plot[Evaluate[Table[x[t]/.%/.C[1]→k,
{k,2Range[-2,2]}]],{t,-√3, √3},ImageSize→150,
PlotStyle→{Black,{Black,Dashed},{Black,Dotted}},
PlotRange→{{-√3, √3},{-2,4}},LabelStyle→10,
Ticks→{{-√3, √3},Range[-1,3,2]},
PlotLegends→Placed["Expressions",Right]]
```

See Figure 2.24.

Remark. One remarks that only four graphics appeared because one particular solution has complex values. This fact is made clear by the figure below because the particular solution is the logarithm of a negative function. ☐

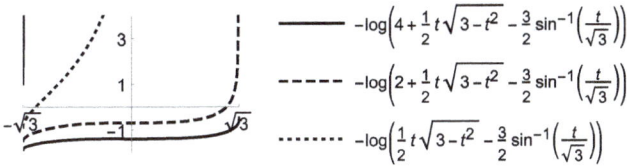

$$-\log\!\left(4+\tfrac12 t\sqrt{3-t^2}\,\right)-\tfrac32\sin^{-1}\!\left(\tfrac{t}{\sqrt3}\right)$$

$$-\log\!\left(2+\tfrac12 t\sqrt{3-t^2}\,\right)-\tfrac32\sin^{-1}\!\left(\tfrac{t}{\sqrt3}\right)$$

$$-\log\!\left(\tfrac12 t\sqrt{3-t^2}\,\right)-\tfrac32\sin^{-1}\!\left(\tfrac{t}{\sqrt3}\right)$$

Figure 2.24: Figure for problem 2.23.

```
Plot[{-2+½ t√3-t²-³⁄₂ ArcSin[t/√3],-4+½ t√3-t²-³⁄₂ArcSin[t/√3]},
{t,-√3, √3}, ImageSize→150],
PlotStyle→{Black,{Black,Dashed},{Black,Dotted}},
Ticks→{{-√3, √3},-2Range[3]},LabelStyle→{Black,[10]}
```

See Figure 2.25.

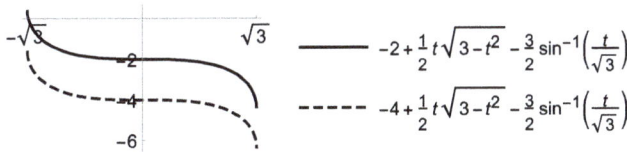

$$-2+\tfrac12 t\sqrt{3-t^2}-\tfrac32\sin^{-1}\!\left(\tfrac{t}{\sqrt3}\right)$$

$$-4+\tfrac12 t\sqrt{3-t^2}-\tfrac32\sin^{-1}\!\left(\tfrac{t}{\sqrt3}\right)$$

Figure 2.25: Negative functions in problem 2.23.

Remark. Other differential equations with separable variables may be found in Section 2.1. ☐

2.6 Homogeneous differential equations

2.24. A first example of a first-order homogeneous ordinary differential equation follows:

$$x'(t) = -\frac{t^2 - 3x^2(t)}{t\,x(t)}, \quad t \in \mathbb{R}\setminus\{0\}.$$

2.24

```
Clear[t,x]
eqn=x'[t]==-t²-3x[t]²/tx[t]; (* The equation *)
sol=DSolve[%,x,t] (* Solve the equation *)
{{x→Function[{t}, -√(t²+2t⁶C[1])/√2]},
{x→Function[{t}, √(t²+2t⁶C[1])/√2]}}

tab=Table[x[t]/.sol/.C[1]→k,{k,{0,3}}]; (* Set four
solutions *)
```

This plots both branches together, showing the complete integral curves $x^2(t)=$ C[1]t^6+t^2/2 for several values of theC[1] constant.

```
Plot[Evaluate[tab],{t,0,1.55},ImageSize→150,Ticks→
{.5Range[3],{-3,3}},TicksStyle→Directive[12],
PlotStyle→{Black,{Black,Dashed},{Black,Dotted},
{Black,DotDashed}},PlotLegends→"Expressions",
LabelStyle→10]
```

See Figure 2.26.

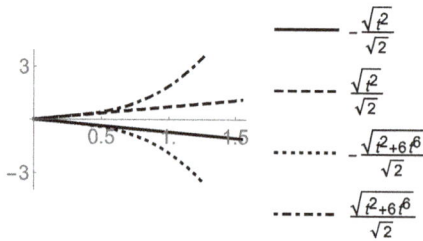

Figure 2.26: Figure for problem 2.24.

We rotate the x(t)/.sol curve around the vertical axis for C[1] equal to 1 and 20, respectively, and put one surface over the other.

```
{x₁,x₂}=x[t]/.sol;
```

```
{RevolutionPlot3D[x₂/.C[1]→#,{t,0,3},Mesh→3,
Ticks→{3Range[-1,1],3Range[-1,1],Automatic},
PlotStyle→Directive[Black,Opacity[.15]],
ImageSize→150]&/@{1,20},
Show[RevolutionPlot3D[x₂/.C[1]→20,{t,0,3},
PlotStyle→Directive[Black,Opacity[.05]],Mesh→3,
ImageSize→175],
RevolutionPlot3D[x₂/.C[1]→1,{t,0,3},ImageSize→175,
PlotStyle→Directive[Black,Opacity[.15]],Mesh→3,
Ticks→{3Range[-1,1],3Range[-1,1],40Range[0,2]}]]}
```

See Figure 2.27.

2.25. Another problem of a first-order homogeneous ordinary differential equation follows [64, p. 16]:

$$x'(t) = \frac{2t\,x(t)}{3t^2 - x^2(t)}, \qquad x(0) = 1, \quad t \in \mathbb{R}.$$

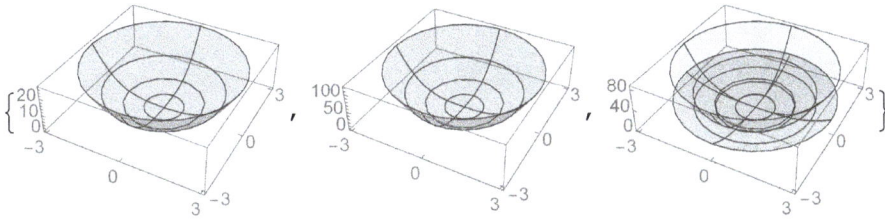

Figure 2.27: Rotate the particular solution for problem 2.24 and overlap the two surfaces.

2.25

```
Clear[t,x,c]
```

We solve the initial value problem.

```
sol=DSolve[{x'[t]==2t x[t]/3t²-x[t]²,x[0]==1},x[t],t]//Flatten
```

$\{x[t] \rightarrow (2\,2^{1/3} + 2(2-27t^2 + 3\sqrt{3}\sqrt{t^2(-4+27t^2)})^{1/3}$
$+2^{2/3}(2-27t^2 + 3\sqrt{3}\sqrt{t^2(-4+27t^2)})^{2/3})/$
$(6(2-27t^2 + 3\sqrt{3}\sqrt{t^2(-4+27t^2)})^{1/3})\}$

This solution is defined on interval $[-.385, .385]$ and visualized below.

```
Plot[x[t]/.sol,{t,-.385,.385},ImageSize→160,
Ticks→{.2 Range[-2,2],Range[.7,1.,.1]},
PlotStyle→Black]
```

See Figure 2.28.

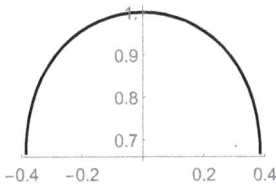

Figure 2.28: A solution for problem 2.25.

Remark. The particular solution is defined only on the interval $[-.385, .385]$ because $\sqrt[3]{\cdot}$ in *Mathematica* 10.3 for negative values of its argument returns complex numbers. We mean that $\sqrt[3]{-1} = .5 + .866025i$. ☐

Now we use the power of the DSolve built-in function to solve the equation.

```
Assuming[t∈Reals,
DSolve[x'[t]==2t x[t]/(3t²-x[t]²),x[t],t]]//Flatten
//FullSimplify;
```

Three solutions are returned. Let us see what they look like.

```
%/.C[1]→0;
tab=Table[x[t]/.%[[k]],k,Length[%]]
```

$$\left\{\frac{1}{3}\left(-1+\frac{2^{1/3}}{(-2+27t^2+3\sqrt{-12t^2+81t^4})^{1/3}}+\frac{(-2+27t^2+3\sqrt{-12t^2+81t^4})^{1/3}}{2^{1/3}}\right),\right.$$

$$\frac{1}{6}\left(-2-\frac{2(-2)^{1/3}}{(-2+27t^2+3\sqrt{-12t^2+81t^4})^{1/3}}+(-2)^{2/3}(-2+27t^2\right.$$

$$\left.+3\sqrt{-12t^2+81t^4})^{1/3}\right),$$

$$-\frac{1}{3}(-\frac{1}{2})^{1/3}(-2+27t^2+3\sqrt{-12t^2+81t^4})^{1/3}$$

$$\left.+\frac{1}{3}\left(-1+\frac{\text{Root}[-2+\#1^3\&,3]}{(-2+27t^2+3\sqrt{-12t^2+81t^4})^{1/3}}\right)\right\}$$

The graph is composed by three arcs.

```
Plot[tab,{t,-1,1},AspectRatio→1,ImageSize→175,
PlotStyle→{Black,{Black,Dashed},{Black,Dotted}},
Ticks→{{-1,1},{-1,.5}}]
```

See Figure 2.29.

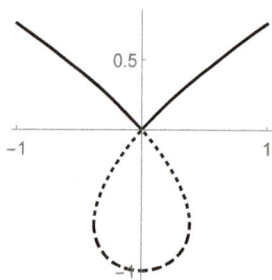

Figure 2.29: Three solutions for problem 2.25.

Indeed, the graph is composed of three arcs, each arc corresponding to a solution supplied by DSolve. We emphasize that the arcs correspond to C[1]=0.

```
Table[(z²-t²==c z³)/.z→tab[[k]]/.t→0//N//Chop,
{k,Length[tab]}]
{True,1.== -1.c,True}
```

Remark. We remark that the first and the last solution satisfy the functional relation $x^2(t) - t^2 = c x^3(t)$ for any real c, whereas the second solution satisfies the same functional relation whenever $c = -1$. ☐

2.26. A new example of a first-order homogeneous ordinary differential equation follows [64, p. 16]:

$$t x^3 d x + (t^4 - 3x^4)d t = 0, \quad t, x \in \mathbb{R}.$$

2.26

```
Clear[t,x]
```

We use the power of DSolve for finding the solutions, assuming that x is a function of t.

```
sol/.C[1]→1;
tab=Table[x[t]/.%[[k]],{k,Length[%]}]; sol/.C[1]→-1;
tab1=Table[x[t]/.%[[k]],{k,Length[%]}]
```

Four solutions are returned. For positive C[1], two of them have real values and the
others have complex values. For negative C[1], two of them have real values and the
others have complex values. The figures follow.

```
{Plot[tab,{t,-2,2},AspectRatio→1,ImageSize→150,
PlotStyle→{Black,{Black,Dashed}},PlotLabel→Style[
"C[1]>0",12,Black],Ticks→{Range[-2,2],4Range[-2,2]}],
Plot[tab,{t,-⁸√1/2, ⁸√1/2},AspectRatio→1,ImageSize→150,
PlotStyle→{Black,{Black,Dashed}},PlotLabel→Style[
"C[1]<0",12,Black],Ticks→{{-⁸√1/2, ⁸√1/2},.65{-1,1}}]]}
```

See Figure 2.30.

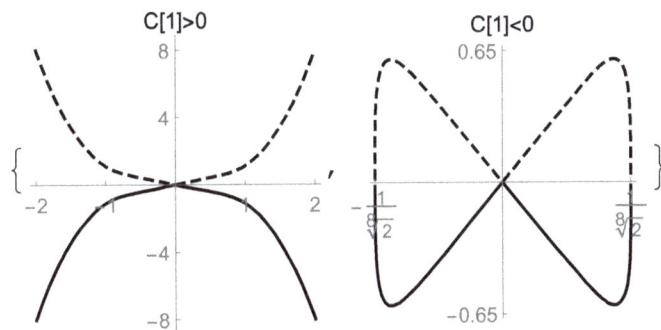

Figure 2.30: Solutions for problem 2.26.

3 First-order ordinary differential equations

The third chapter of this book deals with the next topics: first-order linear differential e-
quation, first-order inverse linear differential equation, Bernoulli differential equation,
Riccati differential equation, exact first-order ordinary differential equation, Lagrange
differential equation, first-order implicit differential equation, and other first-order differential equations.

3.1 First-order linear differential equation

A classification of some first-order ordinary differential equations has been given in
Section 2.1.

3.1. A simple *first-order linear inhomogeneous ordinary differential equation* is introduced here:

$$x'(t) + t\,x(t) = \exp(3t), \quad t \in \mathbb{R}.$$

3.1 We introduce the equation, and look for its solution,

```
Clear[t,x]
sol=DSolve[x'[t]+t x[t]==Exp[3t],x,t]//Flatten
(* Solution as a pure function *)
tab=Table[x[t]/.%/.C[1]→k,{k,4{-1,0,1}}]//Flatten;
{x→Function[{t},e^(-t²/2) C[1]+e^(-9/2-t²/2) √(π/2) Erfi[(3+t)/√2]]}
```

and plot the graphs of the particular solutions contained in the `tab` list.

```
Plot[tab,{t,-2,Sqrt[2]},ImageSize→160,
PlotStyle→{Black,{Black,Dashed},{Black,Dotted}},
Ticks→{Range[-2,1],{-3,5,10,15}}]
```

See Figure 3.1.

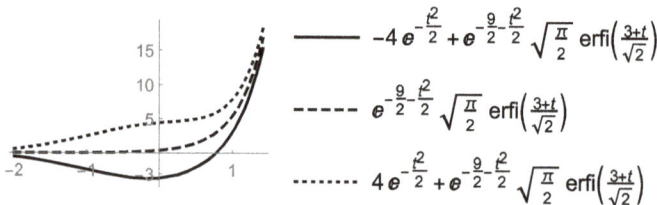

$$\rule{2cm}{0.4pt} \quad -4\,e^{-\frac{t^2}{2}} + e^{-\frac{9}{2}-\frac{t^2}{2}} \sqrt{\frac{\pi}{2}}\ \mathrm{erfi}\!\left(\frac{3+t}{\sqrt{2}}\right)$$

$$- - - - \quad e^{-\frac{9}{2}-\frac{t^2}{2}} \sqrt{\frac{\pi}{2}}\ \mathrm{erfi}\!\left(\frac{3+t}{\sqrt{2}}\right)$$

$$\cdots\cdots \quad 4\,e^{-\frac{t^2}{2}} + e^{-\frac{9}{2}-\frac{t^2}{2}} \sqrt{\frac{\pi}{2}}\ \mathrm{erfi}\!\left(\frac{3+t}{\sqrt{2}}\right)$$

Figure 3.1: Figure for problem 3.1.

https://doi.org/10.1515/9783111411392-004

The behavior of the general solution at $-\infty$ and ∞ is given by the evaluation

```
Limit[x[t]/.sol,t→#]&/@{-∞,∞]}
{{0},{∞}}
```

3.2. Another first-order linear inhomogeneous ordinary differential equation is presented,

$$x'(t) + x(t) = q(t), \quad t \in \mathbb{R}.$$

3.2

```
Clear[t,x,q]
DSolve[x'[t]+x[t]==q[t],x[t],t]//Flatten
{x[t]→ e⁻ᵗC[1]+e⁻ᵗ ∫₁ᵗ eᴷ⁽¹⁾q[K[1]]dK[1]}
```

We consider a particular function q and a particular constant of integration `C[1]` and then plot the resulting solution.

```
sol=Simplify[x[t]/.First[%]/.{q[t_]→Sin[t],C[1]→2}]
½e⁻ᵗ(4+e Cos[1]-eᵗCos[t]-e Sin[1]+eᵗSin[t])
```

```
Plot[sol,{t,-2,10},Ticks→{Range[-2,10,4],{1,3}},
PlotStyle→Black,ImageSize→160]
```

See Figure 3.2.

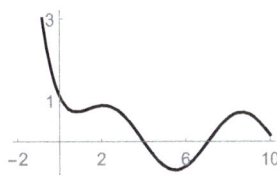

Figure 3.2: Figure for problem 3.2.

Now we find the limits of the particular solution at $-\infty$ and $+\infty$.

```
Limit[sol,t→#]&/@{-∞,∞}
{∞,Interval[{-1,1}]}
```

Remark. The first limit is correct, whereas the second is not. The second limit, i. e., the set of limiting points, is the interval $[-\sqrt{2}/2, \sqrt{2}/2]$. A simple and suggestive argument follows.

```
Plot[{-1,-√2/2,sol,√2/2,1},{t,0,10π},ImageSize→200,
PlotStyle→{{Black,Dotted},{Black,Dashed},Black,
{Black,Dashed},{Black,Dotted}},Ticks→{2π Range[0,5],
{-1,{-√2/2,"-Sqrt[2]/2"},0,{√2/2,"Sqrt[2]/2"},1}}]
```

See Figure 3.3.

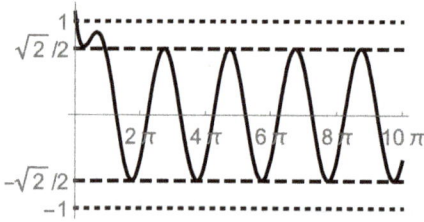

Figure 3.3: A limit in problem 3.2.

Another more rigourous way of finding the same result that the limiting set is the interval $[-\sqrt{2}/2, \sqrt{2}/2]$ can be carried out as we do in the sequel. The midterm in sol tends to zero, so only the first and the last terms count. Then

$$(\sin t - \cos t)/2 \in [-\sqrt{2}/2, \sqrt{2}/2] \iff \sin t - \cos t \in [-\sqrt{2}, \sqrt{2}]$$
$$\iff \sin(t - \pi/4) \in [-1,1] \iff \exists \tau \in [-1,1] \text{ such that } \tau = \sin(t - \pi/4)$$
$$\iff \exists \tau \in [-1,1] \; \forall k \in \mathbb{Z} \text{ such that } t = \pi/4 + (-1)^k \arcsin \tau + k\pi.$$

The revolution figures of Figure 3.2 are introduced.

```
{RevolutionPlot3D[Evaluate[sol],##,Ticks→
{{-1,4.5,10},2{-1,0,1},2{-1,0,1}},RevolutionAxis→"X"],
RevolutionPlot3D[Evaluate[sol],##,Ticks→
{10Range[-1,1],10Range[-1,1],{0,3}}]}&[{t,-1,10},
ImageSize→160,Mesh→2,MeshStyle→Black,
PlotStyle→{Gray,Opacity[.15]}]
```

See Figure 3.4.

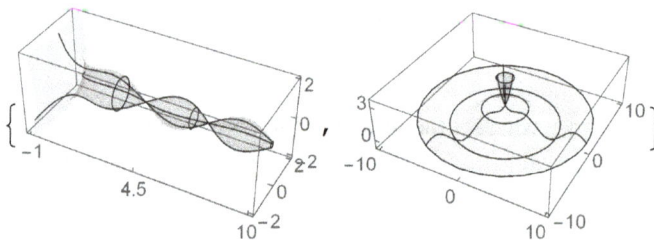

Figure 3.4: Rotated figures in problem 3.2

3.2 First-order inverse linear differential equation

It may happen that a given first-order ordinary differential equation is not linear in $x(t)$, but is linear in its inverse function $t(x)$, if any. In this case, the transformed equation is said to be the *first-order inverse linear* ordinary differential equation.

3.3. Here is our first problem of a differential equation of this kind,

$$x'(t) = \frac{1}{-t\,x(t) + \exp(3x(t))}, \qquad t \in \mathbb{R}.$$

3.3 The equation is not linear and its solution is defined implicitly.

```
Clear[t,x,τ,ξ]
sol=DSolve[x'[t]==     1     ,x[t],t]
                  ─────────────────
                  -t x[t]+Exp[3x[t]]
```
$\text{Solve}[\,t==e^{-\frac{1}{2}x[t]^2}\,C[1]+e^{-\frac{9}{2}-\frac{x[t]^2}{2}}\sqrt{\frac{\pi}{2}}\,\text{Erfi}[\frac{3+x[t]}{\sqrt{2}}]\,,x[t]\,]$

We try using the Reduce built-in function to get an explicit solution.

$\text{Reduce}[\,t==e^{-\frac{1}{2}x[t]^2}\,C[1]+e^{-\frac{9}{2}-\frac{x[t]^2}{2}}\sqrt{\frac{\pi}{2}}\,\text{Erfi}[\frac{3+x[t]}{\sqrt{2}}]\,,x[t]\,]$

The system returns the message "This system cannot be solved with the methods available to Reduce."

We need something else.

`First approach.` We consider the inverse function $\tau = \tau(\xi)$ of $x = x(t)$ and have a new first-order linear differential equation that is instead solvable. Obviously, we suppose that the inverse function $\tau = \tau(\xi)$ exists (see [56, Section 7.3.1]) at least on a nonempty open interval.

```
solinverse=DSolve[ 1  ==       1       ,τ,ξ]//Flatten
                  ────    ───────────────
                  τ'[ξ]   -ξ τ[ξ]+Exp[3ξ]
```
$\{\tau \rightarrow \text{Function}[\{\xi\},e^{-\frac{\xi^2}{2}}\,C[1]+e^{-\frac{9}{2}-\frac{\xi^2}{2}}\sqrt{\frac{\pi}{2}}\,\text{Erfi}[\frac{3+\xi}{\sqrt{2}}]\,]\}$

The inverse function $\tau(\xi)$ is given now:

```
τ[ξ]/.solinverse
```
$e^{-\frac{\xi^2}{2}}\,C[1]+e^{-\frac{9}{2}-\frac{\xi^2}{2}}\sqrt{\frac{\pi}{2}}\,\text{Erfi}[\frac{3+\xi}{\sqrt{2}}]$

Its graph appears in the left-hand side picture in Figure 3.5.

In order to get the graph of the original solution (which solution is still unknown), we symmetrize the graph of the inverse function with respect to the first bisector. Its graph appears in the right-hand side picture in Figure 3.5.

```
{Plot[τ[ξ]/.solinverse/.C[1]→1,##,
Ticks→{Range[-2,2],10{2,3}},
Epilog→{Text[Style["ξ → τ(ξ)",Italic,12],{-1,10}]}],
ParametricPlot[{τ[ξ]/.solinverse/.C[1]→1,ξ},##,
Ticks→{10Range[3],Range[-2,2]},AspectRatio→3/4,
Epilog→{Text[Style["t→x(t)",Italic,12],{20,.5}]}]}
&[{ξ,-2,2},ImageSize→140]
```

See Figure 3.5.

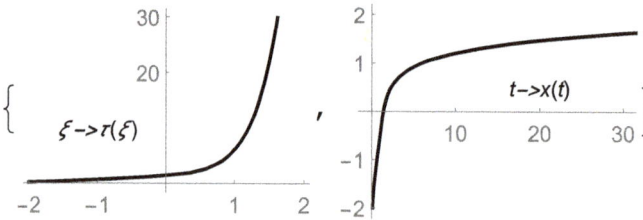

Figure 3.5: Figure for problem 3.3.

Second approach. We study this differential equation in $\mathbb{R}^3$.

```
functx=(Reduce[First[sol],C[1]]/.x[t]→x)//Last
```

$$-\frac{-2e^{\frac{9}{2}+\frac{x^2}{2}}t+\sqrt{2\pi}\,\text{Erfi}[\frac{3+x}{\sqrt{2}}]}{2e^{9/2}}$$

We plot the functx function in $\mathbb{R}^3$.

```
{Plot3D[functx,##,ImageSize→185],
Plot3D[functx,##,PlotRange→All,ImageSize→190]}
&[{t-5,5},{x,-4,2},PlotStyle→{Gray,Opacity[.2]},
Mesh→3]
(* The rightmost figure shows the entire surface *)
```

See Figure 3.6.

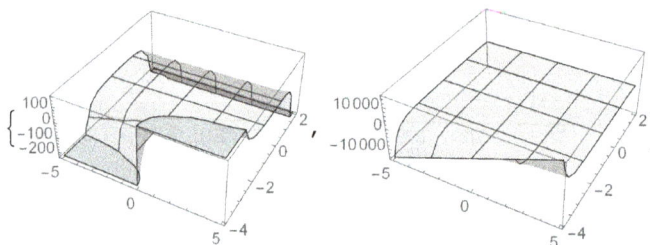

Figure 3.6: Space figures in problem 3.3.

Third approach. Since the function in sol has a particular form, we write t depending on x and represent it.

```
Solve[Last[Reduce[First[sol],C[1]]/.x[t]→x]==1,t]//
Flatten//First//Last
ParametricPlot[{%,x},{x,-2,2},ImageSize→150,
Ticks→{10Range[3],Range[-2,2]},AspectRatio→3/4,
PlotStyle→Black]
```

$\frac{1}{2}e^{-\frac{9}{2}-\frac{x^2}{2}}(2e^{\frac{9}{2}} + \sqrt{2\pi}\,\mathrm{Erfi}[\frac{3+x}{\sqrt{2}}])$

See Figure 3.7.

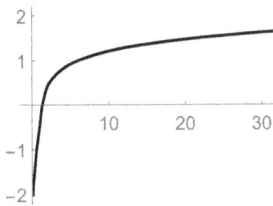

Figure 3.7: Graph of the solution for problem 3.3.

3.3 Bernoulli differential equation

The Bernoulli differential equation has already been introduced in Section 2.1.

Remark. For $\alpha \in \{0,1\}$, the *Bernoulli equation* reduces to a first-order linear differential equation, therefore, we avoid these two cases. □

3.4. We present a simple Bernoulli ordinary differential equation,

$$x'(t) + \frac{x(t)}{t} = t^3 x^3(t), \quad t \in \mathbb{R}\setminus\{0\}.$$

3.4

```
Clear[t,x]
eqn=x'[t]+t⁻¹x[t]==t³x[t]³; (* The equation *)
sol=DSolve[%,x,t]//Flatten (* Solve it *)
```
$\{x\to\mathrm{Function}[\{t\}, -\frac{1}{\sqrt{-t^4+t^2C[1]}}], x\to\mathrm{Function}[\{t\}, \frac{1}{\sqrt{-t^4+t^2C[1]}}]\}$

We have obtained two solutions and check them below:

```
eqn/.sol//Simplify
True
```

and plot the branches of the solution for a particular value of C[1].

```
tab=x[t]/.sol/.C[1]→1
```

$$\left\{-\frac{1}{\sqrt{t^2-t^4}}, \frac{1}{\sqrt{t^2-t^4}}\right\}$$

```
Plot[Evaluate[tab],{t,-1,1},ImageSize→160,
PlotStyle→{Black,{Black,DotDashed}},LabelStyle→10,
Ticks→{{-1,1},3Range[-3,3,2]},PlotLegends→
"Expressions"]
```

See Figure 3.8.

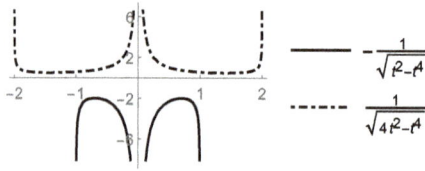

Figure 3.8: Figure for problem 3.4.

3.5. A Bernoulli differential equation with $\alpha = -3$ is introduced,

$$2x'(t)\ln(t) + \frac{x(t)}{t} = \frac{\cos(t)}{x^3(t)}, \quad t > 0.$$

3.5

```
Clear[t,x]
2x'[t]Log[t]+x[t]/t==Cos[t]x[t]^-3;
sol=DSolve[%,x,t] (* Solve the equation *)
Length[%] (* There exist four solutions *)
%%% /.%%//FullSimplify (* All four solutions satisfy
the equation *)
```

$$\left\{\left\{x\rightarrow\text{Function}\left[\{t\}, -\frac{(C[1]+2\text{Log}[t]\,\text{Sin}[t]-2\text{SinIntegral}[t])^{1/4}}{\sqrt{\text{Log}[t]}}\right]\right\},\right.$$
$$\left\{x\rightarrow\text{Function}\left[\{t\}, -\frac{i(C[1]+2\text{Log}[t]\,\text{Sin}[t]-2\text{SinIntegral}[t])^{1/4}}{\sqrt{\text{Log}[t]}}\right]\right\},$$
$$\left\{x\rightarrow\text{Function}\left[\{t\}, \frac{i(C[1]+2\text{Log}[t]\,\text{Sin}[t]-2\text{SinIntegral}[t])^{1/4}}{\sqrt{\text{Log}[t]}}\right]\right\},$$
$$\left\{x\rightarrow\text{Function}\left[\{t\}, \frac{(C[1]+2\text{Log}[t]\,\text{Sin}[t]-2\text{SinIntegral}[t])^{1/4}}{\sqrt{\text{Log}[t]}}\right]\right\}$$

```
4
{True,True,True,True}
```

The differential equation is defined on subsets of the open interval $t > 1$. It is clear that $-2\text{SinIntegral}[t] \rightarrow -\pi$ as $t \rightarrow \infty$ (by (d) in Theorem 1.9) and that $\text{Log}[t]\,\text{Sin}[t]$ oscillates and diverges as $t \rightarrow \infty$. It follows that the term at the 1/4 power oscillates and diverges as $t \rightarrow \infty$. Therefore, the first and the fourth solutions exist only on a union of disjoint intervals in the positive semiaxis.

We select these two solutions (taking C[1]=1)

```
Table[x[t]/.sol[[k]]/.C[1]→1,{k,1,Length[sol]}];
(* Table contains all the solutions *)
solreals=Select[%,FreeQ[#,Complex]&] (* Select the
real solutions *)
```
$$\left\{-\frac{(1+2\mathrm{Log}[t]\,\mathrm{Sin}[t]-2\mathrm{SinIntegral}[t])^{1/4}}{\sqrt{\mathrm{Log}[t]}},\right.$$
$$\left.\frac{(1+2\mathrm{Log}[t]\,\mathrm{Sin}[t]-2\mathrm{SinIntegral}[t])^{1/4}}{\sqrt{\mathrm{Log}[t]}}\right\}$$

and plot them.

```
If[Length[solreals]>0,(* Plot the solutions, if any *)
Plot[Evaluate[solreals],{t,1,25},ImageSize→150,
Ticks→{10Range[2],.5{-1,1}},
PlotStyle→{{Black,Dashed},{Black,DotDashed}},
PlotLegends→"Expressions",LabelStyle→10]]
```

See Figure 3.9.

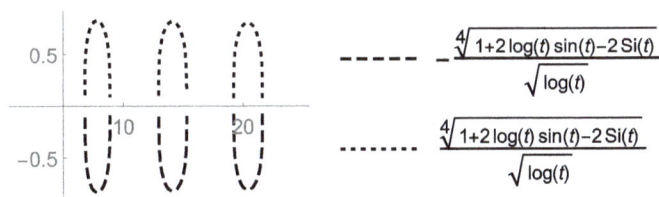

Figure 3.9: Figures for problem 3.5.

Now we rotate the previous figure around the vertical axis.

```
RevolutionPlot3D[Evaluate[solreals[[#]]],{t,1,25},
Ticks→Outer[Times,{20,20,.8},Range[-1,1]],Mesh→1,
ImageSize→160,PlotPoints→45,PlotStyle→{Gray,
Opacity[.15]}]&/@Range[Length[solreals]]
```

See Figure 3.10.

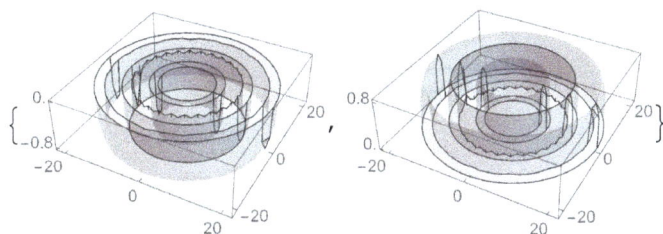

Figure 3.10: Rotated figures for problem 3.5.

If we restrict the above figures to the interval $[1, 10]$, then have that

```
RevolutionPlot3D[Evaluate[solreals[[#]]],{t,1,10},
Mesh→1,ImageSize→160,PlotPoints→45,PlotStyle→
{Black,Opacity[.1]},Ticks→Outer[Times,8{1,1,.1},
Range[-1,1]]]&/@Range[Length[solreals]]
```

See Figure 3.11.

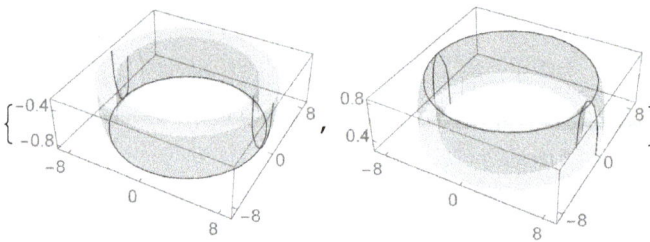

Figure 3.11: Rotated restricted figures for problem 3.5.

We overlap the two surfaces just obtained considering that the functions are defined on the interval $[1, 10]$.

```
RevolutionPlot3D[Evaluate[{{First[solreals]},
{Last[solreals]}}],{t,1,10},ImageSize→160,Mesh→1,
Boxed→False,Axes→False,Ticks→None,PlotStyle→
{{Black,Opacity[.2]},{Black,Opacity[.1]}},
PlotLegends→"Expressions",LabelStyle→10,
PlotPoints→55]
```

See Figure 3.12.

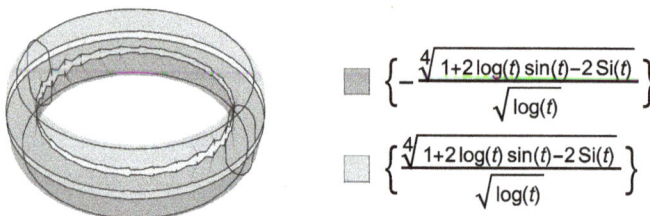

$$\blacksquare \quad \left\{ -\frac{\sqrt[4]{1+2\log(t)\sin(t)-2\,\mathrm{Si}(t)}}{\sqrt{\log(t)}} \right\}$$

$$\blacksquare \quad \left\{ \frac{\sqrt[4]{1+2\log(t)\sin(t)-2\,\mathrm{Si}(t)}}{\sqrt{\log(t)}} \right\}$$

Figure 3.12: Overlapped restricted figure for problem 3.5.

3.6. A Bernoulli equation with $\alpha = \frac{1}{2}$ is introduced now [41, p. 58],

$$x'(t) - 2e^t x(t) = 2\sqrt{e^t x(t)}, \quad t \in \mathbb{R}.$$

3.6

```
Clear[t,x]
x'[t]-2e^t x[t]==2Sqrt[e^t x[t]];  (* The equation *)
sol=DSolve[%,x[t],t]//FullSimplify  (* Solve it *)
Length[%]  (* Number of solutions *)
```
$\{\{x[t]\rightarrow \frac{1}{4}e^{2e^t}\,(C[1]+2\sqrt{\pi}\,\text{Erf}[\sqrt{e}]-2\sqrt{\pi}\,\text{Erf}[e^{t/2}])^2]\},$
$\{x[t]\rightarrow \frac{1}{4}e^{2e^t}\,(C[1]+2\sqrt{\pi}\,\text{Erf}[\sqrt{e}]+2\sqrt{\pi}\,\text{Erf}[e^{t/2}])^2]\}\}$
2

We plot the solutions.

```
Plot[Evaluate[{x[t]/.First[sol],x[t]/.Last[sol]}
/.C[1]→1],{t,-1.5,1.25},ImageSize→160,
PlotStyle→{{Black,Dashed},{Black,Dotted}},
Ticks→{Range[-1.5,1,.5],150Range[3]},
PlotLegends→"Expressions"]
```

See Figure 3.13.

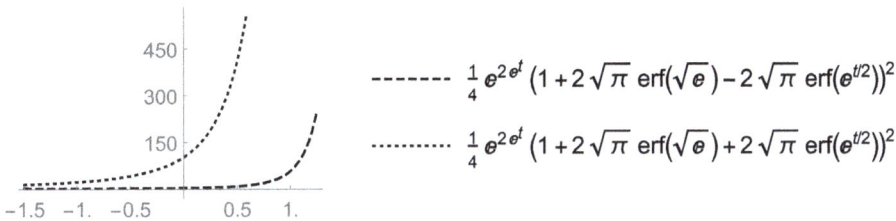

Figure 3.13: Figure for problem 3.6.

3.4 Riccati differential equation

The Riccati ordinary differential equation has already been introduced in Section 2.1.
The Riccati differential equation is of the form:

$$x'(t) = f(t)x^2(t) + g(t)x(t) + h(t), \quad t \in \mathbb{R},$$

with $f(t)\,h(t) \neq 0$.

Given a particular solution $x_0 = x_0(t)$ to the Riccati equation, its general solution can be written as

$$x(t) = x_0(t) + \Phi(t)\left(C - \int f(t)\Phi(t)dt\right)^{-1},$$

where

$$\Phi(t) = \exp\left(\int (2f(t)x_0(t) + g(t))dt\right),$$

with C an arbitrary constant. To the particular solution $x_0(t)$, there corresponds $C = \infty$.

The substitution $u(t) = \exp(-\int f(t)x(t)dt)$ reduces the general Riccati equation to a second-order linear equation

$$f(t)u''(t) - (f'(t) + f(t)g(t))u'(t) + f^2(t)h(t)u(t) = 0,$$

that often can be more easily solved than the original Riccati equation [33, pp. 23–25] and [69, /ode/ode0123.pdf].

3.7. We introduce a solvable Riccati equation,

$$x'(t) + \frac{2}{t^2} - 3x^2(t) = 0, \quad t \in \mathbb{R}\setminus\{0\}.$$

3.7

```
Clear[t,x]
sol=DSolve[x'[t]+2/t²-3x[t]²==0,x[t],t]//First
//Simplify
```
$\{x[t] \to \frac{2C[1]-3t^5}{3tC[1]+3t^6}\}$

We set C[1]=3/2 and plot this particular solution.

```
Plot[x[t]/.sol/.C[1]→3/2,{t,-4,4},ImageSize→160,
Ticks→{{-4,-1,1,4},Range[-2,2]},PlotStyle→Black]
```

See Figure 3.14.

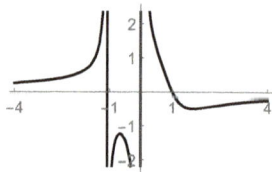

Figure 3.14: Figure for problem 3.7.

3.8. One introduces a Riccati equation so that its solution is expressed by Legendre polynomials (see Section 1.7),

$$x'(t) = -x^2(t) + \frac{2t}{1-t^2}x(t) - \frac{15}{4(1-t^2)}, \quad -1 < t < 1.$$

3.8

```
Clear[t,x]
sol=DSolve[x'[t]==-x[t]²+2t/(1-t²)x[t]-15/4/(1-t²),x[t],t]//Simplify
//Flatten (* Solve the equation *)
{x[t]→5
```

$$-tC[1] LegendreP[\tfrac{3}{2},t]+C[1] LegendreP[\tfrac{5}{2},t]-t LegendreQ[\tfrac{3}{2},t]+LegendreQ[\tfrac{5}{2},t]$$
$$\overline{2(-1+t^2)(C[1]LegendreP[\tfrac{3}{2},t]+LegendreQ[\tfrac{3}{2},t])}$$

```
}
```

We plot the solution for C[1]→1.

```
Plot[Evaluate[x[t]/.sol/.C[1]→1],{t,-1,1},
Ticks→{{-1,{-.6,"-.6"},.75,1},15Range[-2,2]},
PlotStyle→Black,ImageSize→175]
```

See Figure 3.15.

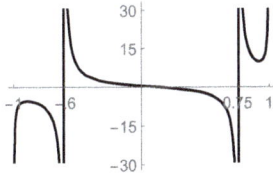

Figure 3.15: Figure for problem 3.8.

We rotate the previous figure around the vertical axis for $t \in [-1, 0]$.

```
RevolutionPlot3D[x[t]/.sol/.C[1]→1,{t,-1,0},
Ticks→Outer[Times,{1,1,20},{-1,0,1}],ImageSize→175,
PlotStyle→{Black,Opacity[.1]},Mesh→3]
```

See Figure 3.16.

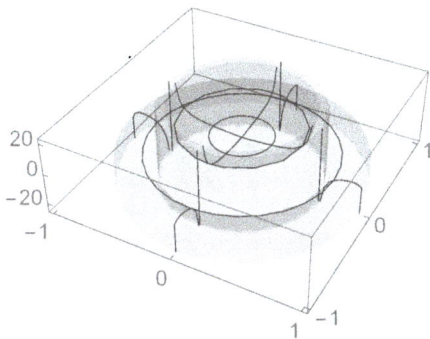

Figure 3.16: Rotated figure in problem 3.8.

3.9. We introduce another Riccati equation [41, p. 75],

$$t\,x'(t) - x^2(t) + (2t+1)x(t) = t^2 + 2t, \quad t \in \mathbb{R}.$$

3.9 First approach. We use the capabilities of *Mathematica*.

```
Clear[t,x]
eqn=t x'[t]-x[t]²+(2t+1)x[t]==t²+2t;  (* The equation *)
sol=DSolve[%,x,t]//Flatten  (* Solve the equation *)
%%/.%//Simplify
```
$\{x\rightarrow Function[\{t\}, -\frac{1}{2}(-\frac{1}{t^2} + \frac{-1-2t}{t^2})t^2 + \frac{1}{-1+\frac{C[1]}{t}}]\}$
```
True
```

```
Plot[x[t]/.sol/.C[1]→1,{t,-3,3},ImageSize→175],
PlotStyle→Black,Ticks→{Range[-3,3,2],4Range[-2,2]}
```

See Figure 3.17.

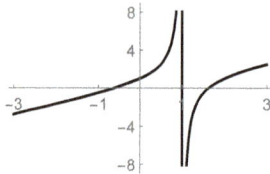

Figure 3.17: Figure for problem 3.9.

Remark. This approach led us to a single solution. □

Second approach. We remark that $x_0(t) = t$ is a particular solution of this Riccati equation. Indeed,

```
eqn/.{x[t]→t,x'[t]→1}//Simplify
```
```
True
```

Then by the substitution $x(t) = x_0(t) + z(t)$ the equation reduces to a Bernoulli equation,

```
Clear[z]
eqnz=eqn/.{x[t]→t+z[t],x'[t]→1+z'[t]}//Simplify
(* This is a Bernoulli equation *)
z[t]²==z[t]+t z'[t]
```

and we solve it.

```
solz=DSolve[eqnz,z,t]//Flatten (* Solve the equation
*)
eqnz/.%//Simplify (* Simplify the solution *)
{z→Function[{t}, 1/(1+e^(1+t)t)]}
True
```

Thus, the general solution of the Riccati equation is

```
xz[t_]:=t+z[t]/.solz
Table[xz[t]/.C[1]→k,{k,{-1,0,1}}] (* We select three
particular solutions of the Riccati equation and plot
them *)
{t + 1/(1+1/e), t + 1/(1+t), t + 1/(1+e t)}
```

```
Plot[Evaluate[%],{t,-4,2},ImageSize→175,
AspectRatio→1,Ticks→{Range[-4,2],4Range[-2,2]},
PlotStyle→{Black,{Black,Dashed},{Black,Dotted}},
PlotLegends→"Expressions"]
```

See Figure 3.18.

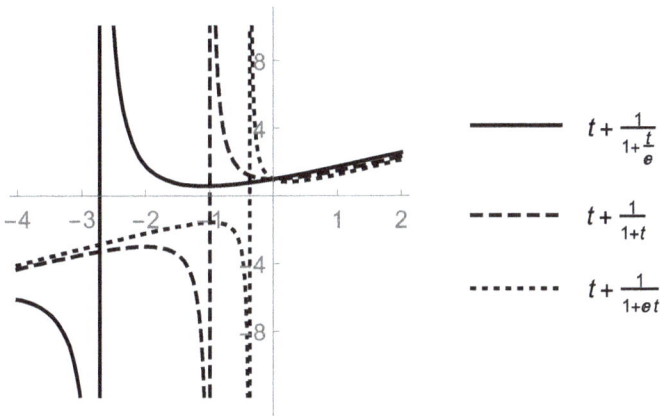

Figure 3.18: Three graphs for problem 3.8.

Remark. We note that the general solution found with the first approach can be obtained from the general solution in the second approach by a convenient transformation of the constant of integration. □

3.10. Our next Riccati equation is of the form [41, p. 75],

$$t^2 x'(t) = t^2 x^2(t) + t\,x(t) + 1, \quad t \in \mathbb{R}.$$

3.10 First approach. We use the capabilities of *Mathematica.*

```
Clear[t,x]
eqn=t²x'[t]==t²x[t]²+t x[t]+1;
sol=DSolve[%,x,t]//Flatten (* Solve the equation *)
%%/.%//Simplify (* Check the solution *)
{x→Function[{t}, (-1-C[1]-Log[t])/(t(C[1]+Log[t]))]}
True
```

The solution requires that $t > 0$, otherwise the logarithm function has complex values.

```
Plot[x[t]/.sol/.C[1]→1,{t,0,2},ImageSize→175,
PlotStyle→Black,Ticks→{{1,2},4Range[-2,1]}]
```

See Figure 3.19.

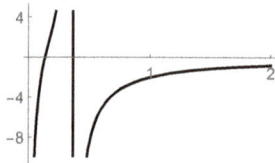

Figure 3.19: Figure of problem 3.10.

Remark. We have got a single solution. ☐

Second approach. We remark that $x_0(t) = -1/t$ is a particular solution of the Riccati equation. Indeed,

```
x₀[t_]:=-1/t
eqn/.{x[t]→x₀[t],x'[t]→x₀'[t]}//Simplify
True
```

Then by the substitution $x(t) = x_0(t)+z(t)$ the Riccati equation reduces to a Bernoulli equation.

```
Clear[z]
eqnz=eqn/.{x[t]→x₀[t]+z[t],x'[t]→x₀'[t]+z'[t]}//
Simplify (* This is a Bernoulli equation *)
t²z[t]²==t (z[t]+t z'[t])

solz=DSolve[eqnz,z,t]//Flatten (* Solve the equation *)
eqnz/.%//Simplify (* Check the solution *)
{z→Function[{t}, 1/(t(C[1]-Log[t]))]}
True
```

Thus, the general solution of the Riccati equation is as follows:

```
xz[t_]:= x₀+z[t]/.solz
Table[xz[t]/.C[1]→k,{k,{-1,0,1}}]//Simplify
```
$$\left\{-\frac{2+\text{Log}[t]}{t+t\,\text{Log}[t]},\ -\frac{1+\text{Log}[t]}{t\,\text{Log}[t]},\ \frac{\text{Log}[t]}{t-t\,\text{Log}[t]}\right\}$$

and requires that $t > 0$.

```
Plot[Evaluate[%],{t,0,2},ImageSize→160,
AspectRatio→1,Ticks→{Range[2],4Range[-2,2]},
PlotStyle→{Black,{Black,Dashed},{Black,Dotted}},
PlotLegends→"Expressions"]
```

See Figure 3.20.

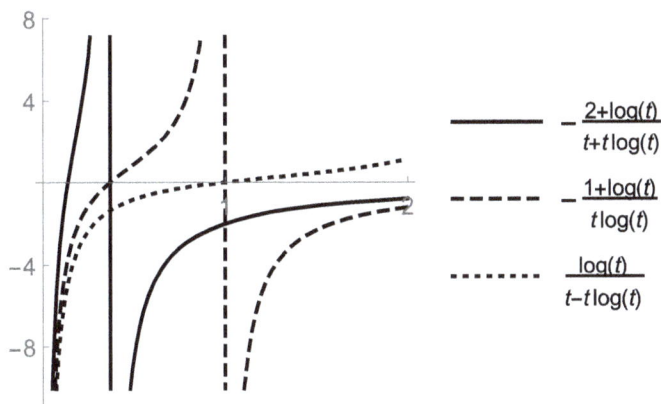

Figure 3.20: Particular solutions for problem 3.10.

Remark. We note that the general solution found with the first approach can be obtained from the general solution in the second approach by a convenient value of the constant of integration.   □

3.11. We introduce another Riccati differential equation,

$$x'(t) = x^2(t) - x(t) - 6, \quad t \in \mathbb{R}.$$

3.11

```
Clear[t,x]
DSolve[x'[t]==x[t]²-x[t]-6,x,t]//Flatten (* Solve
the equation *)
{x→Function[{t}, (-3-2e^{5t+5C[1]})/(-1+e^{5t+5C[1]})]}
```

We plot the solution.

```
Table[x[t]/.%/.C[1]→k,{k,-1,1}];
Plot[Evaluate[%],{t,-2,2},ImageSize→175,Ticks→
{{-2,2},5Range[-2,2]},PlotLegends→"Expressions",
LabelStyle→10]
```

See Figure 3.21.

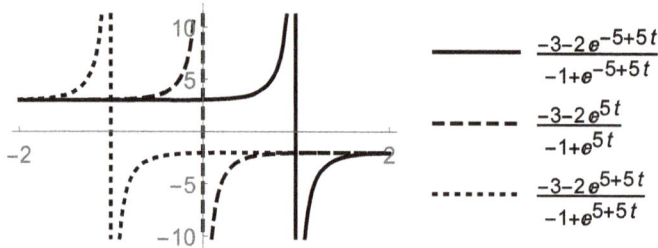

Figure 3.21: Figure for problem 3.11.

Looking at Figure 3.21, it seems that the stripe $\mathbb{R} \times [-2, 3]$ does not contain any part of any trajectory. Let us check if this is true, i. e., we show that a solution exists inside this stripe.

```
Clear[t,x]
Chop[DSolve[{x'[t]==x[t]²-x[t]-6,x[.5]==#},x,t]]
&/@{-2.2,-1.2,1.5,3.2};
Plot[Evaluate[{x[t]/.%[[1,1]],x[t]/.%[[2,1]],
x[t]/.%[[3,1]],x[t]/.%[[4,1]]}],{t,-1.5,2.5},
ImageSize→175,Ticks→{{-1,.5,1,2},{-2,3}},
PlotStyle→{Black,{Black,Dashed},{Black,Dotted},
{Black,DotDashed}},PlotLegends→"Expressions"]
```

See Figure 3.22.

Now we conclude that the stripe $\mathbb{R} \times [-2, 3]$ does contain trajectories.

3.5 Exact first-order ordinary differential equation

The exact first-order ordinary differential equation has already been introduced in Section 2.1.

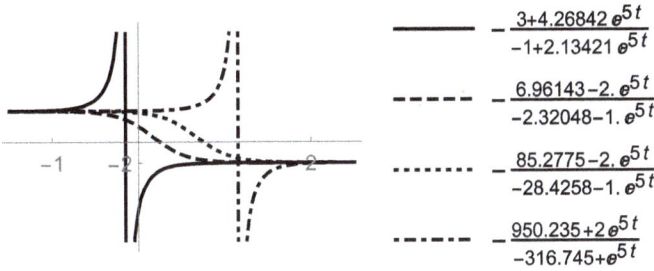

Figure 3.22: Particular solutions for problem 3.11.

We discuss the case of functions of two independent variables and consider the differential form

$$p(x,y)dx + q(x,y)dy = 0, \tag{3.1}$$

where $p(\cdot,\cdot)$, $q(\cdot,\cdot) : D \to \mathbb{R}$ are continuous functions, $p^2(x,y) + q^2(x,y) \neq 0$ on D, and $D \subset \mathbb{R}^2$ is nonempty, open, and connected. The previous differential form is said to be an *exact first-order differential equation* if a differentiable function $u : D \to \mathbb{R}$ there exists so that

$$d\,u(x,y) = p(x,y)dx + q(x,y)dy,$$

for all $(x,y) \in D$.

Theorem 3.12 ([32, Theorem 2.6.1, p. 86]). *Differential form* (3.1) *with* $p, q, \partial p/\partial y$, *and* $\partial q/\partial x$ *continuous on D is an exact first-order ordinary differential equation if and only if*

$$\frac{\partial p(x,y)}{\partial y} = \frac{\partial q(x,y)}{\partial x}$$

for all $(x,y) \in D$.

The initial value problem at a point $(x_0, y_0) \in D$ *associated to* (3.1) *under the above assumptions has a unique solution of the form*

$$u(x,y) = \int_{x_0}^{x} p(s,y)ds + \int_{y_0}^{y} q(x_0,s)ds. \tag{3.2}$$

Remark. An exact first-order differential equation (3.1) has a symmetry of the variables. Therefore, we may consider y a function of x or vice versa. □

3.13. We consider the differential form

$$(2y - x - 1)dx + (2x - y + 1)dy = 0$$

with the initial condition $y(0) = 0$ and integrate it.

3.13

```
Clear[x,y]
p[x_,y_]:=2y-x-1;q[x_,y_]:=2x-y+1;
```

First approach. The differential form satisfies the necessary and sufficient condition in Theorem 3.12 because

```
Simplify[D[p[x,y],y]-D[q[x,y],x]]
0
```

Therefore, by using formula (3.2) one has that

$$\int_0^x p[s,y]ds+\int_0^y q[0,s]ds$$
$$-x-x^2/2+y+2x\,y-y^2/2$$

Thus, the solution of the differential form with initial condition is

$$u(x,y) = (x^2 + y^2 - 4xy + 2x - 2y)/2.$$

If we want an explicit solution, say y depending on x, we write

```
Solve[-x-x²/2+y+2x y-y²/2==0,y]//Flatten
{y→1+2x- √1+2x+3x² , y→1+2x+√1+2x+3x²}
```

Only the first element of the list satisfies the initial condition, thus the solution of the initial value problem is the function

$$y(x) = 1 + 2x - \sqrt{1 + 2x + 3x^2}, \quad x \in \mathbb{R}.$$

Its graph is shown below.

```
Plot[1+2x-√1+2x+3x²,{x,-4,4},ImageSize→175,
Ticks→{2Range[-2,2],{-12,-6,1}},PlotStyle->Black]
```

See Figure 3.23.

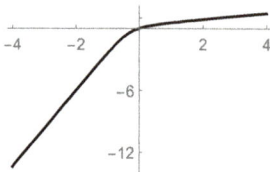

Figure 3.23: Figure for problem 3.13.

Second approach. Taking x as the independent variable, we have the first-order differential equation $y'(x) = -(2y - x - 1)/(2x - y + 1)$ with the initial condition $y(0) = 0$ and solve it.

```
Clear[x,y]
eqn={y'[x]==-p[x,y[x]]/q[x,y[x]],y[0]==0};
sol=DSolve[eqn,y,x]//FullSimplify//Flatten
{y→Function[{x},1+2x+i√-1-2x-3x²]}
```

Because the solution is given in terms of complex numbers, we check it.

```
eqn/.sol//Simplify
{True,True}
```

The domain of definition of this branch is the set $\{x \in \mathbb{R} \mid 3x^2 + 2x + 1 \geq 0\}$.

```
Reduce[3x²+2x+1≥0,x]
True
```

We check this result using a different way. This branch yields the domain of definition, i.e., the set $\{x \in \mathbb{R} \mid 3x^2 + 2x + 1 \geq 0\}$.

```
FindMinimum[3x²+2x+1,x]
{0.666667,{x→-0.333333}}
```

We conclude that the solution of this initial value problem is $y(x) = 1+2x - \sqrt{1 + 2x + 3x^2}$, $x \in \mathbb{R}$, and its graph coincides with the graph in Figure 3.23.

3.14. We consider the differential form

$$(x^2 - y^2 + 1)dx + 2xy\,dy = 0$$

with the initial condition $y(1) = \sqrt{2}$ and integrate it.

3.14

```
Clear[x,y]
p[x_,y_]:=x²-y²+1;q[x_,y_]:=2x y;
```

The necessary and sufficient condition in Theorem 3.12 is not satisfied because

```
Simplify[D[p[x,y],y]-D[q[x,y],x]]
-4y
```

Therefore, we transform it into a first-order differential equation and try to solve it.

```
eqn={y'[x]==-p[x,y[x]]/q[x,y[x]],y[1]==√2};
sol=DSolve[%,y,x]//Flatten
{y→Function[{x}, √1+2x-x²]}
```

We check this solution.

```
eqn/.sol//Simplify
{True,True}
```

The domain of definition of this branch is the set $\{x \in \mathbb{R} \mid -x^2 + 2x + 1 \geq 0\}$.

```
Reduce[-x²+2x+1≥0,x]
1-√2 ≤x≤ 1+√2
```

We conclude that the solution to this initial value problem is $y(x) = \sqrt{1 + 2x - x^2}$ defined for $1 - \sqrt{2} \leq x \leq 1 + \sqrt{2}$, i. e., is the upper-half of the circle $(x - 1)^2 + y^2 = 2$.

```
Plot[y[x]/.sol,{x,1-√2,1+√2},ImageSize→175,
Ticks→{{1-√2,1,1+√2},{1, √2}},PlotStyle→Black]
```

See Figure 3.24.

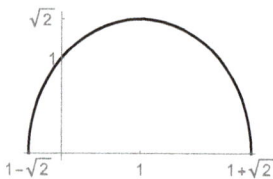

Figure 3.24: Figure for problem 3.14.

3.15. We consider the differential form

$$\left(\frac{1}{x^2} + \frac{1}{y^2}\right)dx + \left(\frac{ax}{y^3} + \frac{1}{y^3}\right)dy = 0,$$

and want to find the real constant a so that it becomes an exact differential form.

3.15 Then we write

```
Clear[a,x,y]
p[x_,y_]:=1/x²+1/y²;q[x_,y_]:=a x/y³+1/y³;
```

First approach. The differential form does not satisfy the necessary and sufficient condition in Theorem 3.12 because

```
Simplify[D[p[x,y],y]-D[q[x,y],x]]
-2+a/y³
```

Taking $a = -2$, we apply formula (3.2) and have

```
sol=Assuming[{s,x,x₀,y,y₀}∈Reals&&x₀≠0&&y₀≠0,
∫ˣ_{x₀}(1/s²+1/y²)ds+∫ʸ_{y₀}(-2x₀/s³+1/s³)ds//Simplify]
ConditionalExpression[1/x₀+1/2(-2/x - 1/y² + 2x/y² + 1/y₀²) - x₀/y₀² ,x>x₀,
y>y₀]
```

Under the above conditions, the solution of the exact differential form is

```
u[x_,y_]:=sol//First;u[x,y]
```
$$\frac{1}{x_0}+\frac{1}{2}\left(-\frac{2}{x} - \frac{1}{y^2} + \frac{2x}{y^2} + \frac{1}{y_0^2}\right) - \frac{x_0}{y_0^2}$$

On the other hand, taking $a = -2$, the differential form is transformed into an exact differential form and we integrate it stepwise:

$$0 = \left(\frac{1}{x^2} + \frac{1}{y^2}\right)dx + \left(\frac{-2x}{y^3} + \frac{1}{y^3}\right)dy = d\left(\frac{-1}{x} - \frac{1}{2y^2}\right) + d\left(\frac{x}{y^2}\right) \Longrightarrow$$

$$0 = d\left(\frac{-1}{x} - \frac{1}{2y^2} + \frac{x}{y^2}\right) \Longrightarrow 2y^2(c\,x + 1) = x(2x - 1), \quad x \in \mathbb{R}.$$

```
sol=Solve[2y²(cx+1)==x(2x-1),y]//Flatten
{y→-√x√(-1+2x)/√(2+2cx) ,y→ √x√(-1+2x)/√(2+2cx) }
```

We have two branches lying symmetrically with respect to the horizontal axis and plot the solutions.

```
tab=Table[sol[[k]]/.c→1,{k,2}];
Plot[Evaluate[tab[[1,2]],tab[[2,2]]],{x,1/2,3/2},
ImageSize→175,AxesOrigin→{0,0},PlotLegends→
"Expressions",Ticks→.5{Range[3],Range[-1,1]},
PlotStyle→{Black,{Black,Dashed}}]
```

See Figure 3.25.

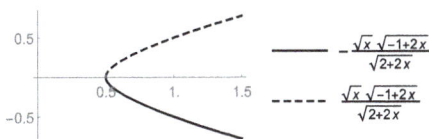

Figure 3.25: Figure for problem 3.15.

Second approach. We transform the differential form into a first-order differential equation.

```
Clear[x,y]
eqn={y'[x]==-p[x,y[x]]/q[x,y[x]]/.a→-2};
sol=DSolve[eqn,y,x]//Simplify (* Solve the
equation *)
```

$$\{\{y \to Function[\{x\}, -\frac{i\sqrt{x}\sqrt{-1+2x}}{\sqrt{-2-2x}\,C[1]}]\},$$
$$\{y \to Function[\{x\}, \frac{i\sqrt{x}\sqrt{-1+2x}}{\sqrt{-2-2x}\,C[1]}]\}\}$$

We check these solutions.

```
eqn/.sol//Simplify
{{True},{True}}
```

A plot of solutions follows below.

```
Table[y[x]/.sol[[k]]/.C[1]→1,{k,2}];
Plot[Evaluate[%],{x,1/2,3/2},ImageSize→175,
AxesOrigin→{0,0},PlotLegends→"Expressions",
Ticks→.5{Range[3],Range[-1,1]},
PlotStyle→{Black,{Black,Dashed}}]
```

This coincides with Figure 3.25.

3.6 Lagrange differential equation

The Lagrange differential equation has already been introduced in Section 2.1.

The first-order ordinary differential is said to be a *Lagrange equation* if it has the form of

$$x(t) = tf(x'(t)) + g(x'(t)),$$

where $t \in I \subset \mathbb{R}$ is the independent variable, x is a function of t, and f and g are arbitrary real functions of a real variable.

If $f(x') = x'$, then this equation is said to be a *Clairaut equation*.

Theorem 3.16. *Let f and g be twice differentiable functions and $f(x') \neq x'$. Then the set of solutions to Lagrange differential equation is represented by the solutions of the system of parametric equations*

$$\begin{cases} t = h(p,c), \\ x = f(p)h(p,c) + g(p), \end{cases}$$

where t is the general solution of the first-order differential equation

$$\frac{dt}{dp} = \frac{f'(p)}{p - f(p)}t + \frac{g'(p)}{p - f(p)},$$

and c is the constant of integration.

If p_1 is a real root of the equation $f(p) - p = 0$, then $x = tp_1 + g(p_1)$ is a solution of the Lagrange equation.

Remark. The solution $x = t\,p_1 + g(p_1)$ is *singular* or *particular*. □

In the case of the Clairaut equation, i. e., $x(t) = tx' + g(x')$, its general solution is $x(t) = ct + g(c)$, where c is the constant of integration. Denote this general solution as $\Phi(t, x, c) = 0$. The *singular solution* of the Clairaut equation follows by eliminating c from the system

$$\begin{cases} \Phi(t, x, c) = 0, \\ \frac{\partial \Phi(t,x,c)}{\partial c} = 0. \end{cases}$$

3.17. We now focus on the Lagrange differential equation,

$$x(t) = \frac{-1 - 4t\, x'(t)}{4x'^2(t)}, \qquad t \in \mathbb{R}.$$

3.17 First approach. One has $f(p) = -1/p$ and $g(p) = -1/(4p^2)$. The equation $f(p) - p = 0$ has no real root. Then

```
Clear[t,x]
f[p_]:=-1/p;g[p_]:=-1/(4p²);
sol=DSolve[t'[p]==f'[p] t[p]/(p-f[p])+g'[p]/(p-f[p]),
t[p],p]//Simplify;

τ[p_]:=t[p]/.First[sol] (* The solutions *)
ξ[p_]:=f[p](t[p]/.First[sol])+g[p]
```

The parametric representation of the solution is of the form

```
param={τ[p],ξ[p]}
```
$$\left\{ \frac{-\sqrt{1+p^2}+p^2(4C[1]-\text{Log}[p]+\text{Log}[1+\sqrt{1+p^2}])}{4\sqrt{1+p^2}}, \frac{-4C[1]+\text{Log}[p]-\text{Log}[1+\sqrt{1+p^2}]}{4\sqrt{1+p^2}} \right\}$$

This solution requires that $p > 0$ and is plotted as follows:

```
ParametricPlot[Evaluate[param/.C[1]→1],{p,.1,100},
AxesOrigin→{0,0},ImageSize→175,AspectRatio→.5,
PlotPoints→75,PlotRange→{{-.45,1.1},{-1.6,.1}},
PlotStyle→Black,Ticks→{{.5,1},{-1.4,-.5}}]
```

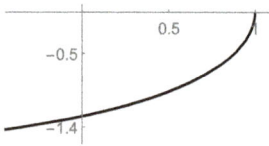

Figure 3.26: Parametric representation of the solution of problem 3.17.

See Figure 3.26.

Taking into account the transformation $p \to \tan(\phi)$, one gets the plot of the solution:

```
σ[φ_]:=Assuming[0≤ φ ≤ π/2,First[param]/.p→Tan[φ]]
ζ[φ_]:=Assuming[0≤ φ ≤ π/2,Last[param]/.p→Tan[φ]]
```

```
ParametricPlot[{σ[φ],ζ[φ]}/.C[1]→1,{φ,.001,π/2-.001},
AxesOrigin→{0,0},PlotPoints→75,ImageSize→175,
AspectRatio→.5,Ticks→{{.5,1},{-1.4,-.5}},
PlotStyle→Black,PlotRange→{{-.45,1.1},{-1.6,.1}}]
```

See Figure 3.27.

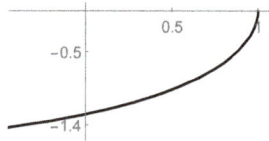

Figure 3.27: Another parametric representation of the solution of problem 3.17.

Second approach. We study the Lagrange differential equation $x = -\frac{1+4t\,x'}{4x'^2}$ directly by using the DSolve built-in function.

```
Clear[t,x]
eqn=x[t]==-(1+4t x'[t])/(4x'[t]²);
DSolve[eqn,x[t],t]; (* We did not get any answer *)
```

Therefore, we decompose the differential equation into two simpler equations.

```
soleq=Solve[eqn,x'[t]]//Flatten;
```

```
DSolve[x'[t]==Last[First[soleq]],x,t];
DSolve[x'[t]==Last[Last[soleq]],x,t];
```

No closed-form solution is returned and we conclude that this approach fails.

Third approach. We try finding a numerical solution and, therefore, set an initial condition.

```
nsol=NDSolve[{eqn,x[1]==-.0001},x,{t,-5,1}];
Length[nsol] (* Number of solutions *)
2
```

```
tab=Table[x[t]/.nsol[[k]],{k,Length[nsol]}];
```

```
Plot[tab,{t,-5,1},ImageSize→175,PlotStyle→
{Black,{Black,Dotted}},Ticks→{Range[-5,1,2],{-2,-1}}]
```

See Figure 3.28.

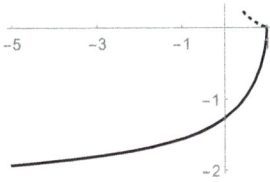

Figure 3.28: Numerical solution for problem 3.17.

Remark. This approach suggests that there exists a second arc lying above the horizontal axis. Indeed, Last[tab] generates the dashed curve. The dashed curve is lying in the region defined by $0 \leq x(t) \leq t^2$. □

```
{Plot[x[t]/.First[nsol],##,PlotStyle→Black],
Plot[{x[t]/.Last[nsol],t²},##,
PlotStyle→{{Black,Dashed},Black}]}]}&[{t,0,1},
ImageSize→140,Ticks→{{0,1},{-1,.2}}]
```

See Figure 3.29.

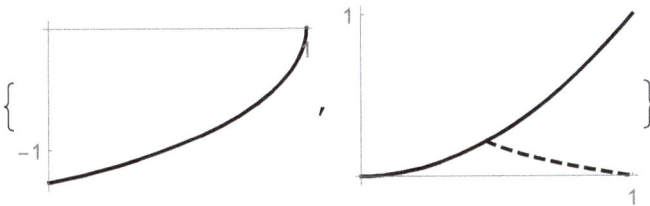

Figure 3.29: The two arcs of the numerical solution for problem 3.17.

3.18. We study the next Clairaut differential equation,

$$x(t) = t\,x'(t) + \sin(x'(t)), \quad t \in \mathbb{R}.$$

3.18

```
Clear[c,t,x,ξ,sol]
sol[t_,x_]:=DSolve[x[t]==t x'[t]+Sin[x'[t]],x,t]/.C[1]→
c//First//First//Last (* Solve the equation *)
t c+Sin[c]
```

We have obtained the general solution and we are looking for its singular solution, if any.

```
φ[t_,x_,c_]:=x-sol[t,c]
red=Reduce[φ[t,x,c]==0&&∂_cφ[t,x,c]==0&&Abs[t]<1,c,
Reals]/.C[1]→0
```
$(-1 < t < 0 \&\&$
$((x == -\sqrt{1-t^2} - 2t \, \text{ArcTan}[\sqrt{\frac{-1-t}{-1+t}}] \&\& c == -2 \, \text{ArcTan}[\sqrt{\frac{-1-t}{-1+t}}]) \, || $
$(x == \sqrt{1-t^2} \&\& c == 2 \, \text{ArcTan}[\sqrt{\frac{-1-t}{-1+t}}]))) \, || $
$(0 < t < 1 \&\&$
$((x == \sqrt{1-t^2} - 2t \, \text{ArcTan}[\sqrt{\frac{-1-t}{-1+t}}] \&\& c == 2 \, \text{ArcTan}[\sqrt{\frac{-1-t}{-1+t}}]) \, || $
$(x == -\sqrt{1-t^2} - 2t \, \text{ArcTan}[\sqrt{\frac{-1-t}{-1+t}}] \&\& c == -2 \text{ArcTan}[\sqrt{\frac{-1-t}{-1+t}}]))) \, || $
$(t == 0 \&\& ((x == -1 \&\& c == -(\text{Pi}/2)) \, || \, (x == 1 \&\& c == \text{Pi}/2)))$

Thus, the singular solution is composed by the arcs:

$$\xi(t) = \begin{cases} \begin{cases} -\sqrt{1-t^2} - 2t \arctan\sqrt{\frac{-1-t}{-1+t}}, \\ \sqrt{1-t^2} + 2t \arctan\sqrt{\frac{-1-t}{-1+t}}, \end{cases} & -1 < t < 0 \\ \begin{cases} \sqrt{1-t^2} + 2t \arctan\sqrt{\frac{-1-t}{-1+t}}, \\ -\sqrt{1-t^2} - 2t \arctan\sqrt{\frac{-1-t}{-1+t}}, \end{cases} & 0 < t < 1 \end{cases}$$

$$= \begin{cases} -\sqrt{1-t^2} - 2t \arctan\sqrt{\frac{-1-t}{-1+t}}, \\ \sqrt{1-t^2} + 2t \arctan\sqrt{\frac{-1-t}{-1+t}}, \end{cases} \quad -1 < t < 1.$$

Because the positive arcs and the negative arcs are continuously differentiable at $t - 0$, we concatenate them.

Thus, the singular solution is composed by two arcs

$\xi[t_]:=\{-\sqrt{1-t^2} - 2t \, \text{ArcTan}[\sqrt{\frac{-1-t}{-1+t}}], \sqrt{1-t^2} + 2t \, \text{ArcTan}[\sqrt{\frac{-1-t}{-1+t}}]\}$

This solution is represented below, where the positive and the negative arcs are distinguished.

```
Plot[Evaluate[ξ[t]],{t,-1,1},ImageSize→175,
PlotStyle→{Black,{Black,Dashed}},
Ticks→Outer[Times,{1,3},{-1,1}]]
```

See Figure 3.30.

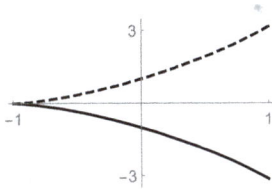

Figure 3.30: Figure of the solution of problem 3.18.

We suggest a dynamical representation of the solution.

```
Animate[ (* Dynamics of the solutions *)
Plot[Evaluate[ξ[t]],c t+Sin[c],{t,-1,1},
PlotStyle→{Black,{Black,Dashed},{Black,Dotted}},Ticks→
{{-1,1},3Range[-1,1]},PlotRange→{{-1,1},{-3,3}},
ImageSize→175],{c,-3π/4,.8π},SaveDefinitions→True,
DefaultDuration→20,AnimationRunning→False,
AnimationDirection→ForwardBackward]
```

See Figure 3.31.

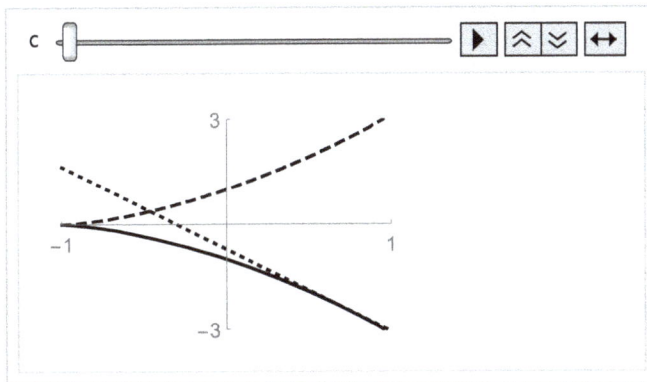

Figure 3.31: Dynamical figure of the solution of problem 3.18.

3.7 First-order implicit differential equation

A first-order *implicit ordinary differential equation* is of the form of $f(t, x, x') = 0$, where the unknown function is $x = x(t)$ depending on the independent variable t. We have already studied some differential equations of this kind, namely Lagrange differential equations.

Theorem 3.19. *Let $f(t, x, x') = 0$ be a first-order ordinary differential equation, with f defined on a nonempty open interval $D \subset \mathbb{R}^3$ centered at (t_0, x_0, x_0'), where x is a real solution of the equation $f(t_0, x_0, x') = 0$. If*

$$\begin{cases} f(t, x, x'), \ \frac{\partial f}{\partial x}(t, x, x'), \text{ and } \frac{\partial f}{\partial x'}(t, x, x') \text{ are continuous on D and} \\ \frac{\partial f}{\partial x'}(t_0, x_0, x_0') \neq 0, \end{cases}$$

then there exists a solution $x = x(t)$ of the differential equation $f(t, x, x') = 0$ defined on a neighborhood of t_0, satisfying $x(t_0) = x_0$ and $x'(t_0) = x_0'$, such that this solution is unique.

3.20. We study and solve the first-order implicit ordinary differential equation with an initial condition,

$$t x'^{\,2}(t) - 2x(t)x'(t) + 4t = 0, \quad x(3) = 10, \quad t \in \mathbb{R}.$$

3.20 First approach. We use the DSolve built-in function.

```
Clear[t,x]
eqn=t x'[t]²-2 x[t] x'[t]+4 t==0; (* The equation *)
sol=DSolve[{eqn,x[3]==10},x[t],t]//TrigToExp
//Simplify (* Solve the equation *)
{{x[t]→1+t²},{x[t]→9+t²/9}}
```

Two solutions are returned, each one being a parabola and we plot them.

```
pair=x[t]/.sol;
Plot[pair,{t,-3.1,3.1},ImageSize→150,
PlotStyle→{Black,{Black,Dashed}},Ticks→{{-3,3},
{2,6,10}},PlotLegends→"Expressions"]
```

See Figure 3.32.

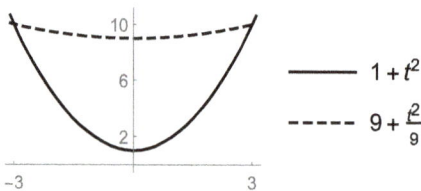

Figure 3.32: Figure for problem 3.20.

Second approach. The differential equation is considered as a second degree equation
in x'.

```
Clear[t,x]
Reduce[eqn,x'[t]];
x'[t]//.{ToRules[%]}
```
$\{\frac{x[t]-\sqrt{-4t^2 + x[t]^2}}{t}, \frac{x[t]+\sqrt{-4t^2 + x[t]^2}}{t}, 0, \text{Derivative}[1][x][0]\}$

For both solutions, it is necessary that $x^2(t) - 4t^2 \geq 0$.
We discuss both differential equations.

```
sol1=DSolve[x'[t]==First[%],x[3]==10,x[t],t]//
FullSimplify//Flatten
sol2=DSolve[x'[t]==Last[%%],x[3]==10,x[t],t]//
FullSimplify//Flatten
```
$\{x[t]\rightarrow 1+t^2, x[t]\rightarrow 9+t^2/9\}$
$\{x[t]\rightarrow 9+t^2/9, x[t]\rightarrow 1+t^2\}$

```
union=Union[{x[t]/.sol1,x[t]/.sol2}]
```
$\{9+t^2/9, 1+t^2\}$

We plot the solutions given by the union list. Thus, we have the following:

```
Plot[union,{t,-3.1,3.1}, ImageSize→150,
PlotStyle→{{Black,Dashed},Black},
Ticks→{{-3,3},2{1,3,5}},PlotLegends→"Expressions"]
```

The graph of this picture coincides with the graph in Figure 3.32.

3.21. We study the first-order implicit ordinary differential equation

$$x(t) = \frac{t\,x'(t)}{2} + \frac{x'^{\,2}(t)}{t^2}, \quad t \in \mathbb{R}\setminus\{0\}.$$

3.21 First approach. We use the DSolve built-in function.

```
Clear[a,t,x]
eqn=x[t]==t x'[t]/2+x'[t]^2/t^2;  (* The equation *)
DSolve[%,x,t];  (* No closed-form solution is returned
*)
```

We try a numerical approach and consider an initial condition $x(0) = 1$.

```
nsol=NDSolve[{eqn,x[0]==1},x,{t,-4,4}]
{{x→InterpolatingFunction[{{-2.12938,2.12895}},<>]},
{x→InterpolatingFunction[{{-4.,4.}},<>]}}
```

We have two solutions whose plots follow.

```
tab=Table[x[t]/.nsol[[k]],{k,Length[nsol]}];
Plot[tab,{t,-4,4},ImageSize→150,PlotStyle→
{Black,{Black,Dashed}},Ticks→{2Range[-2,2],4Range[2]}]
```

See Figure 3.33.

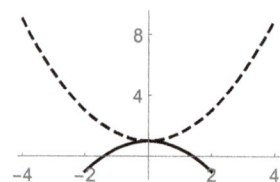

Figure 3.33: Particular numerical solutions for problem 3.21.

Second approach. We try solving it as a polynomial equation of second degree in $x'(t)$.

```
Reduce[eqn,x'[t]];
{x'[t]}/.{ToRules[%]}//Flatten
{eq1=x'[t]==First[%],eq2=x'[t]==Last[%]}//Simplify;
{DSolve[eq1,x,t],DSolve[eq2,x,t]};
{¼(-t³ − √(t⁶+16t²x[t]) ),¼(-t³ + √(t⁶+16t²x[t]) )}
```

No closed-form solution is returned.

Third approach. We introduce a real parameter $p = x'$ and differentiate the equation.

```
D[x[t]=t x'[t]/2+x'[t]²/t²,t]/.{x'[t]→p[t],
x''[t]→p'[t]}//Simplify
```
$$-\frac{(t^3+4p[t])(-p[t]+t\,\mathrm{Derivative}[1][p][t])}{2t^3}$$

The solutions are as follows:

▷ $t\,\mathrm{d}\,p(t) - p(t)\mathrm{d}\,t = 0$, $p(t)t \neq 0 \Longrightarrow p(t) = a\,t \Longrightarrow x(t) = a\,t^2/2 + b$. This is a parabola whenever $a \neq 0$. Otherwise, it is a horizontal straight line. This function is a solution of the given equation whenever $b = a^2$. Thus, a solution is of the form $x(t) = a\,t^2/2 + a^2$, where a is a real constant. ▷ $4p(t)+t^3 = 0 \Longrightarrow x(t) = -t^4/16+c$, where c is a real constant. This function is a solution of the given equation whenever $c = 0$. Thus, a solution is the function $x(t) = -t^4/16$.

Both solutions satisfy the equation.

```
sols={a t²/2+a²,-t⁴/16};
solst=∂ₜsols;
sols==t solst/2+solst²/t²
```
True

We plot the general and singular solutions.

```
Plot[Evaluate[sols/.a→2],{t,-5,5},Ticks→
{2Range[-2,2],10Range[-3,3,2]},ImageSize→150,
PlotStyle→{Black,{Black,Dashed}}]
```

See Figure 3.34.

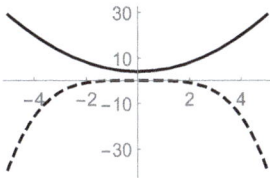

Figure 3.34: General and singular solutions in problem 3.21.

Remark. We conclude the equation $x(t) = t\,x'(t)/2 + x'^2(t)/t^2$ has two solutions, namely one general and one singular:

$$\begin{cases} x(t) = a\,t^2/2 + a^2, & a \in \mathbb{R}, \\ x(t) = -t^4/16. \end{cases}$$

3.22. We try solving the next implicit ordinary differential equation,

$$x'(t)x(t) - \frac{x(t)}{x'(t)} + 2t = 0, \quad t \in \mathbb{R}.$$

3.22 First approach. We study it with the DSolve built-in function.

```
Clear[t,x]
eqn=x'[t]x[t]-x[t]/x'[t]+2t==0;
sol=DSolve[eqn,x[t],t] (* There are two solutions *)
{{x[t]→-e^{C[1]/2}√(e^{C[1]}-2t)},{x[t]→ e^{C[1]/2}√(e^{C[1]}-2t}}}
```

We plot the solutions for C[1]→1.

```
Table[x[t]/.sol[[k]]/.C[1]→1,{k,Length[sol]}];
Plot[Evaluate[%],{t,-2,E/2},ImageSize→150,
PlotStyle→{Black,{Black,Dashed}},Ticks→{{-2,-1,1},
2Range[-2,2]},PlotLegends→"Expressions"]
```

See Figure 3.35.

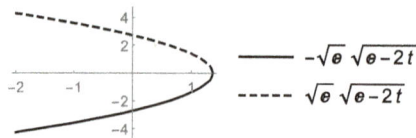

Second approach. We try an approach using the Reduce built-in function.

```
Clear[t,x]
Reduce[eqn,x'[t]];
rules=Drop[x'[t]/.{ToRules[%]},-1]
```
$$\{\frac{-t-\sqrt{t^2+x[t]^2}}{x[t]},\frac{-t+\sqrt{t^2+x[t]^2}}{x[t]}\}$$

```
sol={DSolve[x'[t]==First[rules],x[t],t],
DSolve[x'[t]==Last[rules],x[t],t]}//Flatten
```
$$\{x[t]\to -e^{C[1]/2}\sqrt{e^{C[1]}-2t},x[t]\to e^{C[1]/2}\sqrt{e^{C[1]}-2t},$$
$$x[t]\to -e^{C[1]/2}\sqrt{e^{C[1]}-2t},x[t]\to e^{C[1]/2}\sqrt{e^{C[1]}-2t}\}$$

We eliminate the duplicates,

```
sol1=DeleteDuplicates[sol];
tab=Table[x[t]/.sol1[[k]],{k,Length[sol1]}]
```
$$\{-e^{C[1]/2}\sqrt{e^{C[1]}-2t},e^{C[1]/2}\sqrt{e^{C[1]}-2t}\}$$

and plot the solutions for C[1]→1.

```
Plot[Evaluate[tab/.C[1]→1],{t,-2,e/2},
Ticks→{{-2,-1,1},2Range[-2,2]},ImageSize→150,
PlotStyle→{Black,{Black,Dashed}},
PlotLegends→"Expressions"]
```

See Figure 3.36.

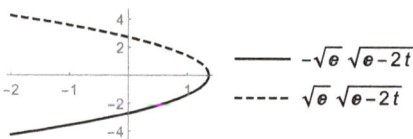

Third approach. We try an approach based on the ContourPlot built-in function.

```
Clear[a]
tab=Table[x[t]/.sol[[k]]/.C[1]→2Log[a],
{k,Length[sol]}]//Simplify//Flatten;
tab=DeleteDuplicates[tab,First#1==First#2&]
```
$$\{-a\sqrt{a^2-2t},a\sqrt{a^2-2t}\}$$

```
Animate[
ContourPlot[{x==First[tab],x==Last[tab]},{t,-4,a²/2},
{x,-5a,5a},ImageSize→150],{a,1,4},
SaveDefinitions→True,DefaultDuration→25,
AnimationDirection→ForwardBackward,
AnimationRunning→False]
```

See Figure 3.37.

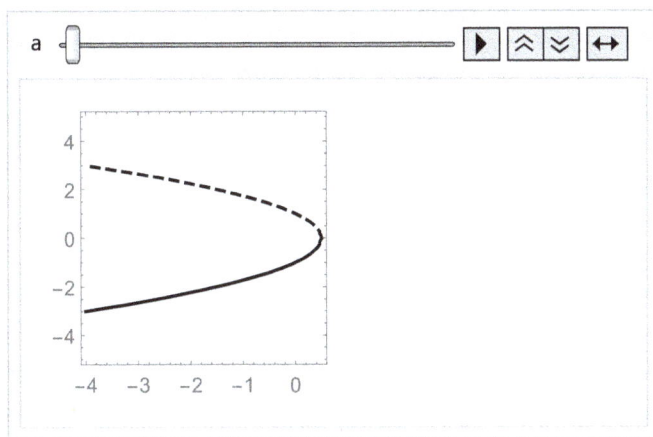

Figure 3.37: Dynamical solutions of problem 3.22.

Fourth approach. We study two initial value problems for this differential equation.

```
Clear[a]
sol=DSolve[{eqn,x[-1]==#},x[t],t]&/@{-2,2}//Flatten;
list=Table[x[t]/.sol[[k]],{k,Length[sol]}]//Sort
```
$$\{-i\sqrt{1+\sqrt{5}}\sqrt{-1-\sqrt{5}-2t}, i\sqrt{1+\sqrt{5}}\sqrt{-1-\sqrt{5}-2t},$$
$$-\sqrt{-1+\sqrt{5}}\sqrt{-1+\sqrt{5}-2t}, \sqrt{-1+\sqrt{5}}\sqrt{-1+\sqrt{5}-2t}\}$$

Remark. Because of

```
{list[[1]]==-list[[2]],list[[3]]==-list[[4]]}
{True,True}
```

the graphs are symmetrical with respect to the horizontal axis. The continuous and the dashed arcs below are defined on the interval $[-\text{GoldenRatio}, +\infty[$, whereas the dotted and the dot-dashed arcs below are defined on the interval $]-\infty, 1/\text{GoldenRatio}]$. ☐

```
Show[
{Plot[Evaluate[{list[[3]],list[[4]]}],##,PlotStyle→
{Black,{Black,Dashed}}],
Plot[Evaluate[{list[[1]],list[[2]]}],##,PlotStyle→
{{Black,Dotted},{Black,DotDashed}}]}&[{t,-GoldenRatio,
1/GoldenRatio},ImageSize→150,PlotRange→All,
Ticks→{{-GoldenRatio,1/GoldenRatio},2Range[-2,2]}],
Graphics[{Black,-1,PointSize[.025],
Point[{{-1,2},{-1,-2}}]}]]
```

See Figure 3.38.

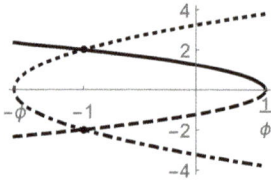

Figure 3.38: Analytical solutions of problem 3.22.

`Fifth approach.` We study the same initial value problems using the NDSolve built-in function.

```
Clear[t,x]
sol1=NDSolve[{eqn,x[-1]==2},x,{t,-GoldenRatio,
1/GoldenRatio}];
sol2=NDSolve[{eqn,x[-1]==-2},x,{t,-GoldenRatio,
1/GoldenRatio}];
```

```
Show[
{Plot[{x[t]/.First[sol1],x[t]/.First[sol2]},
{t,-1.618,.618},PlotStyle→{{Black,Dashed},Black},##],
Plot[{x[t]/.Last[sol1],x[t]/.Last[sol2]},
{t,-1.618,.618},PlotStyle→{{Black,Dotted},
{Black,DotDashed}},##]}&[Point[{{-1,2},{-1,-2}}],
Ticks→{{-1.6,-1,.6},{-4,-2,2,4}},ImageSize→150],
Graphics[{Black,PointSize[.025],PlotRange→All}]]
```

See Figure 3.39.

Remark. The upper arcs are similar to the upper arcs in the fourth approach and the lower arcs are similar to the lower arcs in the fourth approach. ☐

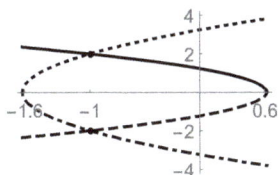

Figure 3.39: Numerical solutions of problem 3.22.

3.8 Other first-order differential equations

Throughout this section, we introduce some *first-order nonlinear* ordinary differential equations.

3.23. We introduce the real *Löwner differential equation* [99, p. 164],

$$x'(t) = -x(t)\frac{1 + k(t)x(t)}{1 - k(t)x(t)}, \quad t \in \mathbb{R}.$$

3.23 We study it by the DSolve built-in function.

```
Clear[t,x,k]
eqn=x'[t]==-x[t] 1+k[t]x[t]/1-k[t]x[t] ;
sol=DSolve[eqn,x[t],t]; (* No closed-form solution
is returned *)
```

Because the equation has no closed-form solution for an arbitrary function k, we consider a constant k, and in this case, we have a solution.

```
Clear[t,x,k]
eqn=x'[t]==-x[t] 1+k x[t]/1-k x[t] ;
sol=DSolve[eqn,x[t],t];
```

The dynamics of the solution is studied by the Animate command.

```
Animate[
Plot[Evaluate[x[t]/.sol/.{C[1]→0,k→j}],{t,2.4,3.4},
ImageSize→150,PlotStyle→{Black,{Black,Dashed}},
Ticks→{Range[2.4,3.4,.2],Automatic}],{j,3,7.4},
DefaultDuration→15,AnimationRunning→False,
AnimationDirection→ForwardBackward]
```

See Figure 3.40.

3.24. We introduce a first-order nonlinear differential equation such that its solution contains the incomplete elliptic integral of the first kind (see Section 1.10 or [99, p. 164]):

$$x'(t) = \sqrt{(1 - x^2(t))(1 - k^2 x^2(t))}, \quad k, t \in \mathbb{R}.$$

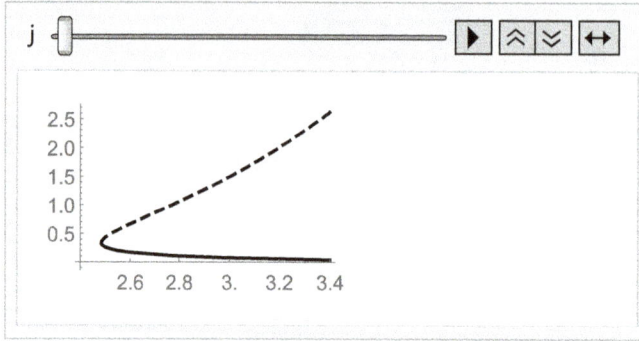

Figure 3.40: Dynamical solution of problem 3.23.

3.24 We try to study it by the DSolve built-in function.

```
Clear[t,x,k]
eqn=x'[t]==√((1-x[t]²)(1-k²x[t]²));
sol=DSolve[eqn,x[t],t];
```

We plot a particular solution.

```
Animate[
Plot[Evaluate[x[t]/.sol/.{C[1]→0,k→j}],{t,-1.5,1.5},
ImageSize→150,Ticks→{{-1,1},Range[-1,1,.5]},
PlotStyle→Black],{j,0,1},DefaultDuration→15,
AnimationDirection→ForwardBackward,
AnimationRunning→False]
```

See Figure 3.41.

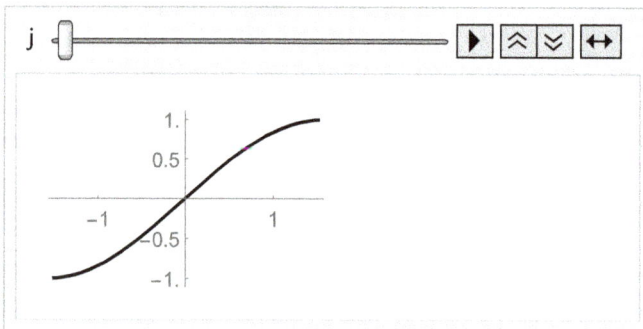

Figure 3.41: Dynamical solution for problem 3.24.

3.25. We now study the *Weierstrass differential equation* [99, p. 164],

$$x'(t) = \sqrt{4\,x^3(t) - p\,x(t) - q}, \quad p, q, t \in \mathbb{R}.$$

3.25 We study it by the DSolve built-in function.

```
Clear[t,x,p,q]
eqn=x'[t]==√(4x[t]³-p x[t]-q);
sol=DSolve[eqn,x[t],t]; (* No closed-form solution is
returned *)
```

We look for a numerical solution with some particular constants p and q and an initial condition.

```
nsol=Block[{p=-3,q=-1},
NDSolve[{eqn,x[1]==1},x[t],{t,-.1,1.9}]]//FullSimplify;
Plot[x[t]/.nsol,{t,-.1,1.9},ImageSize→180,
Ticks→Outer[Times,{.5,5},Range[3]],PlotStyle→Black]
```

See Figure 3.42.

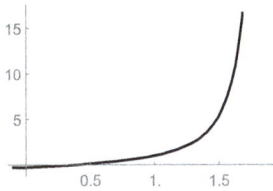

Figure 3.42: Numerical solution for problem 3.25.

4 Higher order and systems of ordinary differential equations

The present chapter deals with the next topics: second-order linear differential equations, Bessel differential equation, Legendre differential equation, Mathieu differential equation, second-order nonlinear differential equations, other higher- order differential equations, and systems of differential equations. Applications of the nonlinear differential equations may be found in many works; we only cite a few of them: [41, 64, 57, 59, 58, 60], and [61].

4.1 Second-order linear differential equations

4.1. Let us find the solution of the following second-order linear inhomogeneous differential equation with constant coefficients [64, p. 42],

$$x''(t) - x(t) = 2\sin(t) - 4\cos(t), \quad t \in \mathbb{R}.$$

4.1

```
Clear[a,b,t,x] (* Clears the values and definitions *)
eqn=x''[t]-x[t]==2Sin[t]-4Cos[t]; (* The equation *)
sol=Flatten[DSolve[eqn,x[t],t]] (* Solve the equation
*)
{x[t]→ e^t C[1]+e^-t C[2]+2 Cos[t]-Sin[t]}
```

Its graph is given below.

```
Animate[
Plot[Evaluate[x[t]/.sol/.{C[1]→a,C[2]→b}],{t,-π,π},
ImageSize→140,AspectRatio→.75,PlotStyle→Black,
Ticks→{π{-1,1},20Range[-3,3]}],{a,-1,1},{b,-3,3},
SaveDefinitions→True,AnimationRunning→False,
DefaultDuration→20]
```

See Figure 4.1.

The general solution shows a possible divergence. If $C[1]^2+C[2]^2 \neq 0$, the solution is unbounded. Otherwise, the solution is oscillatory, and thus bounded.

```
Limit[x[t]/.sol,t→#]&/@{-∞,∞}
{C[2]∞,C[1]∞}
```

We rotate a particular solution around the horizontal axis.

https://doi.org/10.1515/9783111411392-005

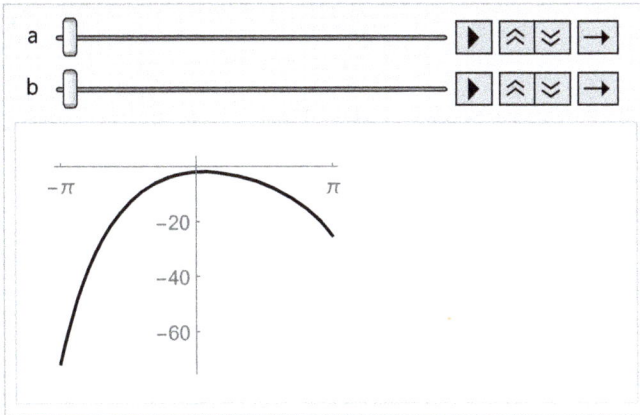

Figure 4.1: Dynamical solution of problem 4.1.

```
RevolutionPlot3D[x[t]/.sol/.{C[1]→-1,C[2]→3},{t,π/6,π/3},
Ticks→Outer[Times,{1,1.6,1.6},{-1,0,1}],Mesh→2,
PlotStyle→{Black,Opacity[.15]},ImageSize→125,
RevolutionAxis→"X"]
```

See Figure 4.2.

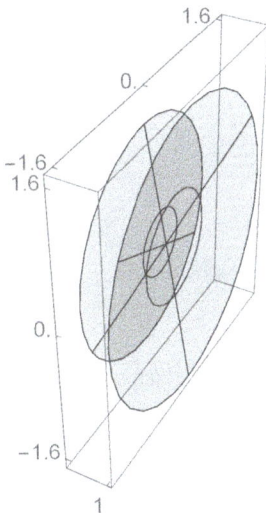

Figure 4.2: Rotated solution of problem 4.1.

4.2. Let us find the solution of the following second-order linear inhomogeneous differential equation with constant coefficients and periodic right-hand side [64, p. 42],

$$x''(t) + x(t) = 6\sin(2t), \quad t \in \mathbb{R}.$$

4.2

```
Clear[t,x]
sol=DSolve[x''[t]+x[t]==6Sin[2t],x[t],t]//Simplify
(* Solve the equation *)
{{x[t]→Cos[t](C[1]-4Sin[t])+C[2]Sin[t]}}

D[x[t]/.sol,{t,2}]+x[t]/.sol//Simplify
(* Check the solution *)
```

We consider a particular solution

```
plot[t_]:=x[t]/.sol/.{C[1]→1,C[2]→3}
```

and plot it on two intervals

```
{Plot[plot[t],{t,-π,π},ImageSize→140,Ticks→{{-π,π},
2 Range[-2,2]}],
Plot[plot[t],{t,-5π,5π},ImageSize→175,
Ticks→{πRange[-5,5,2],2Range[-2,2]}]}&[AspectRatio→
3/4,PlotStyle→Black]
```

See Figure 4.3.

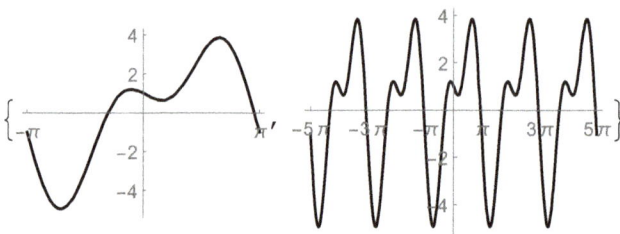

Figure 4.3: Particular solution of problem 4.2.

It appears that the particular solution is periodic. This happens because all the terms are 2π-periodic.

```
Limit[plot[t],t→#]&/@{-∞,∞}
{{Interval[{-6,6}]},{Interval[{-6,6}]}}
```

4.3. We study the solution of the following second-order linear inhomogeneous differential equation with constant coefficients [64, p. 41],

$$x''(t) + x(t) = \frac{1}{\cos^2(t)}, \quad t \in \mathbb{R} \setminus \left\{ (2k + 1)\frac{\pi}{2} \mid k \in \mathbb{Z} \right\}.$$

4.3

```
Clear[t,x] (* Clear the values and definitions *)
sol=DSolve[x''[t]+x[t]==1/Cos[t]²,x[t],t]//Flatten
(* Solve the equation *)
{x[t]→-1+C[1]Cos[t]+2ArcTanh[Tan[t/2]]Sin[t]
+C[2]Sin[t]}
```

The general solution is 2π-periodic because

```
(x[t]/.sol/.{t→t+2π})-(x[t]/.sol)//Simplify
0
```

The solution is defined whenever $|\tan(t/2)| < 1$. We try solving this inequality by the Reduce built-in function. Then

```
Reduce[Abs[Tan[t/2]]<1,t]
Reduce[Abs[Tan[t/2]]<1,t]
```

and we get the message "This system cannot be solved with the methods available to Reduce. >> ."

Therefore, we try discussing this inequality by elementary means:

$$|\tan(t/2)| < 1 \Longleftrightarrow t \in \,]-\pi/2, \pi/2[\, + 2k\pi, \quad k \in \mathbb{Z} \Longleftrightarrow$$
$$t \in \dots]-5\pi/2, -3\pi/2[\, \cup \,]-\pi/2, \pi/2[\, \cup \,]3\pi/2, 5\pi/2[\dots .$$

This result can be justified also by

```
Plot[{Abs[Tan[t/2]],1},{t,-5π/2,5π/2},
Ticks→{π/2Range[-5,5,2],{1,2}},ImageSize→175,
PlotStyle→{Black,{Black,Dotted}}]
```

See Figure 4.4.

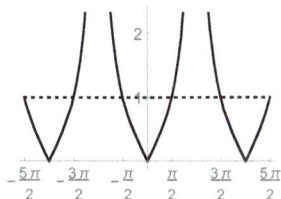

Figure 4.4: Justification of an inequality in the solution of problem 4.3.

The graph of the solution for some particular values of the constants of integration is given below:

```
ξ[t_]:=x[t]/.sol/.{C[1]→1,C[2]→1}
Plot[ξ[t],{t,-3π,3π},ImageSize→150,AspectRatio→.75,
Ticks→{4Range[-2,2],2Range[2]},PlotStyle→Black]
```

See Figure 4.5.

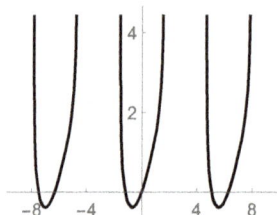

Figure 4.5: Particular solution of problem 4.3.

Figure 4.4 shows that the limits at ±∞ do not exist. Indeed,

```
Limit[ξ[t],t→#]&/@{-∞,∞}
{{Limit[-1+Cos[t]+Sin[t]+2ArcTanh[Tan[t/2]]Sin[t],
t→-∞]},
{Limit[-1+Cos[t]+Sin[t]+2ArcTanh[Tan[t/2]]Sin[t],
t→∞]}}
```

It happens so because the particular solution is defined on a union of disjoint intervals. The graph on the $[-\frac{\pi}{2},\frac{\pi}{2}]$ interval looks like:

```
Plot[ξ[t],{t,-π/2,π/2},ImageSize→140,AspectRatio→ 3/4,
Ticks→{π/2 Range[-1,1],2Range[2]},PlotStyle→Black]
```

See Figure 4.6.

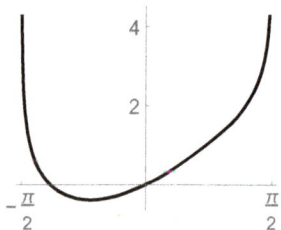

Figure 4.6: Solution of problem 4.3 on an interval.

```
FindMinimum[ξ[t],{t,-π/4}]
{-0.386406,{t→-.691071}}
```

Based on this minimum and the periodicity of the solution, we conclude that the range of the ξ function is the interval $[-0.386406, +\infty[$.

4.4. Let us find the solutions of the following second-order linear inhomogeneous differential equation with constant coefficients [64, p. 42],

$$x''(t) + 4x'(t) + 8x(t) = \frac{4}{\exp(2t)\sin(2t)}, \quad t \in \mathbb{R} \setminus \left\{ \frac{k\pi}{2} \mid k \in \mathbb{Z} \right\}.$$

4.4

```
Clear[t,x] (* Clears the values and definitions *)
sol=DSolve[x''[t]+4x'[t]+8x[t]==Exp[2t]Sin[2t]/4,x[t],t]
//Flatten//FullSimplify (* Solve the equation *)
{x[t]→ e⁻²ᵗ((-2t+C[2])Cos[2t]+(C[1]+Log[Sin[2t]])
Sin[2t])}
```

We set a particular solution.

```
ξ[t_]:=x[t]/.sol/.{C[1]→1,C[2]→1}//FullSimplify
```

Its graph follows.

```
Plot[ξ[t],{t,-3π,3π},ImageSize→175,PlotRange→
{3π{-1,1},{-2,2}},Ticks→{πRange[-3,3],Range[-2,2]},
PlotStyle→Black]
```

See Figure 4.7.

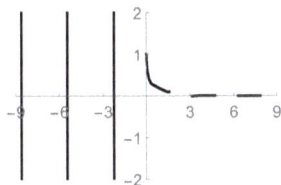

Figure 4.7: Particular solution for problem 4.4.

Remark. The domain of definition of the solution is given by the condition $\sin(2t)>0$.

$\square$

4.5. This is a second-order linear ordinary differential equation of *Euler* type whose solution depends on two real parameters [64, p. 43],

$$t^2 x''(t) + t x'(t) - a^2 x(t) = t^b, \quad a, b \in \mathbb{R}, \quad t > 0.$$

4.5 We note that this equation is of Euler type and the system returns a complex valued solution depending on a, b, and the integration constants. Therefore, we consider certain cases.

```
Clear[a,b,t,x] (* Clear the values and definitions *)
eqn=t²x''[t]+t x'[t]-a²x[t]==t^b;
```

We distinguish several cases.

▷ $a \neq b,\ a \neq 0$.

```
Assuming[{a,b}∈Reals&&a≠b&&a≠0&&t>0,
sol=DSolve[eqn,x[t],t]]//TrigToExp//Simplify//
Flatten
```
$\{x[t] \rightarrow \frac{1}{2}(-\frac{2t^b}{a^2-b^2} + t^{-a}(C[1]-iC[2]) + t^a(C[1]+iC[2]))\}$

Considering certain substitutions,

```
x₁[t_]:=x[t]/.sol/.{C[1]→k1+k2,C[2]→(k1-k2)i}
//FullSimplify
```

the solution just obtained looks like:

```
x₁[t]
```
$\{k1\ t^{-a}+k2\ t^a-\frac{t^b}{a^2-b^2}\}$

▷ $a = b,\ a \neq 0$.

```
Clear[a,k1,k2,t,x]
Assuming[a∈Reals&&a≠0&&t>0,
sol=DSolve[t²x''[t]+t x'[t]-a²x[t]==t^a,x[t],t]]
//TrigToExp//FullSimplify//Flatten
```
$\{x[t] \rightarrow \frac{1}{2}t^{-a}(C[1]-iC[2]) + \frac{t^a(-1+2a^2(C[1]+iC[2])+2a\text{Log}[t])}{4a^2}\}$

Considering certain substitutions,

```
x₂[t_]:=x[t]/.sol/.{a Log[t]→Log[t^a],C[1]→k1+k2,
C[2]→(k1-k2)I}//FullSimplify
```

the solution just obtained looks like:

```
x₂[t]
```
$\{k1\ t^{-a}+\frac{t^a(-1+4a^2 k2+2\text{Log}[t^a])}{4a^2}\}$

▷ $a = 0$, $b \neq 0$, $b \neq 1$.

```
Clear[b,t,x]
Assuming[b∈Reals&&b≠0&&b≠1&&t>0,
sol=DSolve[t²x''[t]+t x'[t]==t^b,x[t],t]]//Simplify
//Flatten
```
$\{x[t] \to \frac{t^b}{b^2} + C[2] + C[1] \text{Log}[t]\}$

▷ $a = 0$, $b = 1$.

```
Clear[t,x]
Assuming[t>0,
sol=DSolve[t²x''[t]+t x'[t]==t,x[t],t]]//Flatten
```
$\{x[t] \to t + C[2] + C[1] \text{Log}[t]\}$

▷ $a = b = 0$.

```
Clear[t,x]
Assuming[t>0,
sol=DSolve[t²x''[t]+t x'[t]==1,x[t],t]]//Flatten
```
$\{x[t] \to C[2] + C[1] \text{Log}[t] + \frac{\text{Log}[t]^2}{2}\}$

4.6. We study an initial value problem with varying initial conditions. The coefficients of this equation are rational functions of the independent variable t:

$$x''(t) - \left(\frac{1}{t} - \frac{3}{16t^2} \right) x(t) = 0, \quad x(1) = a, \quad x'(1) = b, \quad a, b, t \in \mathbb{R}, \quad t \neq 0.$$

4.6

```
Clear[a,b,t,x]
sol=DSolve[{x''[t]-(1/t - 3/16t²)x[t]==0,x[1]==a,x'[1]==b},
x[t],t]//Flatten (* Solve the equation *)
```
$\{x[t] \to \frac{1}{8} e^{-2-2\sqrt{t}} (5ae^4 - 4be^4 + 3ae^{4\sqrt{t}} + 4be^{4\sqrt{t}}) t^{1/4}\}$

It is obvious that $t > 0$.

We show the dependence on parameters a and b of the solution by the Manipulate command.

```
Manipulate[
Plot[x[t]/.sol/.{a→ α,b→ β},{t,0,3},Ticks→{Range[3],
Automatic},ImageSize→160,PlotStyle→Black],
{α,-3,3},{β,-1,3},SaveDefinitions→True]
```

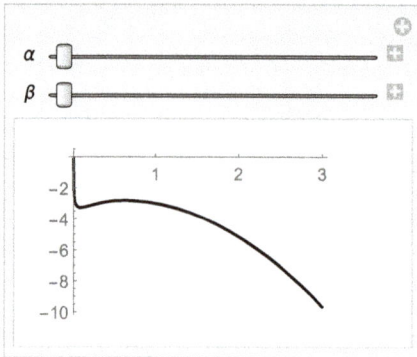

See Figure 4.8.

4.7. We introduce some boundary value problems to second-order linear differential equations and find out how many solutions has each of these problems,

$$x''(t) + x(t) = 0, \quad x(0) = 0, \quad x(\pi) = 0, \quad 0 < t < \pi, \tag{4.1}$$

$$x''(t) + x(t) = 1, \quad x(0) = 0, \quad x(1) = 0, \quad 0 < t < 1, \tag{4.2}$$

$$x''(t) + x(t) = 0, \quad x(0) = 0, \quad x(\pi) = 1, \quad 0 < t < \pi. \tag{4.3}$$

4.7

▷ We discuss problem (4.1) above.

```
Clear[t,x]
DSolve[{x''[t]+x[t]==0,x[0]==x[π]==0},x[t],t]
//Flatten
{x[t]→C[1]Sin[t]}
```

This problem has infinitely many solutions depending on the constant of integration C[1].
We study the problem numerically.

```
nsol=NDSolve[{x''[t]+x[t]==0,x[0]==x[π]==0},x[t],
{t,0,π}]//Flatten;
Plot[x[t]/.nsol,{t,0,π},ImageSize→160,
Ticks→{{1,2,3},Range[-1,1,.5]},PlotStyle→Black]
```

See Figure 4.9.

The numerical approach returns us only the null solution. The null solution belongs to the family of analytical solutions for C[1]= 0.

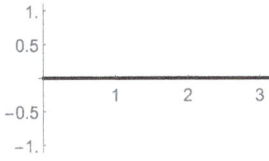

Figure 4.9: Numerical solution to problem 4.1.

▷ We discuss problem (4.2) above.

```
Clear[t,x]
sol=DSolve[{x''[t]+x[t]==1,x[0]==x[1]==0},x[t],t]//
Flatten
{x[t]→1-Cos[t]+Cot[1]Sin[t]-Csc[1]Sin[t]}
```

Problem (4.2) has a unique solution. We check this conclusion by an elementary method. We look for the general solution of the equation,

```
sol=DSolve[x''[t]+x[t]==1,x[t],t]//Flatten
{x[t]→1+C[1]Cos[t]+C[2]Sin[t]}
```

and find the suitable constants to satisfy the boundary conditions:

```
Reduce[(x[t]/.sol/.t→0)==0&&(x[t]/.sol/.t→1)==0,
{C[1],C[2]},Reals]
C[1]==-1&&C[2]==Cot[1]-Csc[1]
```

Thus, this boundary value problem has a unique solution because the constants of integration there exist and are unique.
We also study the problem numerically:

```
nsol=NDSolve[{x''[t]+x[t]==1,x[0]==x[1]==0},x[t],
{t,0,1}]//Flatten;
```

Now we plot the analytical and the numerical solutions.

```
{Plot[x[t]/.sol/.{C[1]→-1,C[2]→Cot[1]-Csc[1]},##],
Plot[x[t]/.nsol,##]}&[{t,0,1},Ticks→{{.5,1},
-{.07,.14}},ImageSize→160,PlotStyle→Black]
```

See Figure 4.10.

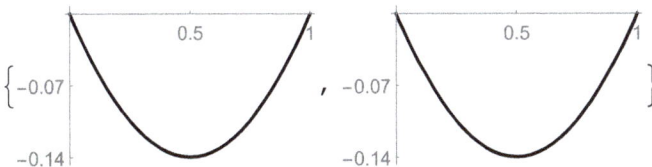

Figure 4.10: Analytical and numerical solutions of problem 4.2.

The pictures strongly suggest similar solutions.

▷ We discuss problem (4.3) above.

```
Clear[t,x]
DSolve[{x''[t]+x[t]==0,x[0]==0,x[π]==1},x[t],t]//
Flatten
{}
```

This result suggests that there is no solution to this problem. We check this result by an elementary means:

```
sol=DSolve[x''[t]+x[t]==0,x[t],t]//Flatten;
Reduce[(x[t]/.sol/.t→0)==0&&(x[t]/.sol/.t→ π)==1,
{C[1],C[2]},Reals]
False
```

We got the same result.
Let us see if the system supplies us with a numerical solution.

```
nsol=NDSolve[{x''[t]+x[t]==0,x[0]==0,x[π]==1},
x[t],{t,0,π}]//Chop//Flatten.
{Plot[x[t]/.nsol,##],Plot[x[t]/.nsol,##,
PlotRange→{{0,π},{0,2.0 10⁶}}]]}&[{t,0,π},ImageSize→
150,Ticks→{{1,2,3},{2.0×10⁷}},PlotStyle→Black]
```

See Figure 4.11.

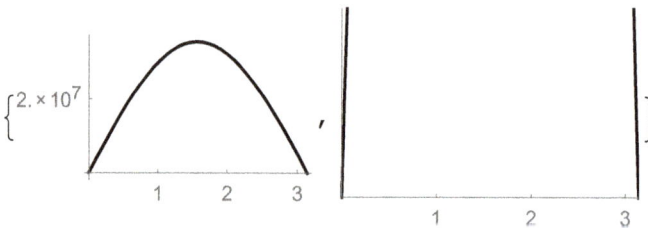

Figure 4.11: Numerical solution to problem 4.3.

4.8. We study whether the following boundary value problem has a solution:

$$x''(t) + \lambda^2 x(t) = 0, \quad x(0) = 0, \quad x(a) = 0, \quad 0 < t < a.$$

4.8

```
Clear[a,λ,t,x]
Assuming[a>0,DSolve[{x''[t]+λ²x[t]==0,x[0]==0,
x[a]==0},x[t],t]]//Flatten
```

$$\left\{ x[t] \rightarrow \begin{bmatrix} C[1]\mathrm{Sin}[t] & \acute{n} \in \mathrm{Integers}\&\&a > 0\&\& \left(\lambda == \frac{2\acute{n}\pi}{a} \,\|\right. \\ & \left. \lambda == -\frac{-\pi-2\acute{n}\pi}{a}\right) \\ 0 & \mathrm{True} \end{bmatrix} \right\}$$

Remark. We note that symbol $\acute{n}$ in *Mathematica* is a letter-like form and is used to represent a formal parameter that will never be assigned a value. □

Remark. The result says that if there exists an integer n so that $\lambda = \frac{2n\pi}{a}$ or $\lambda = \frac{\pi+2n\pi}{a}$, then the problem has infinitely many solutions depending on the constant of integration C[1]. Otherwise, the problem only admits the null solution. □

We solve the equation supposing that $\lambda=2n\pi/a$. Then its solution is of the form

```
sol=DSolve[{x''[t]+((2nπ)/a)²x[t]==0,x[0]==x[a]==0},
x[t],t]/.a→2//Flatten
```

$$\left\{ x[t] \rightarrow \begin{bmatrix} C[1]\mathrm{Sin}[t] & \acute{n} \in \mathrm{Integers}\&\&a > 0\&\& \left(n == \acute{n} \,\|\right. \\ & \left. n == -\frac{-\pi-2\acute{n}\pi}{a}\right) \\ 0 & \mathrm{True} \end{bmatrix} \right\}$$

and we plot it.

```
Manipulate[
Plot[(x[t]/.sol//First//Flatten//First)/.C[1]→1
/.n→ v,{t,0,2},Ticks→{Range[2],.5Range[-2,2]},
ImageSize→160],{v,Range[4]}]
```

See Figure 4.12.

4.9. We introduce some boundary value problems to second-order linear differential equations and want to find out how many solutions each of these problems has:

$$x''(t) + \lambda^2 x(t) = 0, \quad x(0) = 0, \quad x'(a) = 0, \quad 0 < t < a, \tag{4.4}$$

$$x''(t) + \lambda^2 x(t) = 0, \quad x'(0) = 0, \quad x(a) = 0, \quad 0 < t < a, \tag{4.5}$$

$$x''(t) + \lambda^2 x(t) = 0, \quad x'(0) = 0, \quad x'(a) = 0, \quad 0 < t < a. \tag{4.6}$$

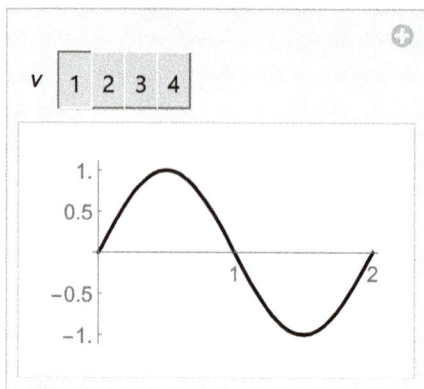

Figure 4.12: Dynamical solution of problem 4.8.

4.9

▷ Problem (4.4).

```
Clear[t,x,a,λ]
Assuming[a>0,DSolve[{x''[t]+λ²x[t]==0,x[0]==
x'[a]==0},x[t],t]]//Flatten
```

$$\left\{x[t]\to\left[\begin{array}{ccc} C[1]\sin[t\lambda] & (a>0\&\&\lambda=0) \| \left(\dot{n}\in Integers\&\&a>0\right. \\ & \&\&(\lambda==\dfrac{-\pi+4\dot{n}\pi}{2a}\|\lambda==\dfrac{\pi+4\dot{n}\pi}{2a})\right) \\ 0 & True \end{array}\right]\right\}$$

The result shows that if there exists an integer n so that $\lambda = \frac{-\pi+4n\pi}{2a}$ or $\lambda = \frac{\pi+4n\pi}{2a}$, then the problem has infinitely many solutions depending on the constant of integration C[1]. Otherwise, the problem only admits the null solution.

▷ Problem (4.5).

```
Clear[t,x,a,λ]
Assuming[a>0,DSolve[{x''[t]+λ²x[t]==0,x'[0]==x[a]
==0},x[t],t]]//Flatten
```

$$\left\{x[t]\to\left[\begin{array}{ccc} C[1]\cos[t\lambda] & (a>0\&\&\lambda=0) \| \left(\dot{n}\in Integers\&\&a>0\right. \\ & \&\&(\lambda==\dfrac{-\pi+4n\pi}{2a}\|\lambda==\dfrac{\pi+4\dot{n}\pi}{2a})\right) \\ 0 & True \end{array}\right]\right\}$$

The result shows that if there exists an integer n so that $\lambda = \frac{-\pi+4n\pi}{2a}$ or $\lambda = \frac{\pi+4n\pi}{2a}$, then the problem has infinitely many solutions depending on the constant of integration C[1]. Otherwise, the problem admits only the null solution.

▷ Problem (4.6).

Let us examine the case of the third problem.

```
Clear[t,x,a,λ]
Assuming[a>0,DSolve[{x''[t]+λ²x'[t]==0,x'[0]==x'[a]
==0},x[t],t]]//Flatten
```

$$x[t] \rightarrow \begin{cases} C[1]\text{Cos}[t\lambda] & (a > 0\&\&\lambda = 0) \parallel \left(\dot{n} \in \text{Integers}\&\&a > 0 \right. \\ & \left. \&\&(\lambda == \frac{2\dot{n}\pi}{a} \parallel \lambda == \frac{\pi+2\dot{n}\pi}{a}) \right) \\ 0 & \text{True} \end{cases}$$

This result shows that if there exists an integer n so that $\lambda = \frac{2n\pi}{a}$ or $\lambda = \frac{\pi+2n\pi}{a}$, then the problem has infinitely many solutions depending on the constant of integration C[1]. Otherwise, the problem admits only the null solution.

4.10. We study whether the following problem with initial and *Robin conditions* has a solution,

$$\begin{cases} x''(t) + \lambda x(t) = 0, & 0 < t < a, \quad \lambda > 0, \\ x(0) = x_0, & h(x(a) - x_a) + \gamma x'(a) = 0. \end{cases}$$

4.10

```
<< Notation`
Symbolize[x_]

Clear[a,h,t,x,γ,λ]

sol=Assuming[a>0&&λ > 0&&{h,x₀,xₐ,γ}∈Reals,
DSolve[{x''[t]+λ x[t]==0,x[0]==x₀,h(x[a]-xₐ)+γ x'[a]
==0},x[t],t]]//Flatten//FullSimplify
```

$$\{x[t] \rightarrow \frac{x_0\gamma \sqrt{\lambda}\text{Cos}[(a-t)\sqrt{\lambda}]+ h\,x_0\text{Sin}[(a-t)\sqrt{\lambda}]+h\,x_a\text{Sin}[t\sqrt{\lambda}]}{\gamma \sqrt{\lambda}\text{Cos}[a\sqrt{\lambda}]+h\,\text{Sin}[a\sqrt{\lambda}]} \}$$

This problem with initial and Robin conditions has a unique solution whenever $\lambda > 0$.

We plot the solution just obtained for some particular values of the parameters.

```
Block[{a=2,λ=3.5,h=1.25,γ=1.5,x₀=5,xₐ=2.2},
Plot[x[t]/.sol,{t,0,a},ImageSize→160,
Ticks→{{1,2},2Range[-2,2]}]]
```

See Figure 4.13.

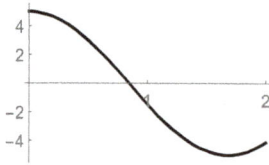

Figure 4.13: Particular solution for problem 4.10.

4.11. We study whether the following problems with initial value conditions have solutions [64, p. 48]:

$$x''(t) + t\,x'(t) + (2n-1)x(t) = 0, \quad x(0) = 0, \quad x'(0) = 1, \quad n \in \mathbb{N}, \tag{4.7}$$

$$x''(t) + t\,x'(t) + 2n\,x(t) = 0, \qquad\quad x(0) = 0, \quad x'(0) = 1, \quad n \in \mathbb{N}. \tag{4.8}$$

4.11

▷ Problem (4.7).

```
Clear[t,x,n]
sol1=Assuming[n≥0&&n∈Integers,
DSolve[{x''[t]+t x'[t]+(2n-1)x[t]==0,x[0]==0,x'[0]
==1},x[t],t]]//Flatten//FullSimplify
{x[t]→t Hypergeometric1F1[n,3/2,-(t²/2)]}
```

▷ Equation (4.8).

```
Clear[t,x,n]
sol2=Assuming[n≥0&&n∈Integers,
DSolve[{x''[t]+t x'[t]+2n x[t]==0,x[0]==0,x'[0]==1},
x[t],t]]//Flatten//FullSimplify
{x[t]→t Hypergeometric1F1[1/2+n,3/2,-(t²t/2)]}
```

Now we plot both solutions setting a particular parameter n.

```
Block[{n=3},
{Plot[x[t]/.sol1,##],Plot[x[t]/.sol2,##]}&[{t,0,5},
ImageSize→200,Ticks→{Range[5],Range[-.1,.4,.1]}],
PlotStyle→Black]
```

See Figure 4.14.

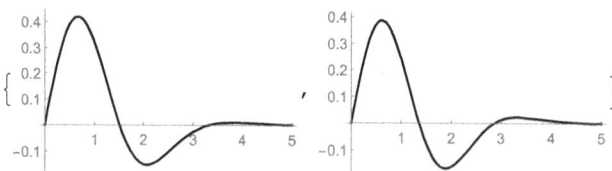

Figure 4.14: Solutions to equations (4.7) and (4.8).

We gather the previous equations.

```
Clear[t,x,n]
sol3=Assuming[n≥0&&n∈Integers,
DSolve[{x''[t]+t x'[t]+n x[t]==0,x[0]==0,x'[0]==1},x[t],
t]]//Flatten//FullSimplify
{x[t]→t Hypergeometric1F1[(1+n)/2,3/2,-(t²/2)]}
```

By the `Manipulate` command, we plot several solutions.

```
Manipulate[
Plot[Evaluate[x[t]/.sol3/.n→ v],{t,0,5},ImageSize→200,
Ticks→{Range[5],Range[-.2,.7,.2]},PlotStyle→Black],
{v,Range[7]}]
```

See Figure 4.15.

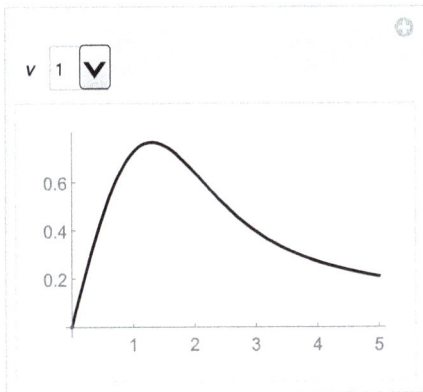

Figure 4.15: Dynamical solution for problem 4.11.

4.2 Bessel differential equation

For details on Bessel differential equations, see Sections 1.6 and 2.1.

4.12. The first Bessel differential equation considered here is of the form,

$$t^2 x''(t) + \left(t^2 + \frac{5}{36} \right) x(t) = 0, \quad t \in \mathbb{R}.$$

We look for its solution.

4.12

```
Clear[t,x]
t²x''[t]+(t²+5/36)x[t]==0;  (* The equation *)
sol=DSolve[%,x,t][[1]]  (* Solve the equation *)
%%/.%//FullSimplify  (* Check the solution *)
ξ[t_]:=x[t]/.sol/.{C[1]→1,C[2]→2}  (* Set a
particular solution *)
{x→Function[{t}, √t BesselJ[1/3,t]C[1]
+√t BesselY[1/3,t]C[2]]}
{True}
```

We plot a particular solution (the rightmost picture), a numerical particular solution
(the mid picture), and the difference between them (the leftmost picture).

```
{Plot[ξ[t],##,Ticks→{5Range[3],Range[-1.5,1.5]}],
Plot[x[t]/.soln,##,Ticks→{5Range[3],Range[-1.5,1.5]}],
Plot[ξ[t]-x[t]/.soln,##,Ticks→{5Range[3],
Automatic}]}&[{t,.1,17},PlotStyle→Black,
ImageSize→165]
```

See Figure 4.16.

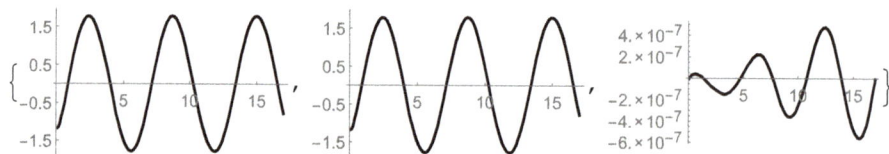

Figure 4.16: Solution for problem 4.12.

4.13. A second-order linear ordinary differential equation is introduced, whose solutions are expressed by Bessel and Γ functions [64, p. 40],

$$t^2x''(t) + a\,t\,x'(t) + (b + c\,t^m)x(t) = 0, \quad a,b,c,m \in \mathbb{R}, \quad t > 0 \quad m \neq 0.$$

4.13 If $m = 0$, then we have an Euler equation. Suppose that $m \neq 0$.

```
Clear[a,b,c,m,t,x]
sol=DSolve[t²x''[t]+a t x'[t]+(b+c tᵐ)x[t]==0,x,t]
//Flatten//FullSimplify
```

(* Solve the equation *)

$$\{x \to \text{Function}[\{t\}, c^{1/(2m)-a/(2m)} m^{-1/m+a/m}(t^m)^{1/(2m)-a/(2m)}$$
$$\text{BesselJ}[-\tfrac{\sqrt{1-2a+a^2-4b}}{m}, \tfrac{2\sqrt{c}\sqrt{t^m}}{m}]C[1]\text{Gamma}[1-\tfrac{\sqrt{1-2a+a^2-4b}}{m}]$$
$$+c^{1/(2m)-a/(2m)} m^{-(1/m)+a/m}(t^m)^{1/(2m)-a/(2m)}$$
$$\text{BesselJ}[\tfrac{\sqrt{1-2a+a^2-4b}}{m}, \tfrac{2\sqrt{c}\sqrt{t^m}}{m}]C[2]\text{Gamma}[1+\tfrac{\sqrt{1-2a+a^2-4b}}{m}]\}$$

We set a particular solution,

```
ξ[t_]:=x[t]/.sol/.{a→1,b→-1,c→1,m→4,C[1]→1,C[2]→2}
```

and plot it.

```
{Plot[ξ[t],{t,0,7},Ticks→{Range[7],Range[-1,5,2]},
ImageSize→160,PlotStyle→Black],
RevolutionPlot3D[ξ[t],{t,0,5},Mesh→2,PlotStyle→
{Black,Opacity[.15]},Ticks→{{0,2.5,5},3{-1,0,1},
2.8{-1,0,1}},ImageSize→160,RevolutionAxis→"X"]}
```

See Figure 4.17.

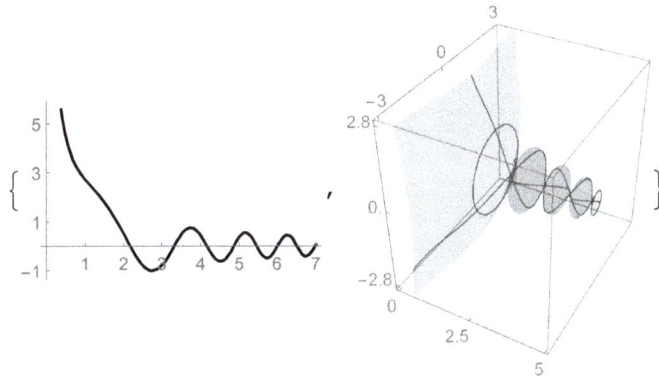

Figure 4.17: Solution and rotated particular solutions for problem 4.13.

4.14. An interesting initial value problem having solutions expressed by Bessel functions is introduced,

$$x''(t) + 3\frac{x'(t)}{t} + 4x(t) = \sin(t), \quad x(1) = 1, \quad x'(1) = 1, \quad t > 0.$$

We look for its solutions, if any.

4.14

```
Clear[t,x]
eqn={x''[t]+3x'[t]/t+4x[t]==Sin[t],x[1]==1,x'[1]==1};
sol=DSolve[eqn,x,t]; (* Solve the equation *)
asol[t_]:=Activate[x[t]/.sol/.{∞ →7}]//FullSimplify (*
Take only a partial sum *)
series[t_]:=Series[asol[t],{t,0,4}] (* Expand in power
series up to order 4 *)
normal[t_]:=Normal[series[t]]
```

To facilitate the plot of the solution, we discretize the data by equidistant nodes.

```
tab=Table[{t,normal[t][[1]]//N},{t,.01,4.01,.1}]
{{0.01,-5804.82},{0.11,-48.4334},{0.21,-12.8862},
{0.31,-5.32051},{0.41,-2.4295},{0.51,-0.991507},
{0.61,-0.166555},{0.71,0.344988},{0.81,0.67311},
{0.91,0.882645},{1.01,1.0097},{1.11,1.07634},
{1.21,1.09728},{1.31,1.08302},{1.41,1.04162},
{1.51,0.979563},{1.61,0.902247},{1.71,0.814312},
{1.81,0.719781},{1.91,0.622149},{2.01,0.524438},
{2.11,0.429221},{2.21,0.338635},{2.31,0.254395},
{2.41,0.177804},{2.51,0.109765},{2.61,0.050808},
{2.71,0.0011053},{2.81,-0.0394897},{2.91,-0.071409},
{3.01,-0.0953285},{3.11,-0.112122},{3.21,-0.122816},
{3.31,-0.128535},{3.41,-0.130462},{3.51,-0.129781},
{3.61,-0.12764},{3.71,-0.125107},{3.81,-0.123134},
{3.91,-0.12253},{4.01,-0.123936}}
```

The graph of the solution is supplied below.

```
ListLinePlot[tab,ImageSize→175,PlotStyle→Black,
Ticks→{Range[4],.5Range[-3,2]}]
```

See Figure 4.18.

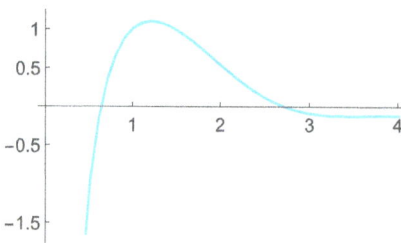

Figure 4.18: Analytical solution for problem 4.14.

For the sake of completeness here, we insert the numerical version of the above approach.

```
Clear[t,x]
NDSolve[eqn,x,{t,.01,4}];
Plot[x[t]/.%,{t,.01,4},Ticks→{Range[4],.5Range[-3,2]},
PlotStyle→Black,ImageSize→175]
```

See Figure 4.19.

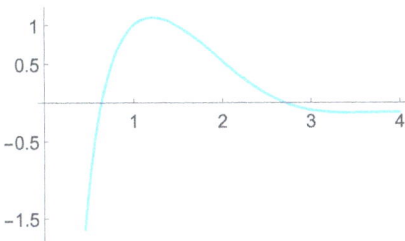

Figure 4.19: Numerical solution for problem 4.14.

Remark. We note that the last two figures are similar.   □

4.3 Legendre differential equation

For details on Legendre differential equations, see Sections 1.7 and 2.1.

4.15. We present a second-order differential equation, whose solutions are given by Legendre functions,

$$(1 - t^2)x''(t) - 2t\, x'(t) + n(n+1)x(t) = 0, \quad n > 0, \quad t \in\,]-1, 1[.$$

For $n \in \mathbb{N}$, one gets solutions expressed by the Legendre polynomials.

4.15 We solve the equation and select the Legendre functions of the first and second kinds.

```
Clear[t,x,n]
eqn=(1-t²)x''[t]-2t x'[t]+n(n+1)x[t]==0;
sol=DSolve[%,x,t]//Flatten (* Solve the equation *)
%%/.%//FullSimplify (* Check the solution *)
x₁[t_]:=x[t]/.sol/.{C[1]→1,C[2]→0} (* Select the
LegendreP[n,t] built-in function *)
x₂[t_]:=x[t]/.sol/.{C[1]→0,C[2]→1} (* Select the
LegendreQ[n,t] built-in function *)
{x→Function[{t},C[1] LegendreP[n,t]
+C[2] LegendreQ[n,t]]}
True
```

Remark. The system returned us Legendre polynomials for the particular constants of integration. These functions are defined on the whole real axis. □

The plot of the P_0, P_1, P_2, and P_3 Legendre polynomials of the first kind have been given by the leftmost picture of Figure 1.21. Here, we exhibit the plot of the P_4, P_5, and P_6 Legendre polynomials of the first kind.

```
Plot[Evaluate[x₁[t]/.n→Range[4,6]//Flatten],
{t,-1.6,1.6},Ticks→{{-1,1},{-1,1}},ImageSize→190,
AspectRatio→.6,PlotLegends→"Expressions",
PlotRange→Outer[Times,1.1{1,1},1.1{-1,1}],
LabelStyle→10]
```

See Figure 4.20.

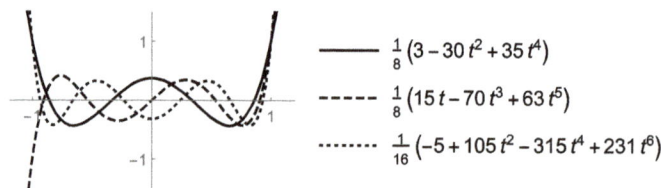

Figure 4.20: The P_4, P_5, and P_6 Legendre polynomials in problem 4.15.

The plot of the Q_0, Q_1, Q_2, and Q_3 Legendre polynomials of the second kind have been given by rightmost picture in Figure 1.21. Here, we exhibit the plot of the Q_4 and Q_5 Legendre polynomials of the second kind.

```
Plot[Evaluate[x₂[t]/.n→Range[4,5]//Flatten],
{t,-1,1},Ticks→{{-1,1},{-1,1}},AspectRatio→.75,
PlotRange→Outer[Times,{1.1,1.2},{-1,1}],
ImageSize→150,PlotLegends→"Expressions",
LabelStyle→10]
```

See Figure 4.21.

4.16. Below there is an associated Legendre differential equation and we plot its solutions for certain values of parameters,

$$(1 - t^2)x''(t) - 2t\,x'(t) + \left(n(n+1) - \frac{m^2}{1 - t^2}\right)x(t) = 0, \quad m \in \mathbb{N}, \quad t \in \,]{-}1, 1[.$$

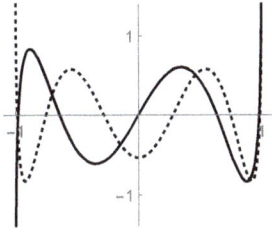

$$-\!\!\!-\!\!\!-\quad \frac{55t}{24}-\frac{35\,t^3}{8}+\frac{1}{8}\left(3-30\,t^2+35\,t^4\right)\left(-\frac{1}{2}\log(1-t)+\frac{1}{2}\log(1+t)\right)$$

$$\cdots\cdots\cdots\quad -\frac{8}{15}+\frac{49\,t^2}{8}-\frac{63\,t^4}{8}+\frac{1}{8}t\left(15-70\,t^2+63\,t^4\right)\left(-\frac{1}{2}\log(1-t)+\frac{1}{2}\log(1+t)\right)$$

Figure 4.21: The Q_4 and Q_5 Legendre polynomials in problem 4.15.

4.16

```
Clear[m,n,p,q,t,x]
sol=DSolve[(1-t²)x''[t]-2t x'[t]+(n(n+1)-m²/(1-t²))x[t]==0,x,
t]//Flatten (* Solve the equation *)
x₁[t_]:=x[t]/.sol/.{C[1]→1,C[2]→0,m→1} (* The first
particular solution *)
x₂[t_]:=x[t]/.sol/.{C[1]→0,C[2]→1,m→1} (* The second
particular solution *)
{x→Function[{t},C[1] LegendreP[n,m,t]
+C[2] LegendreQ[n,m,t]]}
```

We plot $x_1(t)$ for certain values of parameter n.

```
Plot[Evaluate[x₁[t]/.n→{0,2,4}],{t,-1,1},
Ticks→{Range[-1,1,.5],Range[-2,2]},ImageSize→175,
PlotStyle→{Black,{Black,Dashed},{Black,Dotted}},
PlotLegends→"Expressions"]
```

See Figure 4.22.

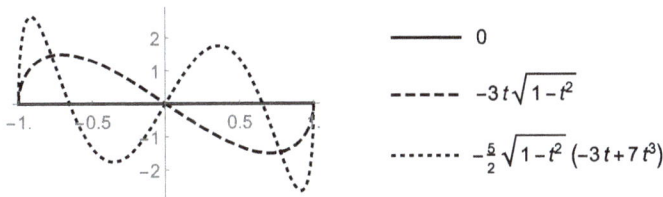

Figure 4.22: First particular solutions for problem 4.16.

We plot $x_2(t)$ for certain values of parameter n.

```
Plot[Evaluate[x₂[t]/.n→2{0,1,2}],{t,-1,1},
Ticks→{.5Range[-2,2],3Range[-2,1]},ImageSize→175,
PlotStyle→{Black,{Black,Dashed},{Black,Dotted}},
PlotLegends→Placed["Expressions",Below]]
```

See Fig. 4.23.

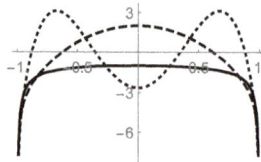

$$-\frac{1}{\sqrt{1-t^2}} \quad \text{-----} \quad \frac{\sqrt{1-t^2}\,(-2+3t^2)}{-1+t^2} - 3t\sqrt{1-t^2}\left(-\frac{1}{2}\log(1-t)+\frac{1}{2}\log(1+t)\right)$$

$$\text{........} \quad \frac{\sqrt{1-t^2}\,(16-115t^2+105t^4)}{6\left(-1+t^2\right)} - \frac{5}{2}\sqrt{1-t^2}\,(-3t+7t^3)\left(-\frac{1}{2}\log(1-t)+\frac{1}{2}\log(1+t)\right)$$

Figure 4.23: Other particular solutions for problem 4.16.

We show the variation of the solution with respect to the parameters n and m.

```
Manipulate[
Plot[x[t]/.sol/.{C[1]→1,C[2]→1,m→p,n→q},{t,-1,1},
Ticks→{.5Range[-2,2],Automatic},ImageSize→175,
PlotStyle→Black],
{{p,0,"m"},0,5,Appearance→"Labeled"},
{{q,1,"n"},1,5,Appearance→"Labeled"},
SaveDefinitions→True]
```

See Figure 4.24.

4.4 Mathieu differential equation

For details on Mathieu differential equations, see Sections 1.9 and 2.1.

4.17. Our first differential equation of Mathieu type is the following:

$$x''(t) + (3\cos(2t) + 5)x(t) = 0, \quad t \in \mathbb{R}.$$

This is the form under which this equation was introduced by Mathieu in 1868 [53].

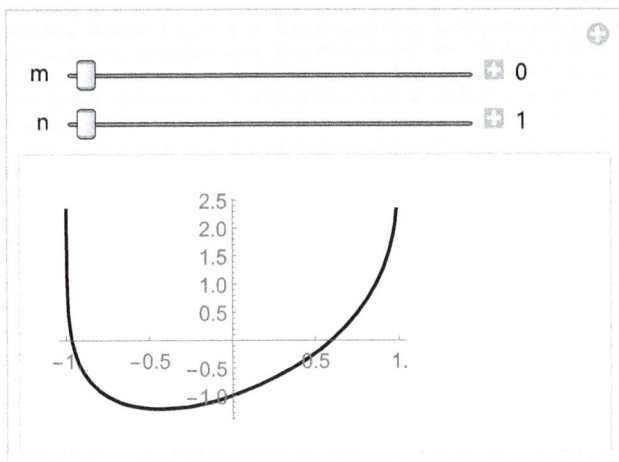

m ▢ 0

n ▢ 1

Figure 4.24: Dynamical solution for problem 4.16.

4.17

```
Clear[a,b,p,t,x]
DSolve[x''[t]+(3Cos[2t]+5)x[t]==0,x,t]//Flatten (*
Solve the equation *)
{x→Function[{t},
C[1]MathieuC[5,-3/2,t]+C[2]MathieuS[5,-3/2,t]]}
```

The surface generated by the product of particular solutions of this equation is given in the sequel:

```
Plot3D[MathieuC[5,-3/2,p]×MathieuS[5,-3/2,q],{p,-π,π},
{q,-π,π},Mesh→7,ImageSize→200,Ticks→Outer[Times,
{π,π,1},{-1,0,1}],PlotStyle→{Black,Opacity[.15]}]
```

See Figure 4.25.

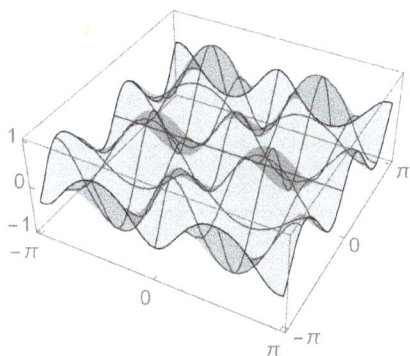

Figure 4.25: Plot of a particular solution for problem 4.17.

We manipulate this figure in order to get the dynamical surface that follows:

```
Animate[
Plot3D[MathieuC[5.,-3/2,a p] MathieuS[5.,-3/2,b q],
{p,-π,π},{q,-π,π},Ticks→Outer[Times,{π,π,1},{-1,0,1}],
ImageSize→175,Mesh→5],
PlotStyle→{Opacity[.15],Black},Boxed→False],
{a,-2.,2.},{b,-1.,3.},AnimationRunning→False,
SaveDefinitions→True,DefaultDuration→50,
AnimationDirection→ForwardBackward]
```

See Figure 4.26.

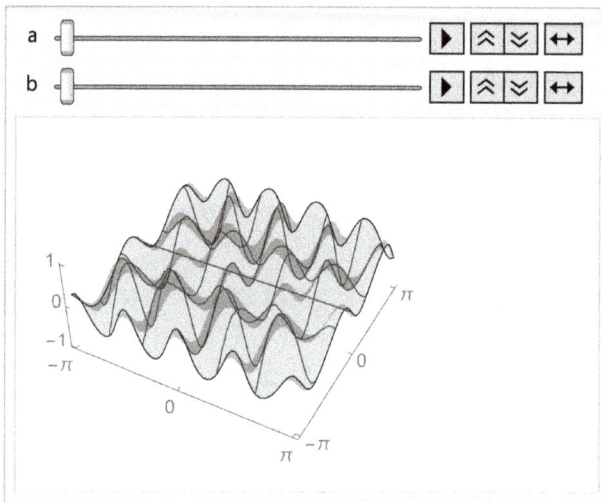

Figure 4.26: Dynamical plot of a particular solution for problem 4.17.

4.18. Below a Mathieu differential equation is introduced and studied,

$$x''(t) + (a - q\cos(2t))x(t) = 0, \quad a, q, t \in \mathbb{R}.$$

4.18

```
Clear[t,x]
With[{a=1,q=1/5}, (* Solve the equation for some
particular values of the parameters and the
constants of integration *)
sol=DSolve[x''[t]+(a-q Cos[2t])x[t]==0,x,t]/.
{C[1]→1,C[2]→-2}]//Flatten
{x→Function[{t},1 MathieuC[1,1/10,t]-2 MathieuS[1,1/10,t]]}
```

We plot the real and the imaginary parts of the solution.

```
Plot[{Evaluate[Re[x[t]/.First[sol]//N]],
Evaluate[Im[x[t]/.First[sol]//N]]},##],
Plot[Evaluate[ReIm[x[t]/.First[sol]//N]],##]
&[{t,-6,6},PlotLegends→"Expressions",Ticks→
{3Range[-2,2],Range[-1.5,1.5,1]},ImageSize→175,
PlotStyle→{Black,{Block,Dashed}}]
```

See Figure 4.27.

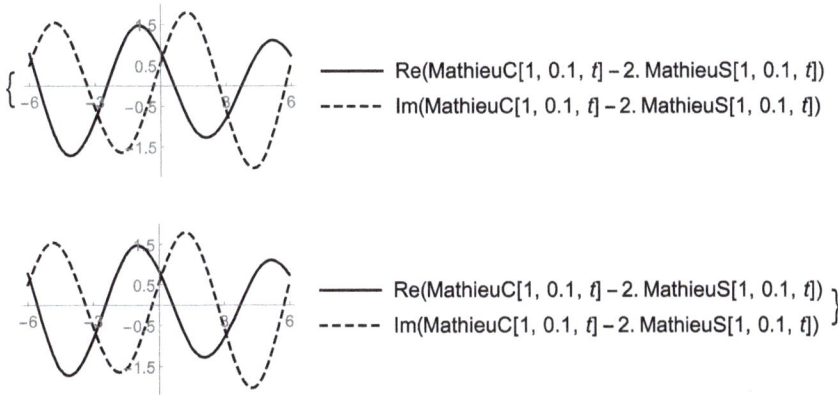

Figure 4.27: Plot of the particular solutions for problem 4.18.

We consider the real part and the imaginary part of the solution and study the dynamics of the trajectory.

```
Clear[a]
With[{time=40.}, (* [-time,time] is the interval
of time where the solution is studied *)
Animate[
ParametricPlot[Evaluate[Through[
{Re,Im}[x[t]/.First[sol]//N]]],{t,-a,a},
ImageSize→160,PlotStyle→Black},
Ticks→Outer[Times,{1,1},2Range[-7,7]],
Epilog→{PointSize[Medium],Black,
Point[Through[{Re,Im}[x[t]/.First[sol]//N]/.t→-a]],
Black,Point[Through[{Re,Im}[x[t]/.First[sol]//N]/.
t→a]]}],{{a,time/50},10^-7.,time},
DefaultDuration→25,AnimationRunning→False,
AnimationDirection→ForwardBackward,
SaveDefinitions→True]]
```

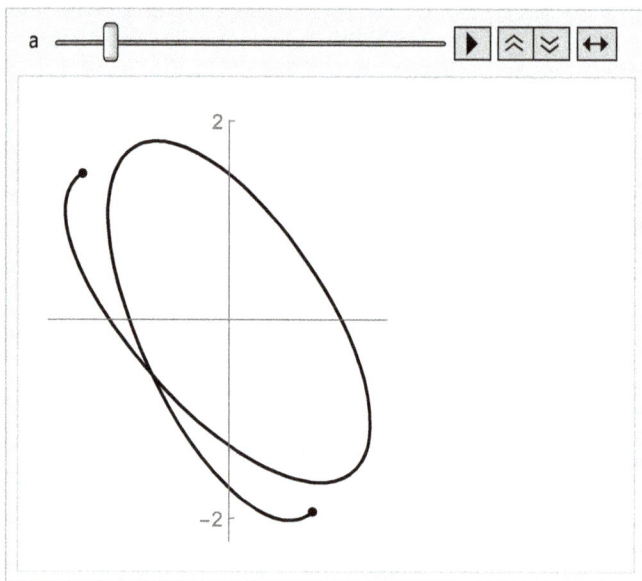

Figure 4.28: Dynamical approach to a particular solution for problem 4.18.

See Figure 4.28.

4.19. The second-order linear ordinary differential equation with rational coefficients exhibited below has a general solution expressed by Mathieu functions,

$$(t^2 - 1)x''(t) + t x'(t) - (2 + 6t^2)x(t) = 0, \quad t \in \ ]{-1},1[.$$

4.19

```
Clear[c,d,p,t,x]
sol=DSolve[(t^2-1)x''[t]+t x'[t]-(2+6t^2)x[t]==0,x[t],t]
//Flatten
{x[t]→C[1]MathieuC[5,-3/2,ArcCos[t]]
+C[2]MathieuS[5,-3/2,ArcCos[t]]}
```

We set a particular solution and plot it.

```
ξ[t_]:=x[t]/.sol/.{C[1]→-1,C[2]→2}
Plot[ξ[t],{t,-1,1},ImageSize→175,PlotStyle→Black,
Ticks→{{-1,1},{-1,1.5}}]
```

See Figure 4.29.

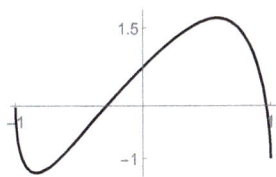

Figure 4.29: Particular solution for problem 4.19.

Now we rotate this figure in order to get the interesting surfaces below.

```
RevolutionPlot3D[ξ[t],{t,-1,1},Mesh→1,
Ticks→Outer[Times,{1,1,1},{-1,0,1}],ImageSize→150,
PlotStyle→Opacity[.1],PlotPoints→25,#]&/@
{{RevolutionAxis→"X",PlotLabel→Style["Around OX
axis",Black]},
{RevolutionAxis→"Y",ViewPoint→{15,-50,-50},
PlotLabel→Style["Around OY axis",Black]},
{RevolutionAxis→"Z",PlotLabel→Style["Around OZ
axis",Black]}}
```

See Figure 4.30.

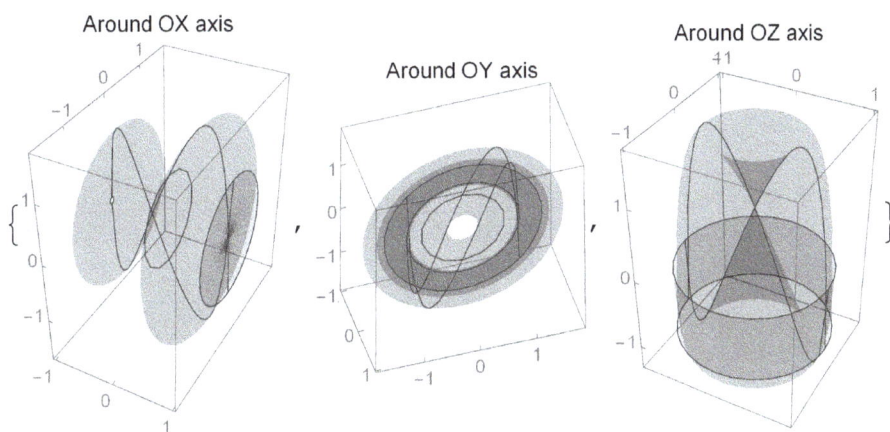

Figure 4.30: Rotating the particular solution for problem 4.19.

Remark. The presence of ArcCos[t] in the previous solution suggests that the equation can be given under a simpler form using trigonometric functions. □

4.5 Equations with discontinuous coefficients or right-hand sides

In this section, we show a few second-order ordinary differential equations having piecewise continuous right-hand sides and/or coefficients.

4.20. We look for the solutions of the next three initial value problems,

$$x''(t) + 2x'(t) + x(t) = 2\{t\}, \qquad\qquad x(0) = 1, \quad x'(0) = 0, \qquad t \in \mathbb{R}, \qquad (4.9)$$
$$x''(t) + 2x'(t) + x(t) = \text{sign}(\sin(5\,t)), \quad x(0) = 1, \quad x'(0) = -1, \qquad t \in \mathbb{R}, \qquad (4.10)$$

where $\{\cdot\}$ denotes the fractional part of a real number [56, p. 20].

4.20

▷ Problem (4.9).
The right-hand side of the equation is discontinuous because of the fractional part function.

```
Clear[t,x]
eqn=x''[t]+2x'[t]+x[t]==2 FractionalPart[t];
sol=DSolve[{eqn,x[0]==1,x'[0]==0},x[t],t]//Flatten//
Simplify (* Solve the equation *)
{x[t]→ e⁻ᵗ(5-2e+3t+t∫₁ᵗ 2eᴷ⁽²⁾FractionalPart[K[2]]dK[2]
+∫₁ᵗ-2eᴷ⁽¹⁾FractionalPart[K[1]]dK[1])}
```

We plot a discretization of the solution.

```
ListLinePlot[Table[x[t]/.sol/.t→k/10,{k,0,40}],
Ticks→{10Range[5],Automatic},ImageSize→150],
PlotStyle→Black
```

See Figure 4.31.

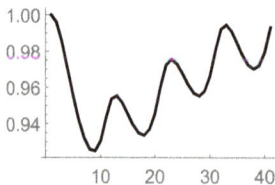

Figure 4.31: Plot of the solution to problem 4.9.

▷ Problem (4.10).

The right-hand side of the equation is discontinuous because of the signum function, Section 1.2.3.

```
Clear[t,x]
x''[t]+2x'[t]+x[t]==Sign[Sin[5t]];
sol=DSolve[{%,x[0]==1,x'[0]==-1},x[t],t]//Flatten//
Simplify (* Solve the equation *)
```
$\{x[t] \rightarrow -e^{-t}t + \frac{1}{Sign[Sin[5t]]}\}$

We plot its solution.

```
Block[{tend=25},
Manipulate[
Plot[x[t]/.sol,{t,0,τ},ImageSize→150,
Ticks→IntegerPart[{τ{1/2,1},{-1,1}}],
AxesOrigin→{0,0},PlotStyle→Black],
{{τ,IntegerPart[tend/2]},10^-5,tend}]]
```

See Figure 4.32.

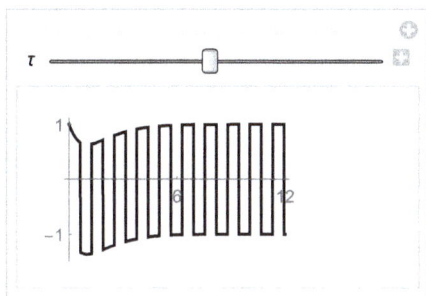

Figure 4.32: Plot of the solution to problem 4.10.

4.21. We look for the solution of an initial value problem having a coefficient and the right-hand side discontinuous,

$$x''(t) + \{t\}x'(t) + x(t) = \lfloor t \rfloor, \quad x(0) = 1, \quad x'(0) = 0, \quad t \in \mathbb{R},$$

where $\lfloor \cdot \rfloor$ denotes the integer part of a real number [56, p. 20].

4.21

```
Clear[t,x]
eqn=x''[t]+FractionalPart[t]x'[t]+x[t]==IntegerPart[t];
DSolve[{%,x[0]==1,x'[0]==0},x[t],t]; (* No
closed-form solution *)
```

We try finding a numerical solution and plot it.

```
Block[{tend=10},
Manipulate[
nsol=NDSolve[{eqn,x[0]==1,x'[0]==0},x[t],{t,0,τ}];
Plot[x[t]/.nsol,{t,0,τ},ImageSize→150,
Ticks→{IntegerPart[τ{1/2,1}],Automatic},
AxesOrigin→{0,0},PlotStyle→Black],
{{τ,IntegerPart[tend/2]},10⁻⁵,tend}]]
```

See Figure 4.33.

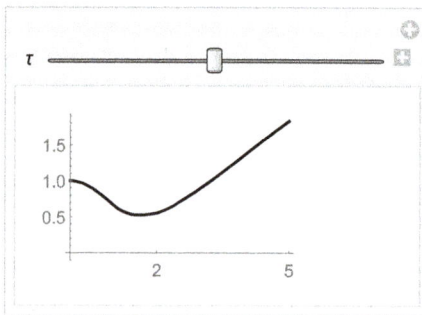

Figure 4.33: Plot of the numerical solution for problem 4.21.

4.22. We look for the solution of another initial value problem having the coefficients and the right-hand side discontinuous,

$$x''(t) + \text{sign}(\sin(5t))x'(t) + \text{clip}(t)x(t) = \delta(t), \quad x(0) = 1, \, x'(0) = 0, \, t \in \mathbb{R},$$

where δ is the Dirac delta function.

4.22

```
Clear[t,x]
eqn=x''[t]+Sign[Sin[5t]]x'[t]+Clip[t]x[t]
==DiracDelta[t];
DSolve[{%,x[0]==1,x'[0]==0},x[t],t]; (* No
closed-form solution is returned *)
```

We try finding a numerical solution and plotting it.

```
Block[{tend=50},
Manipulate[
nsol=NDSolve[{eqn,x[0]==1,x'[0]==0},x[t],{t,0,τ}];
Plot[x[t]/.nsol,{t,0,τ},ImageSize→150,
Ticks→IntegerPart[{τ{1/2,1},{-1,1}}],
AxesOrigin→{0,0},PlotStyle→Black],
{{τ,IntegerPart[tend/2]},10^-5,tend}]]
```

See Figure 4.34.

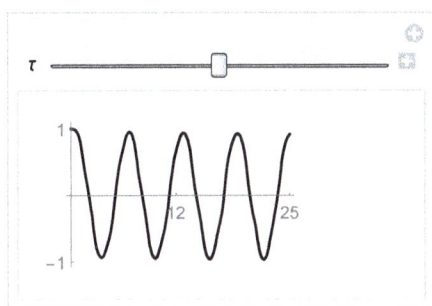

Figure 4.34: Plot of the solution for problem 4.22.

4.6 Second-order nonlinear differential equations

4.23. The equation studied by this problem is a simplified form of the *pendulum equation*, [88]. This is a second-order nonlinear ordinary differential equation that represents the motion of a *circular pendulum*. It is nonlinear because of the term sin($x(t)$),

$$x''(t) + 3\sin(x(t)) = 0, \quad t \in \mathbb{R}.$$

4.23 The Solve::ifun warning message appears because Solve uses JacobiAmplitude (the inverse of EllipticF, the incomplete elliptic integral of the first kind, Section 1.10) to find an expression for $x(t)$.

```
Clear[t,x]
sol=DSolve[x''[t]+3Sin[x[t]]==0,x[t],t]; (* Solve the
equation *)
{{x[t]→-2 JacobiAmplitude[½√(6+C[1])(t+C[2])², 12/(6+C[1])]},
{x[t]→2 JacobiAmplitude[½√(6+C[1])(t+C[2])², 12/(6+C[1])]}}
```

It is obvious that the solutions are symmetrical with respect to the horizontal axis. The solutions intersect at least for $t = 3.19817$, which is a common root for them. Indeed, since

```
x₁[t_]:=x[t]/.First[sol]/.{C[1]→-1,C[2]→3}
x₂[t_]:=x[t]/.Last[sol]/.{C[1]→-1,C[2]→3}
```

and

```
num=FindRoot[x₁[t],{t,3}]//Chop;
m=t/.num
3.19817
```

one has

```
Limit[{x₁[t],x₂[t]},t→m]//N
{0.,0.}
```

We plot the solutions,

```
Plot[{x₁[t],x₂[t]},{t,-5,5},ImageSize→175,
PlotStyle→{{Dashed,Black},{Dotted,Black}},Ticks→
{2Range[-2,2],{-1,1}},PlotLegends→"Expressions"]
```

See Figure 4.35.

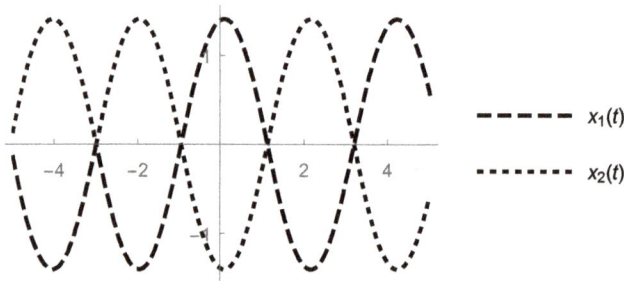

Figure 4.35: Plot of solution for problem 4.23.

4.24. We consider the next second-order nonlinear differential equation,

$$x''(t) + a^2 \sin(x(t)) = 0, \quad x(0) = \pi, \quad x'(0) = 0, \quad a, t \in \mathbb{R}.$$

4.24 First approach.

```
Clear[b,c,t,x]
Block[{a=Sqrt[2]},
{x''[t]+a²Sin[x[t]]==0,x[0]==π,x'[0]==0};
DSolve[%,x[t],t]; (* No closed-form solution
is returned to this initial value problem *)
sol=DSolve[x''[t]+a²Sin[x[t]]==0,x[t],t]] (* This
equation has two general solutions *)
```

$\{\{x[t]\rightarrow-2\,\text{JacobiAmplitude}[\frac{1}{2}\sqrt{(4+C[1])}(t+C[2])^2,\frac{8}{4+C[1]}]\},$

$\{x[t]\rightarrow2\,\text{JacobiAmplitude}[\frac{1}{2}\sqrt{(4+C[1])}(t+C[2])^2,\frac{8}{4+C[1]}]\}\}$

We immediately note that the general solutions are symmetrical with respect to the horizontal axis.

Starting with the first general solution, we try finding the particular solution satisfying the initial conditions.

```
{x₁₁=x[t]/.First[sol]/.{t→0,C[1]→b,C[2]→c}
//FullSimplify,
x₁₂=D[x[t]/.First[sol],t]/.{t→0,C[1]→b,C[2]→c}
//FullSimplify}
```

$\{-2\,\text{JacobiAmplitude}[\frac{1}{2}\sqrt{(4+b)c^2},\frac{8}{4+b}],$

$-\frac{\sqrt{(4+b)c^2}\,\text{JacobiDN}[\frac{1}{2}\sqrt{(4+b)c^2},\frac{8}{4+b}]}{c}\}$

We check whether there exist some constants of integration so that the corresponding particular solution solve the problem.

```
Reduce[x₁₁==π&&x₁₂==0,{b,c},Reals] (* Try to find
the integration constants at t=0 *)
Reduce[x₁₁==π,{c},Reals]
```

$\text{Reduce}[-2\,\text{JacobiAmplitude}[\frac{1}{2}\sqrt{(4+b)c^2},\frac{8}{4+b}]==\pi\&\&$

$-\frac{\sqrt{(4+b)c^2}\,\text{JacobiDN}[\frac{1}{2}\sqrt{(4+b)c^2},\frac{8}{4+b}]}{c}==0,\{b,c\},\text{Reals}]$

$\text{Reduce}[-2\,\text{JacobiAmplitude}[\frac{1}{2}\sqrt{(4+b)c^2},\frac{8}{4+b}]==\pi,\{c\},\text{Reals}]$

We have gotten no constants of integration. Therefore, we try to visually check whether constants of integration satisfying the problem exist.

```
Plot3D[{-2 JacobiAmplitude[½##,8/(4+b)]-π,
-1/c ##JacobiDN[½##,8/(4+b)],0},{b,-10,15},{c,-10,10},
PlotStyle→Opacity[.5],Mesh→5,ImageSize→225,
PlotLegends→"Expressions",PlotPoints→25]&[√((4+b)c²)]
```

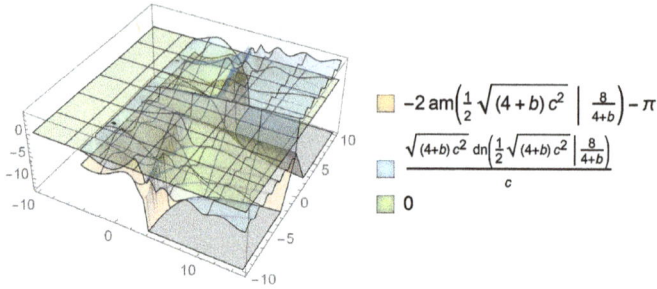

$-2\,\mathrm{am}\!\left(\frac{1}{2}\sqrt{(4+b)\,c^2}\;\middle|\;\frac{8}{4+b}\right)-\pi$

$\dfrac{\sqrt{(4+b)\,c^2}\;\mathrm{dn}\!\left(\frac{1}{2}\sqrt{(4+b)\,c^2}\;\middle|\;\frac{8}{4+b}\right)}{c}$

0

Figure 4.36: Searching the constants of integration for problem 4.24.

See Figure 4.36.

We did not get any indication of the existence of the constants of integration. Second approach. Since the equation is incomplete (no term in $x'(t)$), one reduces its order by 1:

$$x''(t) + a^2 \sin(x(t)) = 0, \quad x(0) = \pi, \quad x'(0) = 0$$
$$x' = y \Longrightarrow y\,dy = -a^2 \sin(x)\,dx \Longrightarrow y^2 = 2a^2 \cos(x) + C, \quad y(0) = 0 \Longrightarrow$$
$$C = 2a^2 \Longrightarrow y = \pm 2a\,\cos(x/2) \Longrightarrow$$
$$\ln\left|(1+\sin(z))/(1-\sin(z))\right| = \pm 2a\,t + D, \quad z(0) = x(0)/2 = \pi/2 \Longrightarrow$$
$$\ln(1/0) = D.$$

The last relation is impossible, and thus we conclude that the initial value problem has no solution.

4.25. We consider a similar to the previous problem, but with initial conditions. We study the existence and the behavior of its solution,

$$x''(t) + \cos(x(t)) = 0, \quad x(0) = 1, \quad x'(0) = 0, \quad t \in \mathbb{R}.$$

4.25

```
Clear[t,x]
eqn={x''[t]+Cos[x[t]]==0,x[0]==1,x'[0]==0};
sol=DSolve[%,x,t];(* No closed-form solution is
returned. *)
```

We try finding a numerical solution.

```
sol=NDSolve[eqn,x,{t,-15,15}];
Plot[x[t]/.sol,{t,-15,15},ImageSize→150,
PlotStyle→Black,Ticks→{5Range[-3,3,2],-4,-2,1}]
```

See Figure 4.37.

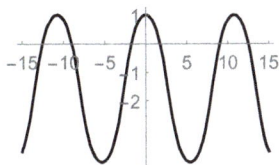

Figure 4.37: Plot of the numerical solution for problem 4.25.

If we are interested in some derivatives of the function, we can proceed as follows:

```
Plot[Evaluate[{x[t],x'[t],x''[t],x'''[t]}/.sol],
{t,-3,3},Ticks→{Range[-3,3],Range[-2,2]},
PlotStyle→{Black,{Black,Dashed},{Black,Dotted},
{Black,DotDashed}},ImageSize→150]
```

See Figure 4.38.

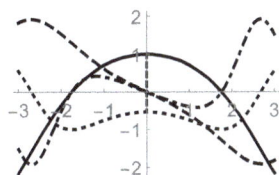

Figure 4.38: Plot of derivatives of solution for problem 4.25.

Suppose we want to study the dependence of the solution on the initial value of the first derivative. Then we write

```
Clear[t,x,p,tm]
Manipulate[
Module[{sol=NDSolve[{x''[t]+Cos[x[t]]==0,
x[0]==First[p],x'[0]==Last[p]},x,{t,0,tm}]},
ParametricPlot[Evaluate[{x[t],x'[t]}/.sol],{t,0,tm},
ImageSize→175,PlotStyle→Black,Ticks→{Range[-5,1],
Range[-2,2]},PlotRange→{{-5,1.5},2.5{-1,1}}]],
{{p,{1,0}},Locator},{{tm,.1},.1,10.6}]
```

See Figure 4.39.

4.26. We introduce a second-order nonlinear ordinary differential equation with several right-hand sides and initial conditions.

$$x''(t) + x'^2(t) = \begin{cases} -1, \\ -1, \\ 1, \\ 1, \end{cases} \quad x(0) = \begin{cases} -2, \\ 3/2, \\ -1/2, \\ 1, \end{cases} \quad x'(0) = -1/2, \quad t \in \mathbb{R}.$$

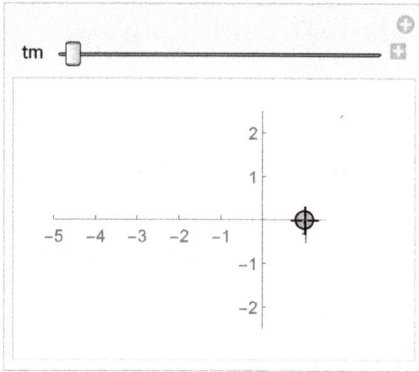

Dynamical solution for Problem 4.25.

4.26 First approach. We find the analytical solutions to this second-order nonlinear ordinary differential equations and try finding a more general method supposing that a list of initial conditions is given. The first component of a pair in the list is equal to a right-hand side of the equation, whereas the second component of that pair is equal to the corresponding initial value.

```
Clear[t,x]
seq={{-1,-2},{-1,3/2},{1,-1/2},{1,1}}; (* List of
pairs *)
nseq=If[ListQ[seq],Length@seq,0]; (* If this is indeed
a list, find the number of its elements *)

sol={}; (* List of solutions to the initial value
problems *)
If[nseq≥1,Do[If[Length[seq[[i]]]==2,(* Check if each
element in seq is indeed a pair; if not, neglect it *)
AppendTo[sol,DSolve[x''[t]+x'[t]²==First[seq[[i]]],
x[0]==Last[seq[[i]]],x'[0]==-1/2,x,t]];
],
{i,nseq}]
];
sol=sol//Flatten;
```

There are Length[sol] solutions:

```
nsol=Length[sol]
```
6

```
Table[x[t]/.sol[[k]],{k,nsol}]; (* List of solutions
*)
solreals=Select[%,FreeQ[#,Complex]&]; (* Select the
real solutions *)
Length[%]
4

If[Length[solreals]>0, (* Plot each real solution *)
Plot[solreals,{t,-2.3,3},ImageSize→150,Ticks→
{{-2,-1,1,3},Range[-5,3,2]},LabelStyle→10,
PlotStyle→{Black,{Black,Dashed},{Black,Dotted},
{Black,DotDashed}},PlotLegends→"Expressions"]]
```

See Figure 4.40.

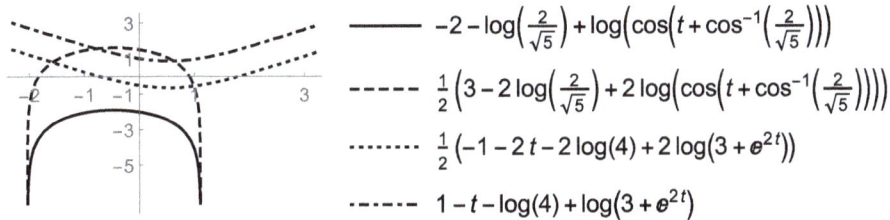

$$-2 - \log\left(\tfrac{2}{\sqrt{5}}\right) + \log\left(\cos\left(t + \cos^{-1}\left(\tfrac{2}{\sqrt{5}}\right)\right)\right)$$

$$\tfrac{1}{2}\left(3 - 2\log\left(\tfrac{2}{\sqrt{5}}\right) + 2\log\left(\cos\left(t + \cos^{-1}\left(\tfrac{2}{\sqrt{5}}\right)\right)\right)\right)$$

$$\tfrac{1}{2}\left(-1 - 2t - 2\log(4) + 2\log(3 + e^{2t})\right)$$

$$1 - t - \log(4) + \log(3 + e^{2t})$$

Figure 4.40: Solutions for problem 4.26.

Second approach. Because the equation is incomplete (the term in $x(t)$ is missing), we reduce its order by 1. Letting $y(t) = x'(t)$, it results

```
Clear[t,x,y]
soly=DSolve[y'[t]+y[t]²==#,y[0]==-1/2,y,t]&/@{-1,1}
//Simplify//Flatten;
Length[soly] (* There are two solutions that are
given below *)
Table[y[t]/.soly[[i]],{i,Length[soly]}];
solyreals=Select[%,FreeQ[#,Complex]&] (* Select the
real solutions *)
Length[%] (* Number of real solutions *)
2
{-Tan[t+ArcTan[1/2]], (-3+e²ᵗ)/(3+e²ᵗ)}
2
```

```
If[Length[solyreals]>0, (* Plot the solutions, if any
*)
Plot[solyreals,{t,-2.3,3.2},ImageSize→150,Ticks→
{{-2,-1,1,3},{-2,3}},PlotLegends→"Expressions",
PlotStyle→{Black,{Black,Dashed}},LabelStyle→11]]
```

See Figure 4.41.

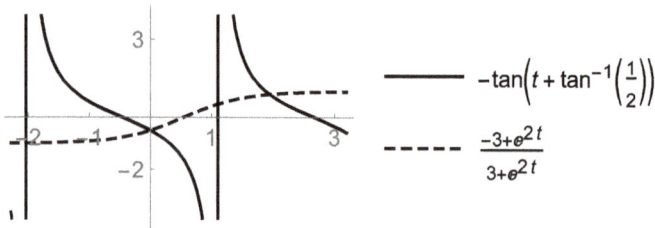

Figure 4.41: Another way of obtaining solutions for problem 4.26.

Now we return to the x variable and consider the next two equations:

```
{solx₁=DSolve[{x'[t]==y[t]/.First[soly],x[0]==#},x,t]
&/@{-2,3/2},
solx₂=DSolve[{x'[t]==y[t]/.Last[soly],x[0]==#},x,t]
&/@{-1/2,1}};
```

and plot them

```
Plot[{Evaluate[x[t]/.First[solx₁]],
Evaluate[x[t]/.Last[solx₁]],
Evaluate[x[t]/.First[solx₂]],
Evaluate[x[t]/.Last[solx₂]]},{t,-2.3,3},
Ticks→{{-2,-1,1,3},Range[-5,3,2]},ImageSize→150,
PlotStyle→{Black,{Black,Dashed},{Black,Dotted},
{Black,DotDashed}},LabelStyle→11,
PlotLegends→"Expressions"]
```

See Figure 4.42.

Because we have four real solutions, we check them.

```
{eqn₁={##,x[0]==-2,x'[0]==-1/2},
eqn₂={##,x[0]==3/2,x'[0]==-1/2}}&[x''[t]+x'[t]²==-1];
{eqn₃={##,x[0]==-1/2,x'[0]==-1/2},
eqn₄={##,x[0]==1,x'[0]==-1/2}}&[x''[t]+x'[t]²==1];
```

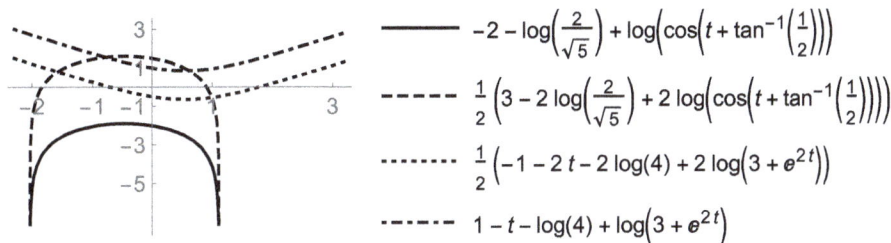

$$-2 - \log\left(\tfrac{2}{\sqrt{5}}\right) + \log\left(\cos\left(t + \tan^{-1}\left(\tfrac{1}{2}\right)\right)\right)$$

$$\tfrac{1}{2}\left(3 - 2\log\left(\tfrac{2}{\sqrt{5}}\right) + 2\log\left(\cos\left(t + \tan^{-1}\left(\tfrac{1}{2}\right)\right)\right)\right)$$

$$\tfrac{1}{2}\left(-1 - 2t - 2\log(4) + 2\log\left(3 + e^{2t}\right)\right)$$

$$1 - t - \log(4) + \log\left(3 + e^{2t}\right)$$

Figure 4.42: Again, the solutions for problem 4.26.

```
{eqn₁/.First[solx₁]//##,eqn₂/.Last[solx₁]//##,
eqn₃/.First[solx₂]//##,eqn₄/.Last[solx₂]//##}
&[FullSimplify] (* Check all solutions *)
{{{True,True,True}},{{True,True,True}},
{{True,True,True}},{{True,True,True}}}
```

We conclude that each of the four solutions satisfies the second-order nonlinear differential equation and the corresponding initial value conditions.

Next, we check if the solutions coincide.

```
(x[t]/.First[solx₁])∪(x[t]/.Last[solx₁])
∪(x[t]/.First[solx₂])∪(x[t]/.Last[solx₂])
==(solreals/.ArcCos[2/√5]→ArcTan[1/2])
∪(solreals/.ArcCos[2/√5]→ArcTan[1/2]) (* These unions
sort the elements of the lists. The Sort command can
be used as well. *)
True
```

Third approach. We take a look at this initial value problem using a numerical path.

```
Clear[t,x]
nsol=NDSolve[{x''[t]+x'[t]²==First[#],x[0]==Last[#],
x'[0]==-1/2},x,{t,-2.3,3.3}]&/@seq;
```

There are Length[nsol] solutions, i. e.,

```
Length[nsol]
4
```

Finally, we plot the solutions.

```
Plot[Evaluate[Table[x[t]/.nsol[[k]],{k,Length[nsol]}]],
{t,-2.033,1.11},ImageSize→130,AspectRatio→1,
PlotStyle→{Black,{Black,Dashed},{Black,Dotted},
{Black,DotDashed}},Ticks→{{-2,-1,1,3},Range[-5,3,2]}]
```

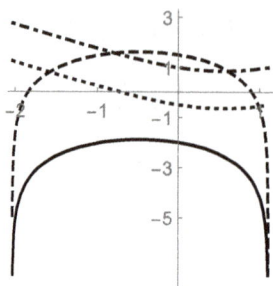

See Figure 4.43.

4.27. Here, we transform a second-order nonlinear differential equation into a system of two first-order differential equations,

$$x''(t) + x'(t)|x'(t)| = t^2, \quad x(0) = 0, \quad x'(1) = -.15, \quad t \in \mathbb{R}.$$

4.27 First approach.

```
Clear[t,x]
eqncond={x''[t]+x'[t] Abs[x'[t]]==t²,x[0]==0,
x'[1]==-.15};
DSolve[eqncond,x,t];  (* No closed-form solution
is returned *)
```

Second approach. Since the equation is incomplete, we reduce its order by one, set $y(t) = x'(t)$, and write

```
soly=DSolve[{y'[t]+y[t] Abs[y[t]]==t²,y[1]==-.15},y,t]
DSolve[Abs[y[t]]y[t]+Derivative[1][y][t]==t²,
y[1]==-.15,y,t]
```

but we get no closed-form solution.
Third approach. Now we use the NDSolve built-in function:

```
Clear[t,x,y]
sol=NDSolve[eqncond,x,{t,-.9,1}];
sol1=NDSolve[{x'[t]==y[t],y'[t]==-y[t] Abs[y[t]]+t²,
x[0]==0,y[1]==-.15},{x,y},{t,-.9,1}];
```

and plot the graphs. Then one has

```
Plot[x[t]/.sol,##],Plot[x[t]/.sol1,##],
ParametricPlot[{x[t],x'[t]}/.sol,##],
ParametricPlot[{x[t],y[t]}/.sol1,##]&[{t,-.9,1},
AspectRatio→1,Ticks→{{-.9,1},{-2,-1,1}},
PlotStyle→Black,ImageSize→125]
```

See Figure 4.44.

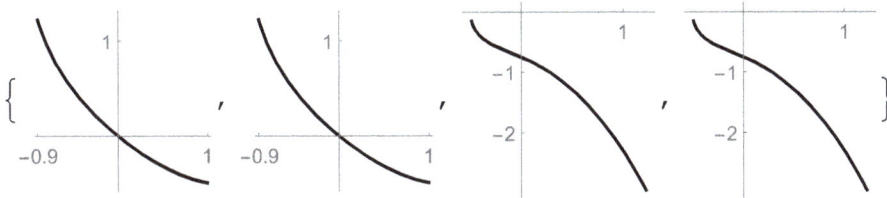

Figure 4.44: Solutions for problem 4.27.

4.28. We introduce a second-order nonlinear differential equation with initial conditions,

$$\exp(x''(t)) - x''^2(t) = t, \quad x(0) = 0, \quad x'(0) = -1, \quad t \in \mathbb{R}.$$

4.28

```
Clear[t,x,y]
DSolve[{e^{x''[t]}-x''[t]^2==t,x[0]==0,x'[0]==-1},x,t];
(* No closed-form solution is returned *)
```

Because the equation is incomplete, we reduce its order by one denoting $y(t) = x'(t)$ and
have

```
DSolve[{t-e^{y'[t]}+y'[t]^2==0,y[0]==-1},y,t]; (* Again no
closed-form solution is returned *)
```

We try a numerical approach and this time we get a solution

```
sol=NDSolve[{t-e^{x''[t]}+x''[t]^2==0,x[0]==-1,x'[0]==-1},
x,{t,0,5}];
```

and plot it.

```
Plot[x[t]/.sol,{t,0,5},ImageSize→150,
PlotStyle→Black,Ticks→{Range[1,5,2],Range[-3,2,2]}]
```

See Figure 4.45.

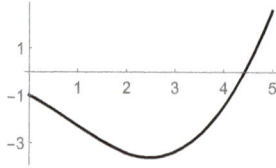

```
t₀=N[IntegerPart[10 FindRoot[x[t]/.sol,{t,4.5}]//First
//Last/10,2] (* The first positive solution *)
x₀=N[IntegerPart[10 FindMinimum[x[t]/.sol,{t,t₀}]
//First]/10,2] (* Minimum of the function on [0,t₀] *)
4.4
-3.6
```

We use the previous results for the plot of the solution on the interval $[0, t_0]$.

```
RevolutionPlot3D[First[x[t]/.sol],{t,0,t₀},Mesh→2,
Ticks→{t₀{-1,0,1},t₀{-1,0,1},{x₀,0}},ImageSize→175,
PlotStyle→{Black,Opacity[.15]}]
```

See Figure 4.46.

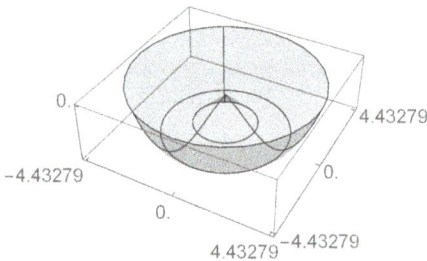

Figure 4.46: Rotated solution for problem 4.28.

4.7 Other higher-order differential equations

Along this section, we focus on differential equations of order at least three.

4.29. We find the general solution of the third-order inhomogeneous ordinary differential equation with constant coefficients,

$$x'''(t) - x''(t) + 4x'(t) - 4x(t) = 3\exp(t^2) - 4\sin(2t), \quad t \in \mathbb{R}.$$

4.29 The roots of the characteristic equation are distinct, one real and two complex conjugate.

```
Clear[t,x]
eqn=x'''[t]-x''[t]+4x'[t]-4x[t]==3e^{t²}-4Sin[2t];
```

Because

```
Reduce[λ³-λ²+4λ-4==0,λ]
λ==-2i||λ==2i||λ==1
```

the characteristic equation has indeed one real root and two complex conjugate roots.

```
sol=DSolve[eqn,x,t]; (* Solve the equation *)
eqn/.%//FullSimplify (* Check the general solution *)
{True}
```

We plot a particular solution.

```
ξ[t_]:=x[t]/.sol/.{C[1]→1,C[2]→1,C[3]→1}
Plot[ξ[t],{t,-³π/4,³π/4},ImageSize→175,
AspectRatio→.8,Ticks→Outer[Times,{³π/4,10},{-1,1}],
PlotStyle→Black]
```

See Figure 4.47.

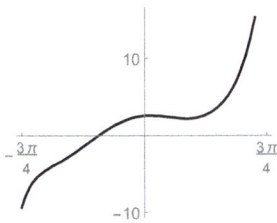

Figure 4.47: Particular solution for problem 4.29.

We find the sided limits of the particular solution.

```
Limit[ξ[t],t→#]&/@{-∞,∞}//FullSimplify//N
{{-∞},{(0.923911,-0.382607i)∞}}
```

4.30. Here is the general solution to a fourth-order ordinary differential equation with constant coefficients,

$$x''''(t) - 16 x(t) = t^2, \quad t \in \mathbb{R}.$$

4.30

```
Clear[a,b,e,d,t,x]
eqn=D[x[t],t,4]-16x[t]==t²;
```

Because

```
Reduce[λ⁴-16==0,λ]
λ==-2||λ==-2i||λ==2i||λ==2
```

the characteristic equation has two real distinct roots and two complex conjugate roots.

```
sol=DSolve[eqn,x,t] (* Solve the equation *)
eqn/.%//FullSimplify (* Check the general solution *)
ξ[t_]:=x[t]/.sol/.{C[1]→.1,C[2]→1,C[3]→1,C[4]→1}
//First (* Consider a particular solution *)
{{x→Function[{t},-t²/16+e²ᵗC[1]+e⁻²ᵗC[3]+C[2]Cos[2t]
+C[4]Sin[2t]]}}
{True}
```

We plot the graph of the particular solution.

```
Plot[ξ[t],{t,-π,π},ImageSize→175,
Ticks→{{-π,-π/2,{π/2,"π/2"},π},40Range[3]},
PlotStyle→Black]
```

See Figure 4.48.

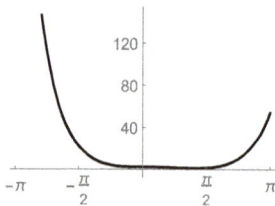

Figure 4.48: Particular solution for problem 4.30.

For the chosen constants of integration the limits of the particular solution at $-\infty$ and ∞ there are

```
Limit[ξ[t],t→#]&/@{-∞,∞}//FullSimplify
{∞,∞}
```

We change the notation of the constants C[·] to the more familiar notation c.

```
sola=DSolve[eqn,x[t],t,GeneratedParameters→
(c_{#1}&)][[1]]
a={.1,1,1,1}; (* List of values of constants *)
Table[c_k →a[[k]],{k,4}];
{x[t]→-t²/16+e²ᵗc₁+Cos[2t]c₂+e⁻²ᵗc₃+Sin[2t]c₄}
```

We plot the particular solution.

```
Plot[x[t]/.sola/.%,{t,-π,π},ImageSize→175,
Ticks→{{-π,-π/2,{π/2,"π/2"},π},40Range[3]},
PlotStyle→Black]
```

See Figure 4.49.

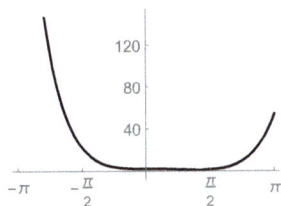

Figure 4.49: Another particular solution for problem 4.30.

Clearly, the two graphs coincide.
 We inspect the variation of the solution with respect to the constants c_{1-4}.

```
val={a,b,e,d};  (* Rename the constants of
integration *)
```

```
Manipulate[
Plot[x[t]/.sola/.{c₁ →a,c₂ →b,c₃ →e,c₄ →d},
{t,-π,π},ImageSize→175,Ticks→{π{-1,1},None},
PlotStyle→Black],{a,-1,1},{b,-2,2},{e,-10,20},{d,-2,2},
SaveDefinitions→True]
```

See Figure 4.50.

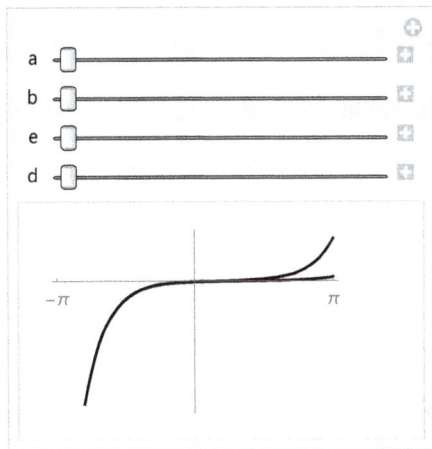

Figure 4.50: Dynamical representations of solution for problem 4.30.

4.31. Next, there is a third-order ordinary differential equation with constant coefficients, but its initial conditions and right-hand sides are distinct,

$$x'''(t) + 3x''(t) - 25x'(t) + 21x(t) = \begin{cases} 0, \\ t^2, \\ t^2 e^{3t}, \end{cases} \quad x(0) = 1, \ x'(0) = 1, \ x''(0) = \begin{cases} -2, \\ 2, \\ 2, \end{cases}$$

with $t \in \mathbb{R}$.

4.31

```
Clear[t,x]
x'''[t]+3x''[t]-25x'[t]+21x[t];
sol=DSolve[{%==First[#],x[0]==1,x'[0]==1,
x''[0]==Last[#]},x[t],t]&/@{{0,-2},{t^2,2},{t^2e^3t,2}};
(* Solve the equations *)
```

We plot the graphs.

```
Plot[Evaluate[Table[x[t]/.sol[[k]],{k,Length[sol]}]],
{t,0,1.5},Ticks→{.5Range[3],4Range[-1,2]},
PlotStyle→{Black,{Black,Dashed},{Black,Dotted}},
ImageSize→150]
```

See Figure 4.51.

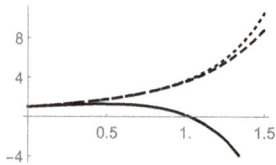

Figure 4.51: Solutions for problem 4.31.

4.32. Here, we introduce a fourth-order nonlinear differential equation with initial conditions,

$$\begin{cases} 2x'''^2(t) = (x''(t) - 1)x''''(t), & t \in \mathbb{R} \\ x(0) = 0, \quad x'(0) = x''(0) = -1, \quad x'''(0) = 3. \end{cases}$$

4.32

```
Clear[t,x]
sol=DSolve[{2x'''[t]²==(x''[t]-1)x''''[t],x[0]==0,
x'[0]==x''[0]==-1,x'''[0]==3},x,t]//Flatten//
Simplify(* The equation is solvable *)
{x→Function[{t}, 1/18 (6t+9t²-16Log[3/2]-24tLog[3/2]
-16Log[2/3+t]-24t Log[2/3+t])]}
```

We simplify the solution

```
ξ[t_]:=Simplify[Together[x[t]/.sol/.{-16Log[3/2]
-24t Log[3/2]-16Log[2/3+t]-24t Log[2/3+t]→
-(16+24t)Log[1+3/2t]}]]
```

and show its representation.

```
ξ[t]
```
$$\frac{1}{18}(2+3t)(3t-8Log[1+\frac{3t}{2}])$$

This solution is defined for $t > -2/3$ and we plot it on the interval by two consecutive roots around the origin but excluding it.

```
tmin=FindRoot[ξ[t],{t,-.64}]//First//Last//Chop
(* The first root at the left-hand side of the origin,
its exact value is -2/3 *)
tmax=FindRoot[ξ[t],t,6]//First//Last (* The first
root at the right-hand side of the origin *)
-0.666667
6.2311
```

We plot the graph of this particular solution between the two roots just found.

```
Plot[ξ[t],{t,-2/3,tmax},PlotStyle→Black,
Ticks→{{-2/3,1,3,5},Range[-3,0]},ImageSize→150]
```

See Figure 4.52.

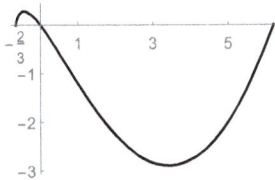

Figure 4.52: Particular solution for problem 4.32 on a given interval.

4.8 Systems of differential equations

4.33. We introduce an elementary differential-algebraic system of equations with initial conditions,

$$\begin{cases} x''(t) = y(t), \\ 2x(t) + y(t) = 4e^t \cos(t), \end{cases} \quad x(0) = 1, \quad x'(0) = 2, \quad t \in \mathbb{R}.$$

4.33 First approach.

```
Clear[t,x,y]
eqns={x''[t]==y[t],2x[t]+y[t]==4e^tCos[t],x[0]==1,
x'[0]==2};
sol=DSolve[%,{x,y},t]//First//Simplify (* Solve the
initial value problem *)
{x→Function[t,e^t(Cos[t]+Sin[t])],
y→Function[t,2e^t(Cos[t]-Sin[t])]}
```

A verification of the solutions of the system immediately can be done.

```
Simplify[eqns/.sol] (* Check the solutions *)
{True,True,True,True}
```

A plot of the solutions shows that their sum satisfies the algebraic relation $2x(t) + y(t) = 4e^t \cos(t)$. The graph of the function $t \rightarrow 2x(t) + y(t) - 4e^t \cos(t)$ is colored in continuous black.

```
Plot[Evaluate[{x[t],y[t],2x[t]+y[t]-4e^tCos[t]}/.sol//
Simplify],{t,-5,7},ImageSize→150,PlotRange→All,
PlotStyle→{{Black,Dashed},{Black,Dotted},Black},
Ticks→{Range[-5,7,4],500Range[3]},
PlotLegends→"Expressions"]
```

See Figure 4.53.

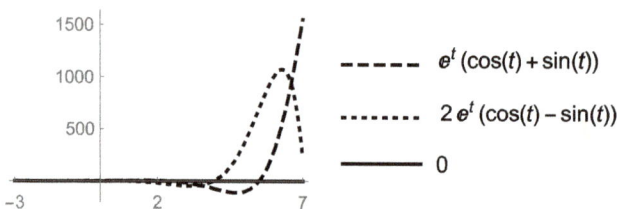

Figure 4.53: Solution for problem 4.33.

The algebraic relation that $2x(t) + y(t) = 4e^t \cos(t)$ can be shown by the next easy sequence of commands, if function $y(t)$ is known.

```
ysol=y[t]/.Last[sol];
xsol=x[t]/.First[sol];
Simplify[2xsol+ysol]
```
$4e^t\text{Cos}[t]$

Second approach. We reduce the system of equations to a simple second-order linear differential equation. Then

```
DSolve[{x''[t]+2x[t]==4 Exp[t] Cos[t],x[0]==1,x'[0]==2},
x,t]/.(Cos[Sqrt[2]t]²+Sin[Sqrt[2]t]²)→1
{{x→Function[t,eᵗ(Cos[t]+Sin[t])]}}
```

which coincides with the result in the First approach.

4.34. Let us see another case of a system of linear ordinary differential equations with constant coefficients so that its characteristic equation has one real root and two real identical roots; see [45, p. 54],

$$\begin{cases} x'(t) = 2x(t) - y(t) - z(t), \\ y'(t) = 3x(t) - 2y(t) - 3z(t), \qquad t \in \mathbb{R}. \\ z'(t) = -x(t) + y(t) + 2z(t), \end{cases}$$

4.34 First approach. We use the power of DSolve.

```
Clear[t,x,y,z]
{x'[t]==2x[t]-y[t]-z[t],y'[t]==3x[t]-2y[t]-3z[t],
z'[t]==-x[t]+y[t]+2z[t]};
DSolve[%,{x,y,z},t]; (* Solve the system *)
%%/.%//Simplify//Flatten (* Check the solutions *)
{True,True,True}
```

We set a particular solution

```
funcn={x[t],y[t],z[t]}/.%%/.{C[1]→1,C[2]→-1,C[3]→-1}
//Simplify; (* Set a particular solution *)
```

and plot it.

```
Plot[funcn,{t,-.22,.45},ImageSize→150,Ticks→
{{{-.2,"-.2"},{.2,".2"},{.4,".4"}},Range[-3,3,2]},
TicksStyle→Directive[Black,13],
PlotStyle→{Black,{Black,Dashed},{Black,Dotted}}
PlotLegends→"Expressions",LabelStyle→10]
```

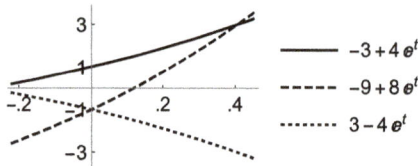

Figure 4.54: Particular solution for problem 4.34.

See Figure 4.54.

We find the limits of the vector solution at -∞ and +∞.

```
Limit[funcn,t→#]&/@{-∞,∞}//Simplify
{{{-3,-9,3}},{{∞,∞,-∞}}}
```

Second approach. We take a look at the matrix approach in the previous system of differential equations.

```
Clear[t,x,y,z]
mat={{2,-1,-1},{3,-2,-3},{-1,1,2}}; (* Coefficients in
the equations *)
Eigenvalues[mat] (* All eigenvalues are real *)
{1,1,0}

v[t_]:={x@t,y@t,z@t}; (* Unknown functions gathered
into a vector *)
system=MapThread[#1==#2&,{v'[t],mat.v[t]}]; (* The
system of differential equations *)

DSolve[system,{x,y,z},t]; (* Solve the system *)
funcn={x[t],y[t],z[t]}/.%/.{C[1]→1,C[2]→-1,C[3]→-1}
//Simplify; (* Set a particular solution *)
```

We plot the particular solution.

```
Plot[funcn,{t,-.22,.42},ImageSize→150,
Ticks→{{{-.2,"-.2"},{.2,".2"},{.4,".4"}},{-5,1}},
TicksStyle→Directive[Black,13],
PlotStyle→{Black,{Black,Dashed},{Black,Dotted}}
PlotLegends→"Expressions",LabelStyle→10]
```

See Figure 4.55.

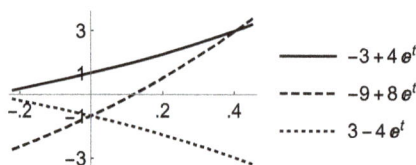

Figure 4.55: Second approach for problem 4.34.

4.35. Let us see a case of a system of linear ordinary differential equations with constant coefficients. Its characteristic equation has one real root and two complex conjugate roots [45, p. 54],

$$\begin{cases} x'(t) = 2x(t) + y(t), \\ y'(t) = x(t) + 3y(t) - z(t), \qquad t \in \mathbb{R}. \\ z'(t) = -x(t) + 2y(t) + 3z(t), \end{cases}$$

4.35 First approach. We use the power of DSolve.

```
Clear[t,x,y,z]
{x'[t]==2x[t]+y[t],y'[t]==x[t]+3y[t]-z[t],
z'[t]==-x[t]+2y[t]+3z[t]};
DSolve[%,{x,y,z},t]; (* Solve the system *)
%%/.%//Simplify (* Check the solutions *)
{{True,True,True}}
```

We set a particular solution

```
funcn={x[t],y[t],z[t]}/.%%/.{C[1]→1,C[2]→-1,C[3]→-1}
//Simplify; (* Set a particular solution *)
```

and plot it.

```
Plot[funcn,{t,-.22,.42},ImageSize→150,
Ticks→{{{-.2,"-.2"},{.2,".2"},{.4,".4"}},{-5,1}},
TicksStyle→Directive[Black,13],
PlotStyle→{Black,{Black,Dashed},{Black,Dotted}}
PlotLegends→"Expressions"]
```

See Figure 4.56.

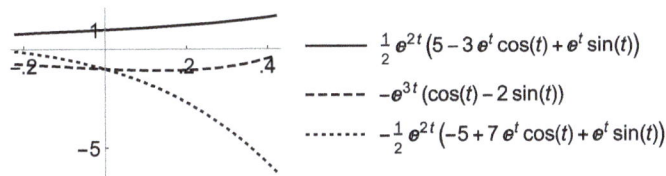

Figure 4.56: Particular solution for problem 4.35.

We find the limits of the vector solution at -∞ and +∞.

```
Limit[funcn,t→#]&/@{-∞,∞}//FullSimplify
{{{0,0,0}},{{ComplexInfinity,Interval[{-∞,∞}],
ComplexInfinity}}}
```

Second approach. We take a look on the matrix approach to this system of differential equations.

```
Clear[t,x,y,z]
mat={{2,1,0},{1,3,-1},{-1,2,3}}; (* Coefficients of
the equations *)
Eigenvalues[mat] (* One eigenvalue is real and two are
complex conjugate. We do not continue in this way,
instead we let DSolve doing its role *)
{3+i,3-i,2}
```

```
v[t_]:={x@t,y@t,z@t}; (* The unknown functions gathered
in a vector *)
system=MapThread[#1==#2&,{v'[t],mat.v[t]}]; (* The
system of differential equations *)
```

We solve the system, set a particular solution,

```
DSolve[system,{x,y,z},t]; (* Solve the system *)
funcn={x[t],y[t],z[t]}/.%/.{C[1]→1,C[2]→-1,C[3]→-1}
//Simplify; (* Set a particular solution *),
```

and plot it.

```
Plot[funcn,{t,-.22,.42},ImageSize→175,
TicksStyle→Directive[Black,13],
Ticks→{{{-.2,"-.2"},{.2,".2"},{.4,".4"}},{-5,1}},
PlotStyle→{Black,{Black,Dashed},{Black,Dotted}}
PlotLegends→"Expressions"]]
```

See Figure 4.57.

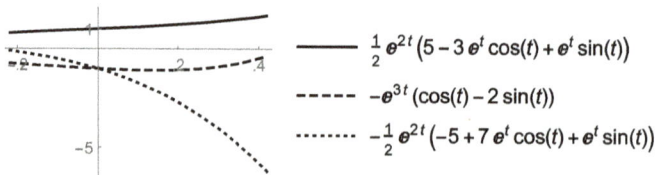

$$\frac{1}{2}e^{2t}\left(5-3\,e^t\cos(t)+e^t\sin(t)\right)$$

$$-e^{3t}\left(\cos(t)-2\sin(t)\right)$$

$$-\frac{1}{2}e^{2t}\left(-5+7\,e^t\cos(t)+e^t\sin(t)\right)$$

Figure 4.57: Second approach for problem 4.35.

Third approach.

```
Clear[c,t,x]
mat={{2,1,0},{1,3,-1},{-1,2,3}}; (* The same system
of coefficients *)
eigenval=Eigenvalues[mat] (* All eigenvalues *)
eigenvec=Eigenvectors[mat] (* All eigenvectors *)
{3+i,3-i,2}
{{2+i,1+3i,5},{2-i,1-3i,5},{1,0,1}}
```

Then the general solution is

```
Array[c,3] (* List of constants *)
{c[1],c[2],c[3]}
```

```
x[t_]:=Sum[c[n] e^eigenval[[n]] t eigenvec[[n]],{n,3}]
(* The general solution *)
x'[t]==mat.x[t]//Simplify (* Check the general
solution *)
True
```

The general solution just obtained contains complex numbers.

```
x[t]
```
$\{(2+i)e^{(3+i)t}c[1]+(2-i)e^{(3-i)t}c[2]+e^{2t}c[3],$
$(1+3i)e^{(3+i)t}c[1]+(1-3i)e^{(3-i)t}c[2],$
$5e^{(3+i)t}c[1]+5e^{(3-i)t}c[2]+e^{2t}c[3]\}$

One can perform some substitutions getting another form of the general solutions. All entries are real.

```
ξ[t_]:=k3 e^2t{1,0,1}
+e^3t(k1{2Cos[t]-Sin[t],Cos[t]-3Sin[t],5Cos[t]}
+k2{Cos[t]+2Sin[t],3Cos[t]+Sin[t],5Sin[t]})
```

```
ξ'[t]==mat.ξ[t]//FullSimplify (* We check the
general solution *)
True
```

4.36. We introduce a nonlinear and inhomogeneous system of three ordinary differential equations and solve it,

$$\begin{cases} x'(t) = \frac{x(t)-y(t)}{z(t)-t}, \\ y'(t) = \frac{x(t)-y(t)}{z(t)-t}, \quad t \in \mathbb{R}. \\ z'(t) = x(t) - y(t) + 1, \end{cases}$$

4.36 First approach.

```
Clear[t,x,y,z]
{x'[t]==(x(t)-y(t))/(z(t)-t) ,y'[t]==(x(t)-y(t))/(z(t)-t),z'[t]==x[t]-y[t]+1};
DSolve[%,x,y,z,t]; (* No closed-form solution is
returned *)
```

Second approach. After a visual inspection, we note that $x'(t) = y'(t)$, and thus $x(t) = y(t)+c_1$. Then $z'(t) = c_1+1$ and $z(t) = (c_1+1)t+c_2$. There follows that $x(t) = \ln(c_1t+c_2)+c_3$ and $y(t) = \ln(c_1t+c_2) - c_1 + c_3$. We check these solutions

```
Clear[t,x,y,z]
x[t_]:=Log[c₁t+c₂]+c₃;
y[t_]:=Log[c₁t+c₂]-c₁+c₃;
z[t_]:=(c₁+1)t+c₂;
{x'[t]==(x[t]-y[t])/(z[t]-t),
y'[t]==(x[t]-y[t])/(z[t]-t),
z'[t]==x[t]-y[t]+1}//Simplify
{True,True,True}
```

We set a particular solution and plot it.

```
Plot[Evaluate[{x[t],y[t],z[t]}/.{c₁ →1,c₂ →2,c₃ →-3}],
{t,0,3}, ImageSize→150,Ticks→{Range[3],3Range[-1,2]},
PlotLegends→"Expressions",LabelStyle→10]
```

See Figure 4.58.

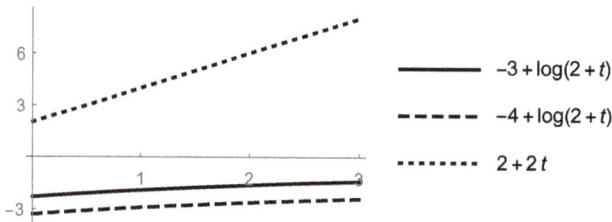

Figure 4.58: Plot a particular solution for problem 4.36.

Third approach. Consider some initial conditions

```
Clear[t,x,y,z] (* Clear the values and definitions *)
{x'[t]==(x[t]-y[t])/(z[t]-t),
y'[t]==(x[t]-y[t])/(z[t]-t),
z'[t]==x[t]-y[t]+1,x[0]==-3+Log[2],y[0]==-4+Log[2],
z[0]==2};
nsol=NDSolve[%,{x,y,z},{t,0,3}];
```

and plot the corresponding solutions.

```
Plot[Evaluate[{x[t],y[t],z[t]}/.nsol],{t,0,3},
PlotStyle→{Black,{Black,Dashed},{Black,Dotted}}
ImageSize→150,Ticks→{Range[3],{-2,2,5,8}}]
```

See Figure 4.59.

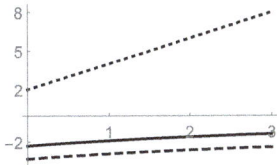

Figure 4.59: Plot another particular solution for problem 4.36.

4.37. We introduce a second-order inhomogeneous system of two differential equations,

$$\begin{cases} x''(t) + y'(t) + x(t) = e^t, \\ y''(t) + x'(t) = 1, \end{cases} \quad t \in \mathbb{R}.$$

4.37 We try finding an analytical solution.

```
Clear[t,x,y] (* Clear the values and definitions *)
{x''[t]+y'[t]+x[t]==e^t,y''[t]+x'[t]==1};
sol=DSolve[%,{x,y},t]//First; (* Solve the system *)
%%/.%//Simplify (* Check the general solution *)
{True,True}
```

We set a particular solution and plot it.

```
Plot[Evaluate[{x[t],y[t]}/.sol/.{C[1]→1,C[2]→-2,
C[3]→2,C[4]→-1}],{t,-3,3},ImageSize→150,
PlotStyle→{Black,{Black,Dashed}},
Ticks→{Range[-3,3,2],5Range[-1,3,2]}]
```

See Figure 4.60.

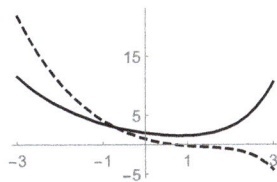

Figure 4.60: Plot the particular solution for problem 4.37.

Now we keep the equations unchanged and impose some boundary conditions.

```
Clear[t,x,y]
{x''[t]+y'[t]+x[t]==e^t,y''[t]+x'[t]==1,x[0]==y[0]==1,
x[2]==y[2]==2};
sol=DSolve[%,{x,y},t]//First;
%%/.%//Simplify
{True,True}
```

```
Plot[Evaluate[{x[t],y[t]}/.sol],{t,-3,3},
ImageSize→150,Ticks→{Range[-3,3,2],4Range[3]},
PlotStyle→{Black,{Black,Dashed}}]
```

See Figure 4.61.

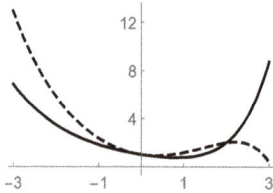

Figure 4.61: Plot another particular solution for problem 4.37.

4.38. We introduce a second-order system of two inhomogeneous differential equations with constant coefficients and initial conditions [64, p. 54],

$$\begin{cases} x''(t) + x'(t) + y'(t) - 2y(t) = 2 - 2t, & x(0) = 2, \quad x'(0) = -1, \\ x'(t) - y'(t) + x(t) = t, & y(0) = \frac{3}{2}, \end{cases} \quad t \in \mathbb{R}.$$

4.38

```
Clear[t,x,y] (* Clear the values and definitions *)
sys={x''[t]+x'[t]+y'[t]-2y[t]==2-2t,
x'[t]-y'[t]+x[t]==t,x[0]==2,x'[0]==-1,y[0]==3/2};
sol=DSolve[%,{x,y},t]//Flatten//Simplify (* Solve the
system *)
{x→Function[{t},1/2 e^{-2t}(2+e^t+e^{3t}+2e^{2t}t)],
y→Function[{t},1/2 e^{-2t}(1+2e^{3t}+2e^{2t}t)]}
```

```
sys/.sol//Simplify//Flatten (* Check the solutions *)
{True,True,True,True,True}
```

We plot the solutions in three ways.

```
{Plot[Evaluate[{x[t],y[t]}/.sol],##],
Plot[{x[t]/.sol,y[t]/.sol},##],
Plot[Plot[Evaluate[{Last@Last@First@sol,
Last@Last@Last@sol}],##]}&[{t,-4,4},
ImageSize→160,PlotStyle→{Black,{Black,Dashed}},
Ticks→{2Range[-2,2],100Range[3]}]
```

See Figure 4.62.

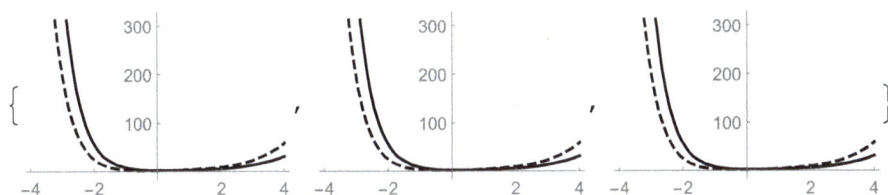

Figure 4.62: Three plots of a particular solutions for problem 4.38.

The limits of the solutions at -∞ and ∞ are given below.

```
Limit[{x[t],y[t]}/.sol,t→#]&/@{-∞,∞}
{{∞,∞},{∞,∞}}
```

We also try solving the system numerically and plot it in two ways.

```
nsol=NDSolve[sys,{x,y},{t,-4,4}];
```

```
{Plot[Evaluate[{x[t],y[t]}/.nsol],##],
Plot[{x[t]/.nsol,y[t]/.nsol},##]}&[{t,-4,4},
ImageSize→160,PlotStyle→{Black,{Black,Dashed}},
Ticks→{2Range[-2,2],100Range[3]}]
```

See Figure 4.63.

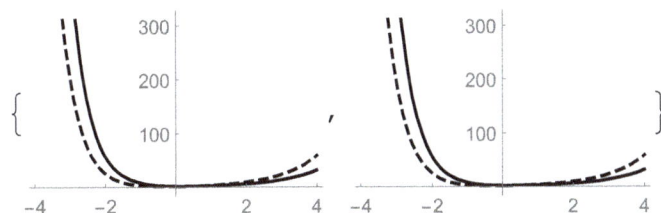

Figure 4.63: Two plots of a numerical particular solution for problem 4.38.

4.39. We introduce a second-order system of two inhomogeneous differential equations with constant coefficients [64, p. 54],

$$\begin{cases} 2x''(t) + 3y''(t) + 2x'(t) + y'(t) + x(t) + y(t) = 2, \\ x''(t) + 3y''(t) + 4x'(t) + 2y'(t) - x(t) - y(t) = -2, \qquad t \in \mathbb{R}. \\ x(0) = 4, \quad x'(0) = 2, \quad y(0) = -6/5, \quad y'(0) = -13/5, \end{cases}$$

4.39

```
Clear[t,x,y] (* Clear the values and definitions *)
sys={2x''[t]+3y''[t]+2x'[t]+y'[t]+x[t]+y[t]==2,
x''[t]+3y''[t]+4x'[t]+2y'[t]-x[t]-y[t]==-2,
x[0]==4,x'[0]==2,y[0]==-6/5,y'[0]==-13/5};
sol=DSolve[%,{x,y},t] (* Solve the initial value
problem *)
{x→Function[{t}, 1/10 (6+11e^t+13Cos[t]+10Cos[t]^2
+9Sin[t]+10Sin[t]^2)],
y→Function[{t}, 1/10 (-6-11e^t-5Cos[t]+10Cos[t]^2
-15Sin[t]+10Sin[t]^2)]}

sys/.sol//Simplify
{True,True,True,True,True,True}
```

We plot the solutions in three ways.

```
Plot[Evaluate[{x[t],y[t]}/.sol],##],
Plot[{x[t]/.sol,y[t]/.sol},##],
Plot[Evaluate[{Last@Last@First@sol,
Last@Last@Last@sol}],##]&[{t,-8,3},ImageSize→160,
Ticks→{2Range[-4,1],5Range[-2,2]},
PlotStyle→{Black,{Black,Dashed}}]
```

See Figure 4.64.

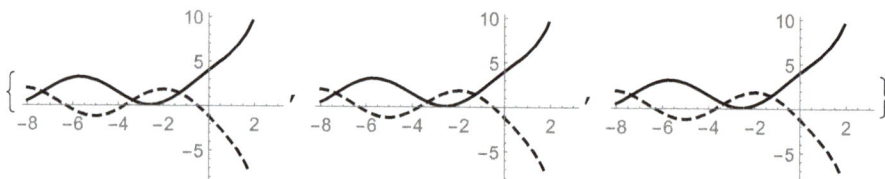

Figure 4.64: Three plots of the solutions for problem 4.39.

```
Limit[{x[t],y[t]}/.sol,t→#]&/@{-∞,∞}
{{Interval[{-(3/5),19/5}],Interval[{-(8/5),12/5}]},
{∞,-∞}}
```

4.40. We introduce a first-order system of two differential equations containing the sign and the unit step functions,

$$
\begin{cases}
(t^2 + 1)x'(t) = -t\,x(t) + y(t) - \text{sign}(t), & x(0) = -\frac{1}{2}, \\
(t^2 + 1)y'(t) = -x(t) - t\,y(t) + t \cdot \begin{cases} 0, & t < 0, \\ 1, & t \geq 0, \end{cases} & y(0) = 2,
\end{cases} \quad t \in \mathbb{R}.
$$

4.40

```
Clear[t,x,y] (* Clear the values and definitions *)
eqn={(t²+1)x'[t]==-t x[t]+y[t]-Sign[t],
(t²+1)y'[t]==-x[t]-t y[t]+t UnitStep[t],x[0]==-1/2,
y[0]==2};
sol=DSolve[%,{x,y},t]//Flatten; (* Solve the system
of equations *)
```

We plot the solutions in three ways.

```
{Plot[Evaluate[{x[t],y[t]}/.sol],##],
Plot[{x[t]/.sol,y[t]/.sol},##],
Plot[Evaluate[{Last@Last@First@sol,
Last@Last@Last@sol}],##]}&[{t,-8,8},
ImageSize→160,PlotStyle→{Black,{Black,Dashed}},
Ticks→Outer[Times,{4,1},Range[-2,2]]]}
```

See Figure 4.65.

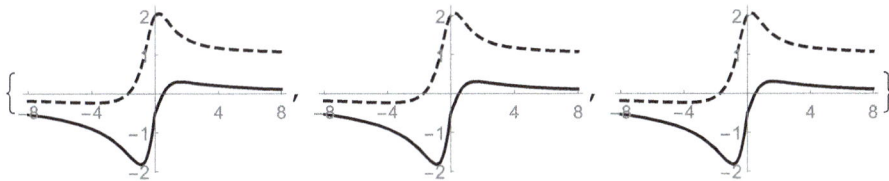

Figure 4.65: Three plots of the solution for problem 4.40.

We also try a numerical approach.

```
nsol=NDSolve[eqn,{x,y},{t,-8,8}];
```

```
{Plot[Evaluate[{x[t],y[t]}/.nsol],##],
Plot[{x[t]/.nsol,y[t]/.nsol},##]}&[{t,-8,8},
ImageSize→160,PlotStyle→{Black,{Black,Dashed}},
Ticks→Outer[Times,{4,1},Range[-2,2]]]
```

See Figure 4.66.

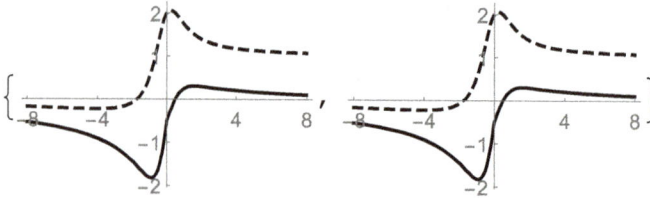

Figure 4.66: Two plots of the numerical solution for problem 4.40.

4.41. We introduce a first-order system of three nonlinear differential equations, which allows predictions of infected peaks and of the number of susceptibles, infected, and recovered during an epidemic outbreak like COVID-19, [72]. The SIR method supposes that we have a large number of homogeneous population, n, which is partitioned into three sets: the susceptibles s, the infected i, and the recovered r.

4.41 The nonlinear system of differential equations generated by this model is

$$\begin{cases} \frac{ds(t)}{dt} = -b\frac{s(t)}{n} i(t), \\ \frac{di(t)}{dt} = b\frac{s(t)}{n} i(t) - k\, i(t), \qquad t \in \mathbb{R}. \\ \frac{dr(t)}{dt} = k\, i(t). \end{cases}$$

We attach some initial conditions: $s(0) = 1$, $i(0) = 10^{-3}$, $r(0) = 0$. b is the *infection rate* whilst k is the *removal rate*.

We solve the system above considering $n = 1$ and several values for b.

```
Block[{k=1},
eqncond:={s'[t]==-b*s[t]*i[t],
i'[t]==b*s[t]*i[t]-k*i[t],r'[t]==k*i[t],s[0]==1,
i[0]==10⁻³,r[0]==0};
sol=ParametricNDSolve[eqncond,{s,i,r},{t,0,17},{b}]]
```

We plot several solutions accordingly to the values of the infection rate.

```
Plot[Evaluate[
Table[{s[b][t],i[b][t],r[b][t]}/.sol,{b,.1,4,.6}]],
{t,0,17},PlotLegends→{"s[t]","i[t]","r[t]"},
PlotStyle→{Black,{Black,Dashed},{Black,Dotted}},
PlotRange→All,ImageSize→275,
Ticks→{5Range[3],.2Range[5]}]
```

See Figure 4.67.

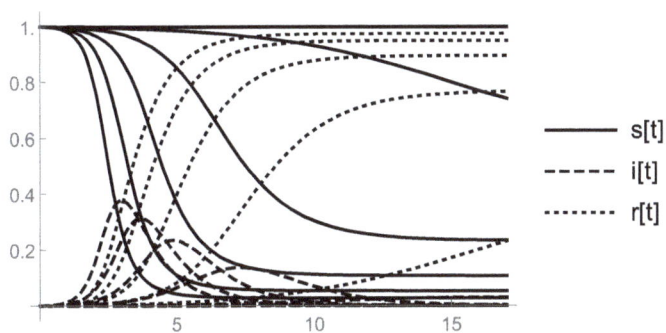

Figure 4.67: Evolution of the SIR epidemic for several values of the infection rate.

The references provided in the second part of this book may be of assistance; see [94, 93, 95, 65, 99, 82, 63], and [71].

A *partial differential equation* is a differential equation that contains unknown functions of several independent variables and their partial derivatives. If we restrict to one unknown function u of three independent variables x, y, and z, it is a relation of the form

$$F(x, y, z, u, \partial_x u, \partial_y u, \partial_z u, \partial_{x,x} u, \partial_{y,y} u, \partial_{z,z} u, \partial_{x,y} u, \partial_{y,z} u, \partial_{z,x} u, \partial_{x,x,x} u, \dots) = 0. \qquad \text{(II-0)}$$

We suppose that function u is real valued, has at least two independent real variables, and is differentiable as many times as it is required.

Sometimes we take into account cases where the unknown function depends on more than two variables. In such cases, there is a more complex problem with the geometrical representation of the solution.

We also admit that the mixed partial derivatives coincide, i. e., $\partial_{x,y} u = \partial_{y,x} u$, etc.

A solution (general or particular) to a partial differential equation is a function that solves the equation and satisfies certain extra conditions, if any.

The general solution of (II-0), if any, depends on one or several integration functions satisfying certain appropriate conditions.

To find a desired particular solution, we impose certain initial conditions, Dirichlet conditions, Neumann conditions, or mixed conditions.

If equation (II-0) is linear in function u as well as in all its partial derivatives, then it is said to be a *linear equation with partial derivatives*. If the highest order of partial derivatives occurring in (II-0) is m, then equation (II-0) is said to be of mth order.

According to [94, p. 67], the DSolve built-in function may find the general solution for certain restricted types of linear homogeneous second-order partial differential equations, namely equations of the form

$$a\frac{\partial^2 u}{\partial x^2} + b\frac{\partial^2 u}{\partial x \partial y} + c\frac{\partial^2 u}{\partial y^2} = 0,$$

where a, b, and c are real constants. Thus DSolve assumes that the equation has constant coefficients and a vanishing nonprincipal part. Actually, DSolve can solve many other types of partial differential equations.

Mathematica for partial differential equations

DSolve[eqn,u[x,y],{x,y}] solves a partial differential equation eqn for u[x,y] in the independent variables x and y.

DSolve[eqn,u,{x,y}] solves a partial differential equation eqn for u in the independent variables x and y. The solution, if any, is given as a "pure function" for x and y.

DSolve[eqn,u[x,y,z],{x,y,z}] solves a partial differential equation eqn for u[x,y,z] in the independent variables x, y, and z.

https://doi.org/10.1515/9783111411392-006

DSolve[eqn,u,{x,y,z}] solves a partial differential equation eqn for $u[x,y,z]$ in the independent variables x, y, and z. The solution, if any, is given as a "pure function" for x, y, and z.

NDSolve[eqn,u[x,y],{x,x_{min},x_{max}}},{y,y_{min},y_{max}}}] solves numerically a partial differential equation eqn for $u[x,y]$ in the independent variables x and y under the constraints that $x \in [x_{min}, x_{max}]$ and $y \in [y_{min}, y_{max}]$.

NDSolve[eqn,u[x,y,z],{x,x_{min},x_{max}},{y,y_{min},y_{max}}, {z,z_{min},z_{max}}] solves numerically a partial differential equation eqn for $u[x,y,z]$ in the independent variables x, y, and z under the constraints that $x \in [x_{min}, x_{max}]$, $y \in [y_{min}, y_{max}]$, and $z \in [z_{min}, z_{max}]$.

5 First-order partial differential equations

This chapter deals with some first-order partial differential equations, linear, quasi-linear, and nonlinear. If equation (II-0) contains only first-order partial derivatives, then it is precisely a first-order partial differential equation. Wolfram documentation on this topic may be found at https://reference.wolfram.com/language/tutorial/DSolveLinearAndQuasiLinearFirstOrderPDEs.html

5.1 First-order linear partial differential equations

A relation of the form

$$f_1(x_1,\ldots,x_n)\partial_{x_1} u(x_1,\ldots,x_n) + \cdots + f_n(x_1,\ldots,x_n)\partial_{x_n} u(x_1,\ldots,x_n) = 0, \qquad (5.1)$$

where $\emptyset \neq D \subset \mathbb{R}^n$, D is nonempty, open, and connected, $u : C^1(D,\mathbb{R})$ is the unknown function, $f_i : C^1(D,\mathbb{R})$, $i = 1,\ldots,n$, and $\sum_{i=1}^n f_i^2 \neq 0$, is said to be a *first-order linear partial differential equation*. If $\sum_{i=1}^n f_i^2 \equiv 0$ on D, then equation (5.1) reduces to an identity.

5.1. We introduce our first problem of a first-order linear partial differential equation,

$$a\,\partial_x u(x,y) + b\,\partial_y u(x,y) = 0, \quad a,b,x,y \in \mathbb{R}.$$

5.1

```
Clear[a,b,x,y,u] (* Clear values and definitions *)
eqn=a D[u[x,y],x]+b D[u[x,y],y]; (* The equation *)
```

We distinguish several cases depending on the constants a and b.

▷ $a = b = 0$.

```
sol₁=Assuming[a==0&&b==0,DSolve[eqn==0,u,{x,y}]]
{{}}
```

We conclude that any differentiable function u of two variables is a solution of the equation.

▷ $a = 0, b \neq 0$.

```
sol₂=Assuming[a==0&&b≠0,DSolve[eqn==0,u,{x,y}]]
//Flatten
{u→Function[{x,y},x]}
```

Any differentiable function u of one variable x is a solution of the equation.

https://doi.org/10.1515/9783111411392-007

▷ $a \neq 0, b = 0$.
This case is similar to the previous and, therefore, we omit it.

▷ $a \neq 0, b \neq 0$.

```
sol₃=Assuming[a≠0&&b≠0,DSolve[eqn==0,u,{x,y}]
//Flatten]
{u→Function[{x,y},C[1][-bx+ay/a]]}
```

Here, C[1] is an arbitrary differentiable function of one real variable $t = \frac{ay-bx}{a}$. We can write this solution under the classical form $u(x,y) = \phi(\frac{ay-bx}{a})$, where ϕ is a differentiable function on its domain of definition, i. e., on an nonempty open interval in $\mathbb{R}$.

The general solution is correct because

```
sol=eqn/.sol₃ (* Check the general solution *)
0
```

We look for certain particular solutions corresponding to constants $a = b = 1$ and some selections of function C[1]. Then we write

```
sol=sol₃/.{a→1,b→1};
```

This solution is correct because

```
eqn/.sol (* Check the solution *)
0
```

We plot some particular solutions corresponding to certain selections of function C[1]. Then we write

```
Plot3D[u[x,y]/.sol/.{C[1][t_]→First@#},{x,-5,5},
{y,-5,5},ImageSize→150,Mesh→2,PlotStyle→{Black,
Opacity[.15]},Ticks→{5{-1,0,1},5{-1,0,1},Automatic},
PlotLabel→Style[#[[1]],Black,12]]&/@{{t,15},
{t²,130},{Sin[t],2},{ArcTan[t],3},{Log[1+Abs[t]],4}}
```

See Figure 5.1.
Below we gather all graphs above into a single one and add in each case the ContourPlot command to the solution.

```
listofsol=u[x,y]/.sol/.{C[1][t_]→#}&/@{t,t²,Sin[t],
ArcTan[t],Log[1+Abs[t]]}; (* List of particular
solutions *)
```

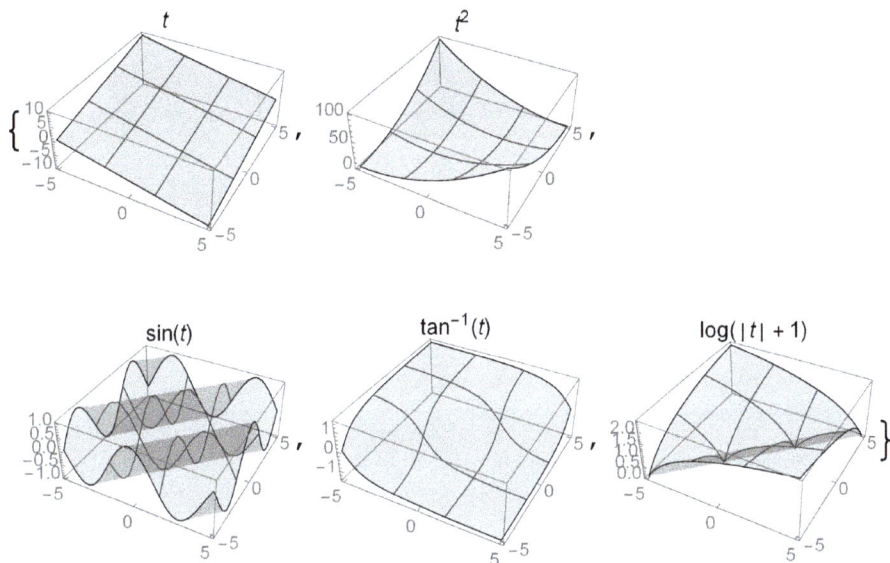

Figure 5.1: Plot of some particular solutions for problem 5.1.

```
icons=Table[lists→Plot3D[lists,{x,-5,5},{y,-5,5},
Axes→False,Boxed→False,Mesh→3,ImageSize→60,
PlotStyle→{Black,Opacity[.15]}],{lists,listofsol}];

Manipulate[
Plot3D[solution,##,ImageSize→150,Mesh→3,
PlotStyle→{Black,Opacity[.15]}]&[{x,-5,5},{y,-5,5},
PlotLabel→Style[solution,Black,12]],{{solution,
icons[[1,1]]},icons,Setter}]
```

See Figure 5.2.

5.2. We consider the *transport equation* [13, Section 2.2] with an initial condition,

$$\partial_x u(x,y) + \partial_y u(x,y) = 0, \quad u(x,0) = x\sin(x) + e^{-1/x^2}\cos(x), \quad x,y \in \mathbb{R}.$$

5.2

```
Clear[x,y,u]
sol=DSolve[{∂ₓu[x,y]+∂ᵧu[x,y]==0,
u[x,0]==x Sin[x]+e^{-1/x²}Cos[x]},u,{x,y}]//Flatten
{u→Function[{x,y},
e^{1/(x-y)²} (Cos[x-y]+e^{1/(x-y)²} x Sin[x-y]-e^{1/(x-y)²} y Sin[x-y])]}
```

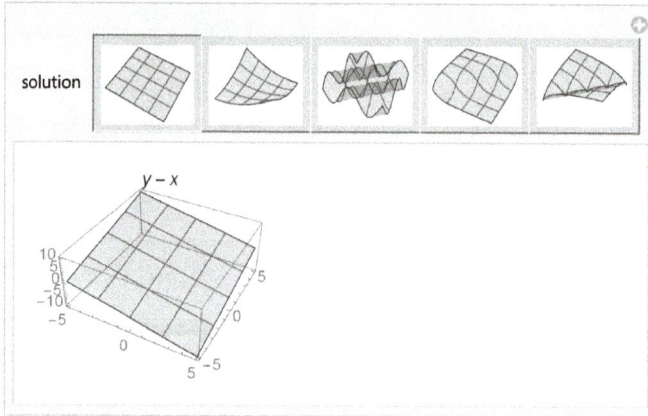

Figure 5.2: Dynamical solutions for problem 5.1.

We plot the solution of this problem with initial condition.

```
{Show[Plot3D[u[x,y]/.sol,{x,-6,6},{y,-5,5},Mesh→3,
ImageSize→160,PlotStyle→{Black,Opacity[.15]},
Ticks→{6{-1,0,1},5{-1,0,1},2{-5,-1,3}}],
ParametricPlot3D[{x,0,x Sin[x]+e^(-1/x^2) Cos[x]},{x,-6,6},
PlotStyle→{Black,Thick}]],
Animate[
Plot[Evaluate[u[x,y]/.sol],{x,-2π,2π},PlotRange→All,
Ticks→{π Range[-2,2],3Range[-1,2]},ImageSize→150,
Filling→Axis],{y,-π,π},AnimationRunning→False,
SaveDefinitions→True,DefaultDuration→20,
AnimationDirection→ForwardBackward]]}
```

See Figure 5.3.

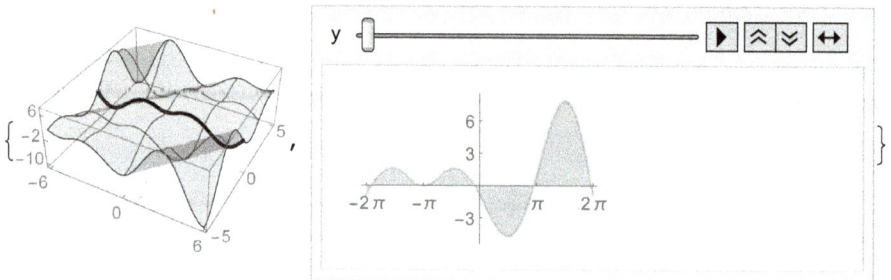

Figure 5.3: Plot of the solution for problem 5.2.

5.3. We introduce another first-order linear homogeneous partial differential equation with nonconstant coefficients, [45, p. 89],

$$x \, \partial_x \, u(x,y) + y \, \partial_y \, u(x,y) = 0, \quad x,y \in \mathbb{R}.$$

5.3

```
Clear[a,b,x,y,u]
eqn=x ∂ₓu[x,y]+y ∂ᵧu[x,y];
sol=DSolve[%==0,u,{x,y}]//Flatten (* Solve the
equation *)
{u→Function[{x,y},C[1][y/x]]}
```

Here, C[1] is an arbitrary differentiable function defined for all $t = y/x \in \mathbb{R}$. We can write this solution under the classical form $u(x,y) = \phi(x/y)$, where ϕ is any differentiable function on its domain of definition.

We immediately verify the solution just gotten by the command

```
eqn/.sol (* Check the general solution *)
0
```

Some particular solutions satisfying conditions similar to problem 5.1 are introduced below. Therefore, we write

```
Plot3D[u[x,y]/.sol/.{C[1][t_]→#},{x,-5,5},{y,-5,5},
ImageSize→150,Mesh→3,PlotStyle→{Black,Opacity[.1]},
Ticks→{5{-1,0,1},5{-1,0,1},Automatic},PlotPoints→50,
PlotLabel→Style["C1[t]→"#,11]]&/@{ArcTan[t],
Log[1+Abs[t]],t+1/t}
```

See Figure 5.4.

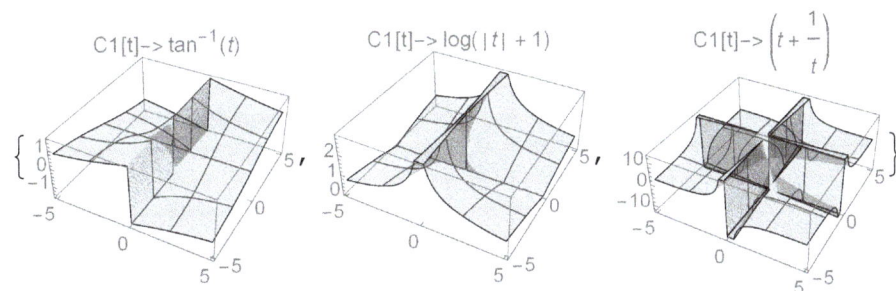

Figure 5.4: Plot of solutions for problem 5.3.

We consider two moving particular solutions.

```
Manipulate[
Plot3D[Evaluate[u[x,y]/.sol/.{C[1][t_]→Sin[t]}],
{x,-a,a},{y,-a,a},PlotRange→All,Ticks→
Outer[Times,{a,a,1},{-1,0,1}],Filling→Axis,
PlotStyle→{Black,Opacity[.15]},ImageSize→160,
PlotLabel→Style["C1[t]→Sin[t]",11],PlotPoints→100],
{a,.01,1,.01},SaveDefinitions→True]
```

See Figure 5.5.

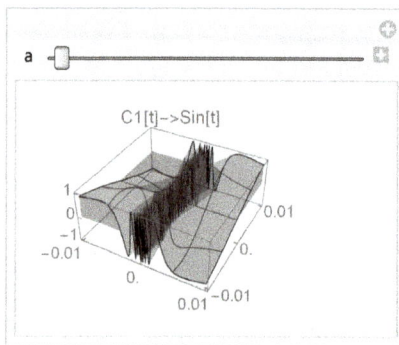

Figure 5.5: Plot of the solution of the initial value problem 5.3.

```
Manipulate[
Plot3D[Evaluate[u[x,y]/.sol/.{C[1][t_]→t²}],
{x,-b,b},{y,-b,b},ImageSize→160,Mesh→3,
PlotLabel→Style["C1[t]→t2",11,Black],
PlotStyle→Opacity[.15],Ticks→{b{-1,0,1},b{-1,0,1},
10{0,1,2}}],{b,.1,10,.01},SaveDefinitions→True]
```

See Figure 5.6.

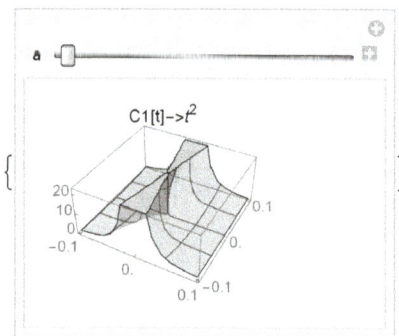

Figure 5.6: Dynamical plot of a particular solution for problem 5.3.

We get the same function as above by the initial condition $u(1, y) = y^2$ instead of $C[1] \to t^2$.

5.4. Another first-order partial differential equation follows [45, p. 89]:

$$y\, \partial_x u(x, y) - x\, \partial_y u(x, y) = 0, \quad x, y \in \mathbb{R}.$$

5.4

```
Clear[x,y,u]
eqn=y ∂ₓu[x,y]-x ∂ᵧu[x,y];
sol=DSolve[%==0,u,{x,y}]//Flatten (* Solve the
equation *)
{u→Function[{x,y},C[1][½(x²+y²)]]}
```

We verify the general solution.

```
%%/.%
0
```

We show that the above partial differential equation is the equation of a paraboloid of rotation, a sphere, and a cone. For it, we select three particular functions.

```
funcs=u[x,y]/.sol/.{C[1][t_]→#}&/@{t,Sqrt[9-t],
Sqrt[t]}
{½(x²+y²), √(9 + ½(-x²-y²)), √(x²+y²)/√2 }
```

```
icons=Table[listf→Plot3D[listf,{x,-5,5},{y,-5,5},
Axes→False,Boxed→False,Mesh→3,ImageSize→66,
PlotStyle→{Black,Opacity[.15]}],{listf,funcs}];
```

```
Manipulate[
 {Plot3D[function,##,ImageSize→150,Mesh→3,
 PlotStyle→{Black,Opacity[.15]}]}&[{x,-5,5},{y,-5,5},
 PlotLabel→Style[function,Black,10]],
 {{function,icons[[1,1]]},icons,Setter}]
```

See Figure 5.7.

We introduce a continuous deformation of the first surface into the second one by a convex combination.

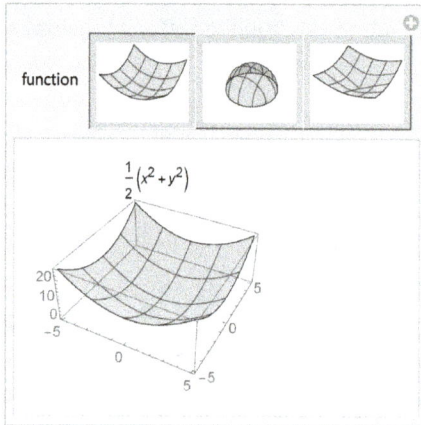

Figure 5.7: Plot of the particular solutions for problem 5.4.

```
Clear[a]
Manipulate[
Plot3D[Evaluate[(1-a)funcs[[1]]+a funcs[[2]]],{x,-5,5},
{y,-5,5},Ticks→{5{-1,0,1},5{-1,0,1},Automatic},
PlotStyle→{Black,Opacity[.15]},ImageSize→160,
Mesh→3],{a,0,1},SaveDefinitions→True]
```

See Figure 5.8.

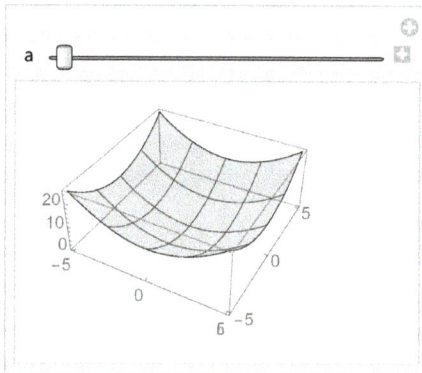

Figure 5.8: Convex combination of the first two surfaces in problem 5.4.

We consider a particular solution satisfying the initial condition $u(x,1) = 2x, x \in \mathbb{R}$.

```
sol=DSolve[eqn==0,u[x,1]==2x,u,{x,y}]//First (* Solve
the initial value problem *)
u→Function[{x,y},-2√(-1+x²+y²)]
```

Remark. We note that for the solution just yielded, the initial value problem is satisfied if and only if $x^2 + y^2 \geq 1$. □

We plot the particular solution.

```
Show[Plot3D[u[x,y]/.sol,{x,-2,2},{y,-2,2},Mesh→1,
ImageSize→160,PlotStyle→{Black,Opacity[.15]},
Ticks→Outer[Times,{2,2,5},{-1,0,1}]],
ParametricPlot3D[{x,1,2x},{x,-2,2},
PlotStyle→{Black,Thick}],ViewPoint→{-4,4,3}]
```

See Figure 5.9.

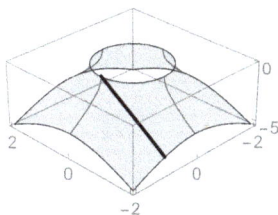

Figure 5.9: Plot of a particular solution in problem 5.4.

Now we want to find the particular solution satisfying $u(x,1) = 2x$, $x \in \mathbb{R}$, by elementary means, [45, 12, p. 89]. Therefore, we write the system of equations

$$\begin{cases} C[1] = x^2 + y^2, \\ y = 1, \\ u(x,1) = 2x \end{cases}$$

characterizing the initial value problem.

From the first and second equations, it follows that $C[1] = x^2 + 1$. From here and the last equation, we get the particular solution

$$u(x,y) = 2\,\text{sgn}(x)\,\sqrt{C[1] - 1} = 2\,\text{sgn}(x)\,\sqrt{x^2 + y^2 - 1}.$$

Obviously, $x^2 + y^2 \geq 1$. We plot this function.

```
Show[Plot3D[2Sign[x]√x² + y² - 1,{x,-2,2},{y,-2,2},
Mesh→3,ImageSize→160,PlotStyle→{Black,Opacity[.15]},
Ticks→Outer[Times,{2,2,5},{-1,0,1}]],
ParametricPlot3D[{x,1,2x},{x,-2,2},
PlotStyle→{Black,Thick}],ViewPoint→{-4,4,3}]
```

See Figure 5.10.

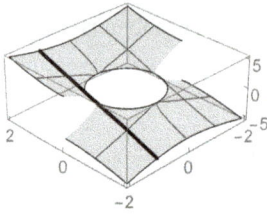

Figure 5.10: Solution to an initial value problem in problem 5.4.

We directly check the solution $u(x,y) = 2\operatorname{sign}(x)\sqrt{x^2+y^2} - 1$. Indeed

```
check=y D[2Sign[x] √x²+y²-1 ,x]-x D[2Sign[x] √x²+y²-1,y]
//FullSimplify
```
$2y\sqrt{x^2+y^2}-1$ Derivative[1][Sign][x]

If $x \neq 0$, then $\operatorname{sign}'(x) = 0$, and so check $= 0$. Otherwise one has

```
Limit[y D[2Sign[x]Sqrt[x²+y²-1],x]
-x D[2Sign[x]Sqrt[x²+y²-1],y],x→0]
0
```

Remark. We conclude that the two functions coincide whenever $x \leq 0$. Only the second function satisfies the initial value condition for all real x. □

5.5. An example of a first-order partial differential equation so that the unknown function depends on three independent variables is supplied now,

$$x\partial_x u(x,y,z) + y\partial_y u(x,y,z) + x y\partial_z u(x,y,z) = 0, \quad x,y,z \in \mathbb{R}.$$

5.5

```
    First approach.

Clear[x,y,z,u] (* Clear the values and definitions *)
eqn=x ∂ₓu[x,y,z]+y ∂ᵧu[x,y,z]+x y∂_zu[x,y,z]==0;
sol=DSolve[%,u,{x,y,z}]//Flatten (* Solve the equation
*)
```
$\{u\rightarrow\text{Function}[\{x,y,z\},C[1][\frac{y}{x},\frac{1}{2}(-x\,y+2z)]]\}$

C[1] is a differentiable function of two real variables, say t and v.

```
u[x,y,z]/.sol/.{C[1][t_,v_]→-2v(t+1/t)}//Simplify
(* Set a particular solution *)
%%%/.%%//FullSimplify (* Check the particular solution
*)
```
$\frac{(x^2+y^2)(x\,y-2z)}{x\,y}$
True

Suppose that we want to find the particular solution satisfying the condition that $u(x, y, 0) = x^2 + y^2$.

Second approach. We use the DSolve built-in function to solve the initial value problem.

```
Clear[x,y,z,u]
eqncond={eqn,u[x,y,0]==x²+y²};
DSolve[%,u,{x,y,z}]; (* No closed-form solution is
returned *)
```

Third approach. Then we try to find a numerical solution

```
nsol=NDSolve[eqncond,u,{x,.1,5},{y,.1,5},{z,0,5},
PrecisionGoal→1]
{{u→InterpolatingFunction[{{0.1,6.},{0.1,5.},
{-5.,5.}},<>]}}
```

and plot it for three values of the ImageSize parameters and the upper bounds of the variables.

```
ContourPlot3D[Evaluate[First[u[x,y,z]/.nsol]],
{x,.1,First[#]},{y,.1,#[[2]]},{z,#[[3]]],5},
ImageSize→Last[#],Contours→3,Mesh→None,
ContourStyle→{{Opacity[.15],Black},{Opacity[.15],
Black},{Opacity[.15],Black}}]&/@{{5,5,0,140},
{.5,.5,4.8,150},{.45,.4,4.98,160}}
```

See Figure 5.11.

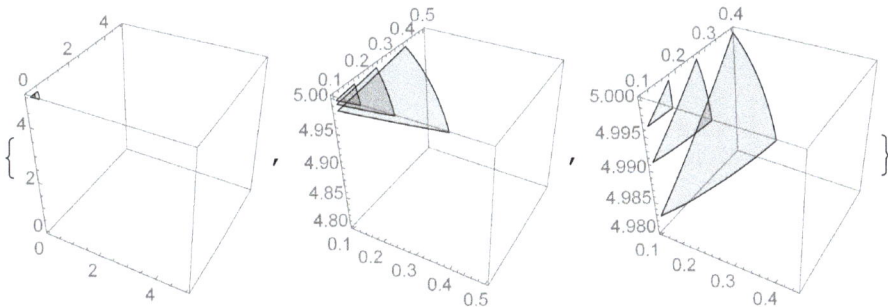

Figure 5.11: Numerical solution for problem 5.5.

Remark. Another accomplishment of the above figure is given by gradually modifying the PlotRange command and the upper bounds of the variables. □

Fourth approach. We follow the classical way and consider the following system of equations:

$$\begin{cases} a = x/y, \\ b = 2z - xy, \\ u = x^2 + y^2, \\ z = 0. \end{cases}$$

From the second and the last equations, it follows that $b = -xy$. From here and the first equation, we get that $x^2 = -ab$ and $y^2 = -b/a$. Now the particular solution follows, which is $u = -b(a + 1/a)$. Thus, the particular solution is

$$u(x, y, z) = \frac{(xy - 2z)(x^2 + y^2)}{xy}.$$

This function satisfies the equation and the initial condition, because setting

```
v[x_,y_,z_]:=(x²+y²)(xy-2z)/xy
eqn/.{u[x,y,0]→v[x,y,0],u^(0,0,1)[x,y,z]→v^(0,0,1)[x,y,z],
u^(0,1,0)[x,y,z]→v^(0,1,0)[x,y,z],u^(1,0,0)[x,y,z]→
v^(1,0,0)[x,y,z]}//Simplify
```
{True,True}

We plot function v using the `ContourPlot3D` built-in function.

```
ContourPlot3D[v[x,y,z],{x,.001,5},{y,.001,5},
{z,0,5},Mesh→3,ImageSize→160,Contours→3,
ContourStyle→Directive[Black,Opacity[.1]],
PlotRange→{{.001,6},{.001,5},5{0,1}}]
```

See Figure 5.12.

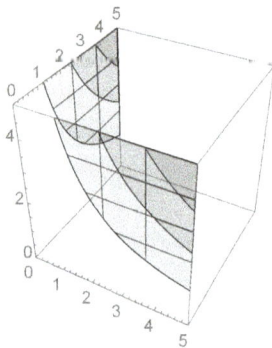

Figure 5.12: Contour plot of the particular solution for problem 5.5.

We enlarge the domain of definition of function v.

```
ContourPlot3D[v[x,y,z],{x,0,6},{y,0,5},{z,-5,5},
Mesh→3,ImageSize→160,Contours→3,
ContourStyle→Directive[Black,Opacity[.1]]]
```

See Figure 5.13.

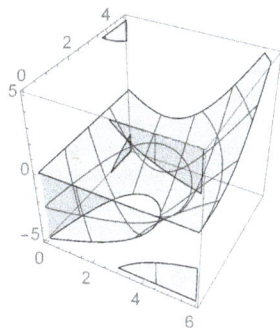

Figure 5.13: Another contour plot of the particular solution for problem 5.5.

Remark. The two `ContourPlot3D` supply similar results. □

Remark. Another possible representation is by considering variable z a parameter. □

5.6. We present a simple first-order linear partial differential equation having three independent variables, which is not handled at all by *Mathematica* 10.3 [45, p. 89],

$$(x - z)\partial_x u(x, y, z) + (y - z)\partial_y u(x, y, z) + 2z\,\partial_z u(x, y, z) = 0, \quad x, y, z \in \mathbb{R}.$$

5.6

```
Clear[x,y,z,u,ϕ,ξ,η,γ]
eqn=(x-z)∂ₓu[x,y,z]+(y-z)∂ᵧu[x,y,z]+2z∂_zu[x,y,z]==0;
sol=DSolve[eqn,u,{x,y,z}]; {(* No closed-form
solution is returned *)}
```

We try finding the general solution by a classical method. The given equation implies that

$$\frac{dx}{x - z} = \frac{dy}{y - z} = \frac{dz}{2z} = \frac{dx + dz}{x + z} = \frac{dy + dz}{y + z}.$$

Then one has two prime integrals

$$C_1 = \frac{x + z}{\sqrt{z}} \quad \text{and} \quad C_2 = \frac{y + z}{\sqrt{z}}$$

and so the general solution is of the form

$$u = \phi(C_1, C_2) = \phi\left(\frac{x+z}{\sqrt{z}}, \frac{y+z}{\sqrt{z}}\right),$$

where ϕ is an arbitrary differentiable function of two real variables. Below we consider two particular functions ϕ.

```
listoffunctions=ϕ[ x+y/√z , y+z/√z ]/.{ϕ[ξ_,η_]→#}&/@{
Log[1+(ξ²+η²)],ηSin[ξ]};
```

and plot them.

```
icons=Table[
listf→ContourPlot3D[listf,{x,-5.,5.},{y,-5.,5.},
{z,1.,5.},Contours→3,Axes→False,Boxed→False,
Mesh→None,ContourStyle→Directive[Black,Opacity[.1]],
ImageSize→60],{listf,listoffunctions}];
```

See Figure 5.14.

$$\left\{\text{Log}\left[1 + \frac{(x+y)^2}{z} + \frac{(y+z)^2}{z}\right] \rightarrow \right. \qquad , \qquad \frac{(y+z)\,\text{Sin}\left[\frac{x+y}{\sqrt{z}}\right]}{\sqrt{z}} \rightarrow \left. \right\}$$

Figure 5.14: Two particular solutions for problem 5.6.

```
Manipulate[
ContourPlot3D[function,{x,-5.,5.},{y,-5.,5.},{z,1.,5.},
PlotLabel→Style[function,Blue,9],ImageSize→150,
Mesh→None,Contours→3,ContourStyle→Directive[Black,
Opacity[.1]],Ticks→{5{-1,0,1},5{-1,0,1},{1,5}}],
{{function,icons[[1,1]]},icons,Setter},
SaveDefinitions→True]
```

See Figure 5.15.

We introduce another function

```
Clear[x,y,z,γ,μ,ν]
```

```
functionga=γ[ x+y/√z , y+z/√z ]/.{γ[μ_,ν_]→(μCos[ν])};
```

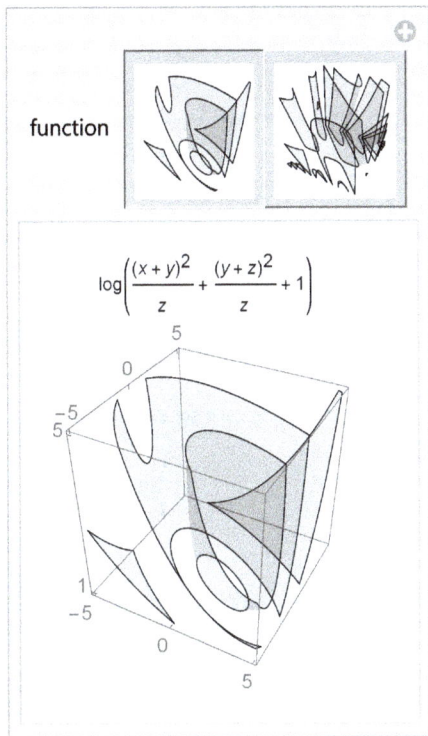

$$\log\left(\frac{(x+y)^2}{z} + \frac{(y+z)^2}{z} + 1\right)$$

Figure 5.15: Dynamical plot of solutions for problem 5.6.

and plot its contours

```
Manipulate[
ContourPlot3D[functionga==c,{x,-5.,5.},{y,-5.,5.},
{z,1.,5.},Contours→3,Axes→False,Mesh→None,
ContourStyle→Directive[Black,Opacity[.1]],
Boxed→False,ImageSize→160],{c,-3.,3.},
SaveDefinitions→True]
```

See Figure 5.16.

5.7. We study the next initial value problem,

$$\begin{cases} x(y-z)\partial_x u(x,y,z) + y(z-x)\partial_y u(x,y,z) + z(x-y)\partial_z u(x,y,z) = 0, \\ u(x,1,z) = \frac{1}{x} + \frac{1}{z}, \quad x,y,z \in \mathbb{R}, \quad x \neq 0, \quad z \neq 0. \end{cases}$$

5.7

```
Clear[x,y,z,u]
```

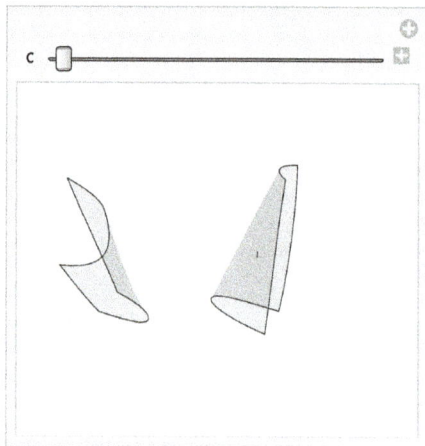

Figure 5.16: Two particular solutions for problem 5.6.

```
eqn=x(y-z)∂ₓu[x,y,z]+y(z-x)∂ᵧu[x,y,z]
+z(x-y)∂_zu[x,y,z]==0;
sol=DSolve[{eqn,u[x,1,z]==1/x+1/z},u,{x,y,z}]; (* Solve
the problem *)
```

Because no closed-form solution is returned, we solve only the equation neglecting for a while the initial condition.

```
sol=DSolve[eqn,u,{x,y,z}]//Flatten; (* Solve the
equation *)
{u→Function[{x,y,z},C[1][-x y z,x+y+z]]}
```

With this general solution, we try finding the particular solution satisfying the initial condition $u(x, 1, z) = 1/x + 1/z$. Therefore, we impose the next condition on function C[1] thus getting the particular solution.

```
fn=u[x,y,z]/.sol/.{C[1][t_,v_]→(v-1)/(-t)}//Simplify
```
$$\frac{-1+x+y+z}{x\,y\,z}$$

The contour plot of the particular solution immediately appears.

```
ContourPlot3D[fn,{x,.1,.13},{y,.1,.13},{z,.1,.13},
Mesh→None,Contours→3,ImageSize→150,
ContourStyle→{LightGray,Opacity[.1]},
Ticks→{{.1,.13},{.1,.13},{.1,.13}}]
```

See Figure 5.17.

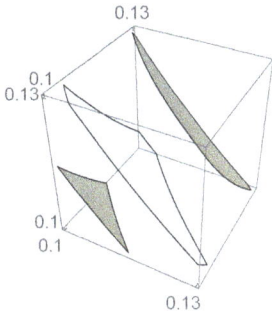

Figure 5.17: Particular solution for problem 5.7.

Now we take a look at a numerical approach.

```
nsol=NDSolve[{eqn,u[x,1,z]==1/x+1/z},u,
{x,.1,.13},{y,.1,.13},{z,.1,.13}]
NDSolve[{(x-y)z u^(0,0,1)[x,y,z]+y(-x+z)u^(0,1,0)[x,y,z]
+x(y-z)u^(1,0,0)[x,y,z]==0,u[x,1,z]==1/x + 1/z},u,
{x,.1,.13},{y,.1,.13},{z,.1,.13}]
```

Obviously, this numerical approach does not yield any solution.

5.2 First-order quasilinear partial differential equations

A relation of the form

$$f_1(x_1,\ldots,x_n,u)\partial_{x_1}u + \cdots + f_n(x_1,\ldots,x_n,u)\partial_{x_n}u = f_{n+1}(x_1,\ldots,x_n,u), \tag{5.2}$$

where $\emptyset \neq D \subset \mathbb{R}^n$, D is nonempty, open, and connected; $u : C^1(D, \mathbb{R})$ is the unknown function, $f_i : C^1(D \times \mathbb{R}, \mathbb{R})$, $i = 1,\ldots,n + 1$, and $\sum_{i=1}^{n+1} f_i^2 \neq 0$ is said to be a *first-order quasilinear partial differential equation*. If $\sum_{i=1}^{n+1} f_i^2 \equiv 0$ on D, then (5.2) is an identity. If $\sum_{i=1}^{n} f_i^2 \equiv 0$ on D and $f_{n+1}(x_1,\ldots,x_n,u) \neq 0$, then (5.2) is an equation.

5.8. Our first quasilinear partial differential equation of the first-order is considered below,

$$x^2\partial_x u(x,y) + y^2\partial_y u(x,y) - (x + y)u(x,y) = 0, \quad x,y \in \mathbb{R}.$$

5.8 We use the power of *Mathematica* to find its general solution.

```
Clear[x,y,u]
eqn=x^2∂ₓu[x,y]+y²∂ᵧu[x,y]-(x+y)u[x,y]==0;  (* Equation
*)
sol=DSolve[%,u,{x,y}]//Flatten (* Solve the equation
*)
{u→Function[{x,y},x y C[1][-x+y/x y]]}
```

Here, C[1] is an arbitrary differentiable function depending on a variable $t = \frac{-x+y}{xy}$. We check the general solution

```
FullSimplify[eqn/.sol]
True
```

One obtains a particular solution to this problem by a specific choice of C[1].

```
partsol=u[x,y]/.sol/.C[1][t_]→Sin[t²]
```
$x\,y\,Sin[\frac{(-x+y)^2}{x^2 y^2}]$

Here is a plot of the surface to this particular solution and its symmetry with respect to the horizontal plane.

```
Plot3D[#,{x,-5,5},{y,-5,5},ImageSize→160,Mesh→8,
PlotStyle→{Black,Opacity[.1]},PlotPoints→75,
Ticks→Outer[Times,5{1,1,1},{-1,0,1}]]&/@{partsol,
-partsol}
```

See Figure 5.18.

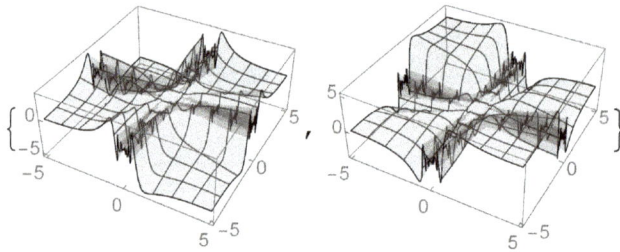

Figure 5.18: A particular solution for problem 5.8.

Remark. This equation also appears in [94, p. 4]. □

5.9. The second example of a first-order quasilinear partial differential equation follows:

$$x\,\partial_x u(x,y) + y\,\partial_y u(x,y) = 2xy, \quad x,y \in \mathbb{R}.$$

5.9

```
Clear[x,y,u]
x∂ₓu[x,y]+y∂ᵧu[x,y]==2x y;
sol=DSolve[%,u,{x,y}]//Flatten (* Solve the equation
*)
%%/.%//Simplify
{u→Function[{x,y},x y C[1][ʸ⁄ₓ]]}
True
```

We vary the figure by using the `PlotRange` built-in function.

```
{Plot3D[##,PlotRange→Outer[Times,{1,1,50},{-1,1}],
Ticks→Outer[Times,{1,1,50},{-1,0,1}],
ImageSize→165],
Plot3D[##,PlotRange→Outer[Times,{1,1,5},{-1,1}],
ImageSize→145,Ticks→Outer[Times,{1,1,5},{-1,0,1}]],
Plot3D[##,PlotRange→Outer[Times,.5{1,1,10},{-1,1}],
Ticks→Outer[Times,.5{1,1,10},{-1,0,1}],
ImageSize→145]}&[{u[x,y]/.sol/.C[1][t_]→t-1},
{x,-1,1},{y,-1,1},Mesh→4,ViewPoint→{2,-3,2},
PlotStyle→{Black,Opacity[.1]},PlotPoints→50]
```

See Figure 5.19.

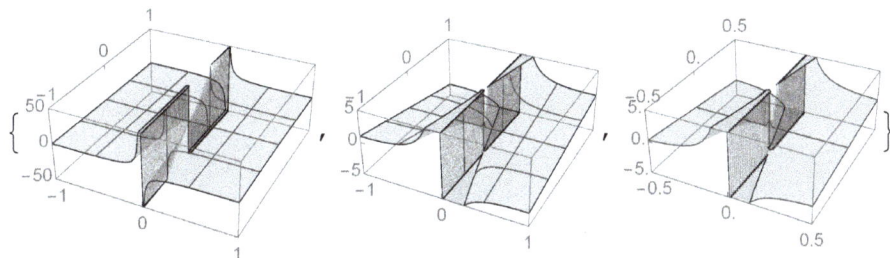

Figure 5.19: A particular solution with several `PlotRange` for problem 5.9.

5.10. Suppose we are interested in solving the next initial value problem,

$$x\, \partial_x u(x,y) + y\, \partial_y u(x,y) = 2xy, \quad u(x,x) = x^2, \quad x,y \in \mathbb{R}.$$

This problem is connected to the previous one and may be found in [45, p. 96].

5.10

```
Clear[x,y,u]
eqn=x∂ₓu[x,y]+y∂ᵧu[x,y]==2x y; (* Equation *)
DSolve[{%,u[x,x]==x²},u,{x,y}]; (* No closed-form
solution is returned since the error message appears:
DSolve::conarg: "The arguments should be ordered
consistently." *)
```

We try finding the general solution of the equation.

```
sol=DSolve[eqn,u,{x,y}]//Flatten
{u→Function[{x,y},x y+C[1][ʸ⁄ₓ]]}
```

Now we impose the initial condition, so

```
(u[x,y]/.sol/.y→x)-x²
C[1][1]
```

Thus, C[1] is an arbitrary differentiable function vanishing for 1. We show some plots to this initial value problem.

```
Plot3D[u[x,y]/.sol/.C[1][t_]→#,{x,-5,5},{y,-5,5},
ImageSize→150,PlotStyle→{Black,Opacity[.1]},Mesh→3,
Ticks→{5{-1,0,1},5{-1,0,1},Automatic},PlotPoints→50,
PlotLabel→Style["C1[t]→"#,Black,10]]&/@{t⁴-1,
Tan[t-1],Log[Abs[t]]}
```

See Figure 5.20.

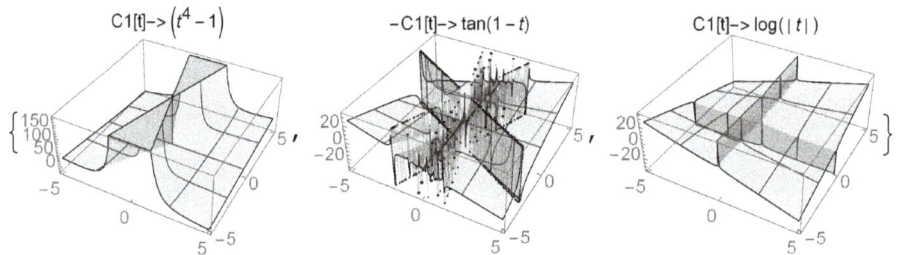

Figure 5.20: The particular solutions for problem 5.10.

5.11. Another example follows below:

$$a x \partial_x u(x,y) + b y \partial_y u(x,y) = c \sin(x), \quad u(1,y) = by, \quad a,b,c,x,y \in \mathbb{R}.$$

This problem may be found in [45, p. 96].

5.11

```
Clear[a,b,c,x,y,u]
sol=DSolve[{a x∂ₓu[x,y]+b y∂ᵧu[x,y]==c Sin[x],
u[1,y]==b y},u[x,y],{x,y}]//Flatten (* Solve the
problem *)
```
$$\{u[x,y] \to -\frac{x^{-\frac{b}{a}}(-c\,\text{SinIntegral}[x]x^{\frac{b}{a}}+c\,\text{SinIntegral}[1]x^{\frac{b}{a}}-ab\,y)}{a}\}$$

```
Manipulate[
Show[Plot3D[Evaluate[u[x,y]/.sol/.{a→ α,b→ β,c→ γ}],
{x,1,5},{y,-5,5},ImageSize→160,PlotRange→All,
Ticks→{{1,3,5},5{-1,0,1},None},PlotStyle→
{Black,Opacity[.2]},Mesh→3],
ParametricPlot3D[{1,y,b y}/.{b→ β},{y,-5,5},
PlotStyle→{Black,Thick}]],{α,.1,3},{β,-5,5},
{γ,-5,8},SaveDefinitions→True]
```

See Figure 5.21.

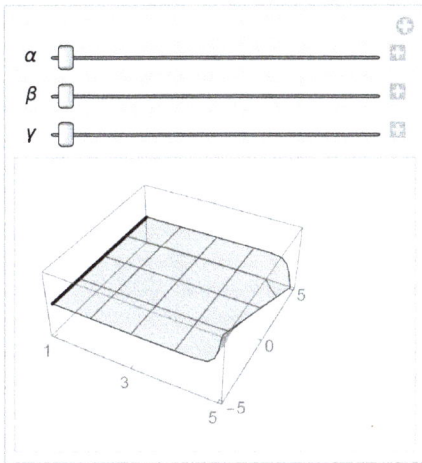

Figure 5.21: Dynamical solution for problem 5.11.

5.12. The fifth example of a first-order quasilinear partial differential equation follows:

$$\frac{1}{x}\partial_x u(x,y) - \frac{1}{y}\partial_y u(x,y) = \frac{1}{x} + \frac{1}{y}, \quad x,y \in \mathbb{R}\setminus\{0\}.$$

5.12

```
Clear[x,y,u]
1/x ∂ₓu[x,y]-1/y ∂ᵧu[x,y]==1/x+1/y  (* Equation *)
sol=DSolve[%,u,{x,y}]  (* Solve the equation *)
%%/.sol//FullSimplify
{{u→Function[{x,y},x- √y²+C[1][½(x²+y²)]]},
{u→Function[{x,y},x+√y²+C[1][½(x²+y²)]]}}
{1/y==1/√y²,1/y+1/√y²==0}
```

Remark. From the verification step, we conclude that the first solution holds for $y > 0$, whereas the second solution holds for $y < 0$. ☐

Ignoring for a while the above remark and supposing that C[1][t]=Log[t] for all $t > 0$, we get the next particular functions to plot.

```
psol=u[x,y]/.sol/.C[1][t_]→Log[t]
{x-√y²+Log[½(x²+y²)],x+√y²+Log[½(x²+y²)]}
tabsol=Table[psol[[k]],{k,Length[psol]}];

Plot3D[tabsol,{x,-3,3},{y,-4,4},ImageSize→175,
PlotStyle→{{Black,Opacity[.2]},{Black,Opacity[.1]}},
Ticks→Outer[Times,{2.9,4,9},{-1,0,1},Mesh→1]]
```

See Figure 5.22.

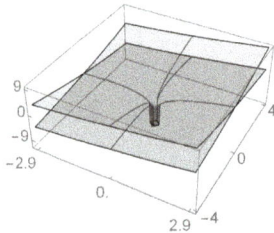

Figure 5.22: The particular solutions for problem 5.12.

5.3 First-order nonlinear partial differential equations

5.13. Our first equation of a first-order nonlinear partial differential equation follows:

$$\partial_x u(x,y) \times \partial_y u(x,y) = u(x,y), \quad x,y \in \mathbb{R}.$$

5.13 We use the power of *Mathematica*.

```
Clear[x,y,u]
sol=DSolve[∂ₓu[x,y]*∂ᵧu[x,y]==u[x,y],u,{x,y}]//Flatten
(* Solve the equation *)
tab=u[x,y]/.%/.{C[1]→2,C[2]→1} (* Set a particular
solution *)
{u→Function[{x,y}, (y²+2 x y C[1]+x²C[1]²+2 y C[2]+2 x C[1] C[2]+C[2]²)/(4C[1]) ]}
{ (1+4x+4x²+2y+4x y+y²)/8 }
```

We plot the particular solution.

```
Plot3D[tab,{x,-5,5},{y,-4,4},ImageSize→160,Mesh→2,
Ticks→{5{-1,0,1},4{-1,0,1},13{0,1,2}},
PlotStyle→{Black,Opacity[.15]}]
```

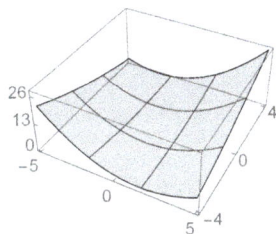

Figure 5.23: Plot a particular solution for problem 5.13.

See Figure 5.23.

For this equation, we consider an initial condition.

```
sol=DSolve[{∂ₓu[x,y]*∂yu[x,y]==u[x,y],u[0,y]==y²},u,
{x,y}]//Flatten (* Solve the problem *)
{u→Function[{x,y}, x²+8x y+16y²/16 ]}
```

and plot the resulting particular solution.

```
Show[Plot3D[u[x,y]/.sol,{x,-5,5},{y,-4,4},ImageSize→
160,Mesh→3,PlotStyle→{Black,Opacity[.15]},
Ticks→{5{-1,0,1},4{-1,0,1},13{0,1,2}}],
ParametricPlot3D[{0,y,y²},{y,-4,4},
PlotStyle→{Black,Thick}]]
```

See Figure 5.24.

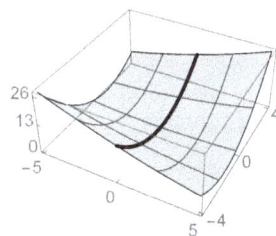

Figure 5.24: Plot the solution of the initial value problem.

The initial condition is drawn by a black thick line.

5.14. We introduce the scalar *conservation law* (*Burgers' inviscid equation*),

$$\partial_x u(x,y) + u(x,y)\partial_y u(x,y) = 0, \quad x,y \in \mathbb{R}.$$

5.14

```
Clear[x,y,u]

eqnconserv=∂ₓu[x,y]+u[x,y]∂yu[x,y]; (* The equation *)
```

```
DSolve[eqnconserv==0,u,{x,y}] (* There is no explicit
solution. The general solution is given under the
form of a functional equation *)
Solve[u[x,y]==C[1][x-  y  ],u[x,y]]
                       u[x,y]
```

Here, we introduce an initial condition, and thus an explicit solution is obtained.

```
sol=DSolve[{eqnconserv==0,u[0,y]==y},u,{x,y}]//First
{u→Function[{x,y},  y  ]}
                   1+x
```

Remark. A classical approach to this initial value problem may be found in https://math. stackexchange.com/questions/305727/solve-burgers-equation. □

```
Plot3D[u[x,y]/.sol,{x,-3,3},{y,-3,3},ImageSize→160,
PlotStyle→{Green,Opacity[.1]},Mesh→5,
Ticks→Outer[Times,{3,3,5},{-1,0,1}]]
```

See Figure 5.25.

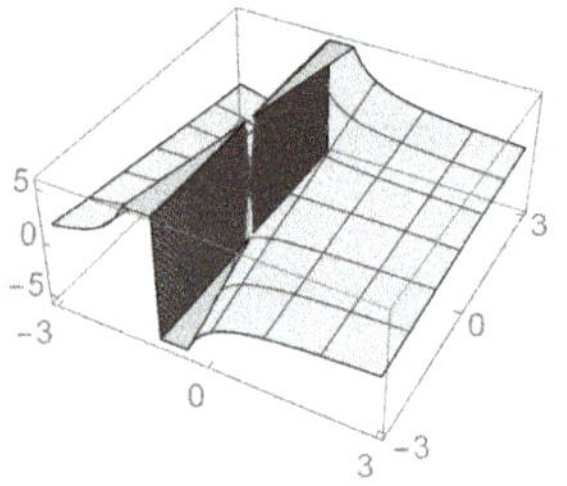

Figure 5.25: Particular solution for problem 5.14.

We change the initial condition and find the solution to this new problem.

```
solp=DSolve[{eqnconserv==0,u[x,0]== 1  },u,{x,y}]
                                    x+1
//Flatten
{u→Function[{x,y},  1+y  ]}
                   1+x
```

This solution also results by the next reasoning:

$$u(x,y) = C[1]\left(x - \frac{y}{u(x,y)}\right) \implies u(x,0) = C[1](x) \implies C[1](x) = \frac{1}{1+x}$$

$$u(x,y) = 1/\left(1 + x - \frac{y}{u(x,y)}\right) \implies \begin{cases} u(x,y) = 0, & \text{impossible} \\ u(x,y) = \frac{1+y}{1+x}. \end{cases}$$

Below there is a numerical approach in which the discontinuity situation is rounded about away.

```
nsolp=NDSolve[{eqnconserv==0,u[x,0]==1/(x+1)},u,
{x,0,5},{y,0,5}]; (* Numerical solution *)
```

```
{Plot3D[u[x,y]/.solp,##,PlotRange→All],
Plot3D[u[x,y]/.nsolp,##]}&[{x,0,5},{y,0,5},Mesh→2,
PlotStyle→{Black,Opacity[.15]},ImageSize→145,
Ticks→{{0,5},{0,5},Automatic}]
```

See Figure 5.26.

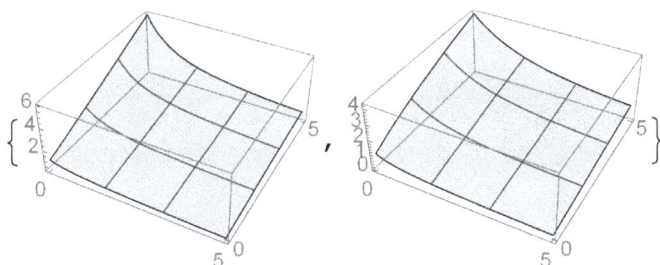

Figure 5.26: Analytical and numerical solutions for problem 5.14.

Remark. The two previous particular solutions are not numerically similar. A simple numerical example illustrates it,

```
{u[0,5]/.solp,u[0,5]/.nsolp[[1]]}
{6,4.2494}
```

□

5.15. Another initial value problem to a first-order quasilinear partial differential equation is introduced below,

$$x\,u(x,y)\partial_x u(x,y) + y\,u(x,y)\partial_y u(x,y) + x\,y = 0, \quad u(x,2) = x, \quad x,y \in \mathbb{R}.$$

5.15

```
Clear[x,y,u]
sol=DSolve[{x u[x,y]∂ₓu[x,y]+y u[x,y]∂ᵧu[x,y]+x y==0,
u[x,2]==x},u,{x,y}]//Flatten (* Solve the problem *)
```
$$\{u\to Function[\{x,y\},-\sqrt{\tfrac{x(4x+4y-y^3)}{y^2}}],$$
$$u\to Function[\{x,y\},\sqrt{\tfrac{x(4x+4y-y^3)}{y^2}}]\}$$

We simultaneously plot both solutions and the straight line $u(x, 2) = x$. The two figures differ by the PlotRange options.

```
{Show[Plot3D[u[x,y]/.First[sol],##],
ParametricPlot3D[{x,2,x},{x,-1,1},
PlotStyle→{Black,Thick}]],
Show[Plot3D[u[x,y]/.Last[sol],##],
ParametricPlot3D[{x,2,x},{x,-1,1},
PlotStyle→{Black,Thick}]]}&[{x,-1,1},{y,-2,2.3},
ImageSize→180,PlotRange→All,PlotStyle→{Black,
Opacity[.15]},Ticks→{{-1,0,1},{-2,0,2.3},Automatic}]
```

See Figure 5.27.

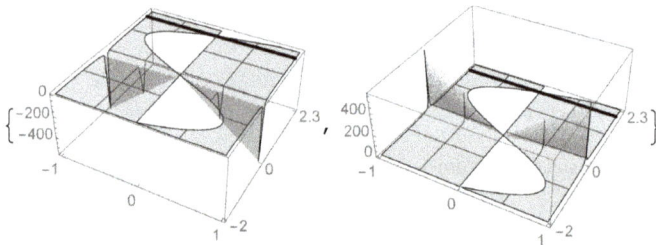

Figure 5.27: Particular solutions for problem 5.15.

5.16. The next equations are similar to the previous one. We successively consider them.

$$\partial_x u(x,y) \times \partial_y u(x,y) + 1 = u(x,y),$$
$$\partial_x u(x,y) \times \partial_y u(x,y) = x + y, \quad x, y \in \mathbb{R}.$$

5.16 We study the first equation.

```
Clear[x,y,u]
sol=DSolve[∂ₓu[x,y]*∂yu[x,y]+1==u[x,y],u,{x,y}]
//Flatten (* Solve the equation *)
```
$$\{u \to \text{Function}[\{x,y\}, \frac{y^2 + 4C[1] + 2x\,y\,C[1] + x^2 C[1]^2 + 2y\,C[2] + 2x\,C[1]\,C[2] + C[2]^2}{4C[1]}]\}$$

Thus, the general solution of the first equation is a second degree polynomial in the variables x and y.

Now we study the second equation.

```
Clear[x,y,u]
DSolve[∂ₓu[x,y]*∂ᵧu[x,y]==x+y,u,{x,y}]; (* Solve the
equation *)
tab=Table[u[x,y]/.%[[k]]/.{C[1]→1,C[2]→5},{k,2}] (*
Set some constants of integration *)
```

$$\left\{ \frac{-15-\sqrt{2}\sqrt{-4-12x-9x^2+6x\,y+9x^2y+3y^2+6x\,y^2+y^3}}{3}, \right.$$

$$\left. \frac{-15+\sqrt{2}\sqrt{-4-12x-9x^2+6x\,y+9x^2y+3y^2+6x\,y^2+y^3}}{3} \right\}$$

It is useful to know whether the expression under the square root is nonnegative. Then we write

```
v[x_,y_]:=(3Last[tab]+15)²/2 (* The expression under
the square root *)
{FindMinimum[##],FindMaximum[##]}&[{v[x,y],
{x,y}∈Reals},{x,y}]
{{-5.44968*10⁶,{x→1.,y→-179.}},
{6.22728*10⁶,{x→1.,y→181.}}}
```

Thus, the expression under the square root has a domain where it is nonnegative. Indeed, we have

```
{FindMinimum[{v[x,y],-7≤x≤4&&2≤y≤4},{x,y}]//Chop
Plot3D[v[x,y],{x,-7,4},{y,2,4},ImageSize→160,Mesh→1,
PlotStyle→{Black,Opacity[.15]},
Ticks→{{-7,-1.5,4},{2,3,4},Automatic}]}
{0,{x→-1.63473,y→2.90418}}
```

See Figure 5.28.

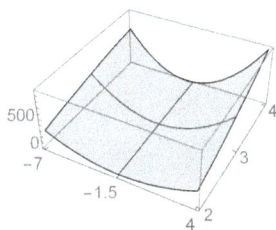

Figure 5.28: Plot of the expression under the square root in the solution of the second equation for problem 5.16.

The solutions of the two equations are represented below.

```
{Plot3D[u[x,y]/.sol/.{C[1]→1,C[2]→5},##,
Ticks→{{-7,-1.5,4},{2,3,4},20{0,1,2}},
PlotStyle→{Black,Opacity[.15]}],
Plot3D[tab,##,Ticks→{{-7,-1.5,4},{2,3,4},5{-4,-1,2}}
PlotStyle→{{Black,Opacity[.1]},{Black,Opacity[.2]}},
&[{x,-7,4},{y,2,4},ImageSize→160,Mesh→1]
```

See Figure 5.29.

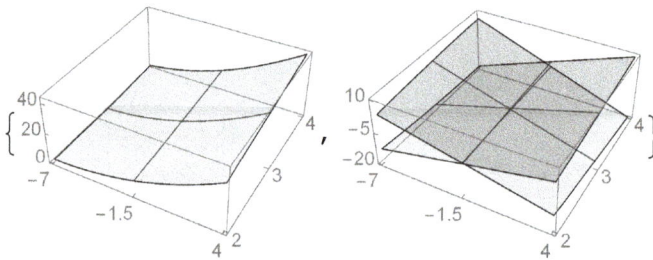

Figure 5.29: The two particular solutions for problem 5.13.

5.17. The next first-order nonlinear partial differential equation has a different form, i. e.,

$$(\partial_x u(x,y))^2 + (\partial_y u(x,y))^2 = 1, \quad x,y \in \mathbb{R}.$$

5.17 First approach. We use the `DSolve` built-in function.

```
Clear[x,y,u]
sol=DSolve[(∂ₓu[x,y])²+(∂ᵧu[x,y])²==1,u,{x,y}];
(* Solve the equation *)
tab=Table[u[x,y]/.%[[k]]]/.{C[1]→1,C[2]→.5},
{k,Length[%]}] (* Set some particular solutions *)
{{u→Function[{x,y},C[1]+yC[2]-x √1-C[2]²]},
{u→Function[{x,y},C[1]+y C[2]+x √1-C[2]²}}
{1-0.866025x+0.5y,1+0.866025x+.5y}
```

The solutions are two planes.

```
Plot3D[tab,{x,-2π,2π},{y,-2π,2π},ImageSize→180,
PlotStyle→{{Black,Opacity[.1]},{Black,Opacity[.2]}},
Ticks→Outer[Times,2{π,π,3},{-1,0,1}],Mesh→1]
```

See Figure 5.30.

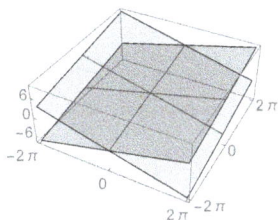

Figure 5.30: Plot of the particular solutions for problem 5.17.

Second approach. We also use a classical way to solve this equation.

$$\partial_x u(x,y) = \cos(a), \quad \partial_y u(x,y) = \sin(a), \text{ for an arbitrary real } a \implies$$

$$u(x,y) = x\cos(a) + f(y) = y\sin(a) + g(x), \text{ for some functions } f \text{ and } g \implies$$

$$y\sin(a) - f(y) = x\cos(a) - g(x) = k, \text{ for a constant } k \implies$$

$$\frac{g(x) - g(0)}{x} = \cos(a) \text{ and } \frac{f(y) - f(0)}{y} = \sin(a) \implies$$

$$g(x) = x\cos(a) + g(0) \text{ and } f(y) = y\sin(a) + f(0) \implies$$

$$u(x,y) = x\cos(a) + y\sin(a) + f(0).$$

Thus, we get the same result.

5.18. Our next initial value problem for a first-order nonlinear partial differential equation is of the form,

$$\partial_x u(x,y) + (\partial_y u(x,y))^2 = x + y, \quad u(x,0) = x^2/2, \quad x, y \in \mathbb{R}.$$

5.18

```
Clear[x,y,u]
sol=DSolve[{∂ₓu[x,y]+(∂ᵧu[x,y])²==x+y,u[x,0]==x²/2},u,
{x,y}]//Flatten (* Solve the problem *)
{u→Function[{x,y}, (3x²-4y^(3/2))/6],u→Function[{x,y}, (3x²+4y^(3/2))/6]}
```

We plot the solution to this problem. The solution consists in two surfaces that intersect at the initial condition. The surfaces are defined whenever $y \geq 0$.

```
Show[Plot3D[Evaluate[u[x,y]/.sol],{x,-5,5},{y,0,5},
PlotStyle→{{Black,Opacity[.1]},{Black,Opacity[.2]}},
Ticks→{5{-1,0,1},{1,3,5},10{0,1,2}},Mesh→1,
ImageSize→175],
ParametricPlot3D[{x,0,x²/2},{x,-5,5},
PlotStyle→{Black,Thick}]]
```

See Figure 5.31.

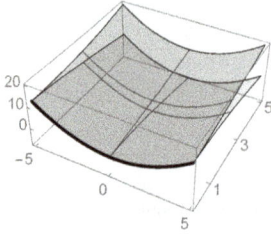

The initial condition is drawn by a black thick curve.

5.19. Here, we introduce a *Clairaut equation* in $\mathbb{R}^2$ and study it,

$$u(x,y) = x\,\partial_x u(x,y) + y\,\partial_y u(x,y) + \frac{(\partial_x u(x,y))^2 + (\partial_y u(x,y))^2}{2}, \quad x,y \in \mathbb{R}.$$

5.19 We introduce the problem

```
Clear[a,b,x,y]
eqn=u[x,y]==x ∂ₓu[x,y]+y ∂ᵧu[x,y]+(∂ₓu[x,y])²+(∂ᵧu[x,y])²/2 ;
```

and solve it by the DSolve built-in function.

```
sol=DSolve[eqn,u,{x,y}];
{u→Function[{x,y},x C[1]+y C[2]+½ (C[1]²+C[2]²)]}
```

The general solution of the equation is written as

```
ϕ[x_,y_,a_,b_]:=u[x,y]/.sol/.{C[1]→a,C[2]→b}
```

Now we look for the singular solution, if any,

```
Reduce[ϕ[x,y,a,b]==0&&∂ₐϕ[x,y,a,b]==0&&
∂_bϕ[x,y,a,b]==0,{a,b}]
(x==-i y||x==i y)&&a==-x&&b=-y
```

Then

```
usingular[x_,y_]:=ϕ[x,y,a,b]/.{a→-x,b→-y}//Simplify
```

We plot the figure.

```
Manipulate[
Plot3D[Evaluate[{ϕ[x,y,a,b],usingular[x,y]}],
{x,-5,10},{y,-5,10},ImageSize→175,Mesh→1,
PerformanceGoal→"Quality",MeshStyle→{{Black,Dashed},
{Black,Dotted}},Boxed→False,Axes→None,
PlotStyle→{{Black,Opacity[.1]},{Black,Opacity[.2]}},
PlotPoints→50],{a,-5,5},{b,-9,5}]
```

See Figure 5.32.

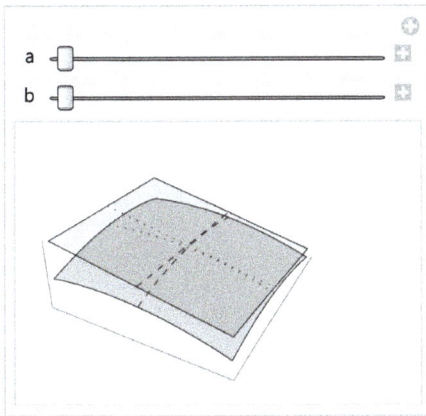

Figure 5.32: Dynamical solution of Clairaut's problem 5.19.

6 Linear hyperbolic partial differential equations

The present chapter deals with the following topics: linear hyperbolic partial differential equations, linear hyperbolic equations with variable coefficients, linear hyperbolic equations on curvilinear domains, linear hyperbolic equations in solid space, and Klein–Gordon equation.

According to [90, p. 2], a *wave* is any recognizable signal that is transferred from one part of the medium to another with a recognizable velocity of transfer. We also can say that a wave is a partial differential equation that models a time evolution phenomenon.

Modern introductions to this topic may be found in many references; we only mention [65, 39], and [13].

6.1 Hyperbolic equations with constant coefficients

A *linear hyperbolic differential equation* is a second-order partial differential equation so that its principal part is of the form

$$a\frac{\partial^2 u}{\partial x^2} + b\frac{\partial^2 u}{\partial x \partial y} + c\frac{\partial^2 u}{\partial y^2} = 0 \quad \text{with } b^2 - 4ac > 0.$$

The dynamics of an inextensible infinite vibrating string is modeled by a hyperbolic differential equation. This is an 1D *wave equation*, i. e., a partial differential equation of the second order whose reduced form is given by

$$\partial_{t,t} u(t,x) = c^2 \partial_{x,x} u(t,x), \quad c,t,x \in \mathbb{R}, \quad c \neq 0.$$

Its general, the solution is of the form

$$u(t,x) = f(x - ct) + g(x + ct),$$

where f and g are arbitrary twice differentiable functions [68].

The initial value problem for this hyperbolic differential equation is of the form

$$\begin{cases} \partial_{t,t} u(t,x) = c^2 \partial_{x,x} u(t,x), & c \in \mathbb{R} \setminus \{0\}, \quad t,x \in \mathbb{R}, \\ u(0,x) = f(x), \quad \partial_t u(0,x) = g(x), & x \in \mathbb{R}, \quad \text{(Initial conditions)}. \end{cases}$$

Here, function f describes the *initial shape* of the string while function g gives the *initial speed* of the string. Its solution is given below [68],

$$u(t,x) = \frac{1}{2}(f(x - ct) + f(x + ct)) + \frac{1}{2c}\int_{x-ct}^{x+ct} g(s)\mathrm{d}s.$$

https://doi.org/10.1515/9783111411392-008

The dynamics of an inextensible *finite* vibrating string is modeled by a hyperbolic differential equation with initial and boundary conditions of the form

$$\begin{cases} \partial_{t,t}u(t,x) = c^2\partial_{x,x}u(t,x), & c \in \mathbb{R}\setminus\{0\}, \quad x \in [0,l], \quad t,l > 0, \\ u(0,x) = f(x), \quad \partial_t u(0,x) = g(x), & \text{(Initial conditions)} \\ u(t,0) = 0, \quad u(t,l) = 0 & \text{(Boundary or Dirichlet conditions).} \end{cases}$$

The consistency conditions of the finite vibrating string problem read

$$f(0) = f(l) = 0 \text{ and } g(0) = g(l) = 0.$$

The solution of the finite vibrating string problem is of the form

$$u(t,x) = \sum_{n=1}^{\infty}\left(a_n \cos\left(\frac{c\,n\pi}{l}t\right) + b_n \sin\left(\frac{c\,n\pi}{l}t\right)\right)\sin\left(\frac{n\pi}{l}x\right),$$

$$a_n = \frac{2}{l}\int_0^l f(x)\sin\left(\frac{n\pi}{l}x\right)dx, \quad b_n = \frac{2}{c\,n\pi}\int_0^l g(x)\sin\left(\frac{n\pi}{l}x\right)dx,$$

supposing that all the operations above are valid.

6.1. Below there is the simplest example of the one-dimensional linear hyperbolic equation with constant coefficients,

$$\partial_{t,t}u(t,x) - c^2\partial_{x,x}u(t,x) = 0, \quad t,x \in \mathbb{R}, \quad c > 0.$$

6.1

```
Clear[c,t,x] (* Clear values and definitions *)
sol=Assuming[c>0,
DSolve[∂t,tu[t,x]-c²∂x,xu[t,x]==0,u,{t,x}]//Flatten]
(* Solve the equation *)
{u→Function[{t,x},C[1][-c t+x]+C[2][c t+x]]}
```

That is the general solution of the reduced hyperbolic equation, where C[1] and C[2] are arbitrary twice differentiable functions [56, Chapter 7].
We propose two particular functions C[1] and C[2]. For the domain of this equation, we take the square $[-5,5]\times[-5,5]$. We vary the plot by the PlotStyle built-in function.

```
fn=u[t,x]/.sol/.{C[1][p_]→Sin[p²],C[2][p_]→Cos[p]}
(* Particular solution *)
Cos[c t+x]+Sin[(-c t+x)²]
```

We plot the particular solution.

```
Manipulate[
Plot3D[fn/.c→1/4,{t,-5,5},{x,-5,5},ImageSize→175,
Ticks→Outer[Times,{5,5,2},{-1,0,1}],Mesh→5,
PlotStyle→{Black,Opacity[.1k],PlotPoints→20],
{k,Range[0,2]},SaveDefinitions→True]
```

See Figure 6.1.

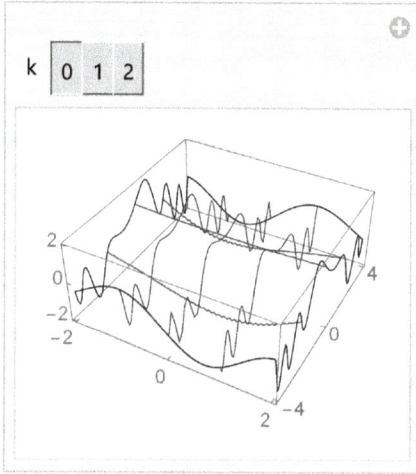

Figure 6.1: Plot of a particular solution for problem 6.1.

We exhibit three figures for corresponding values of constant c and some functions
C[1] and C[2].

```
sollist=
Table[u[t,x]/.DSolve[∂t,tu[t,x]-c²∂x,xu[t,x]==0,u,
{t,x}],{c,1,5,2}]/.{C[1][t_]→Sin[t],C[2][t_]→Cos[t]}
//Flatten
{Cos[t+x]-Sin[t-x],Cos[3t+x]-Sin[3t-x],
Cos[5t+x]-Sin[5t-x]}
```

The corresponding plots now follow.

```
Manipulate[
Plot3D[sollist[[k]],{t,-5,5},{x,-5,5},Mesh→{4,7},
PlotStyle→{Black,Opacity[.15(k-1)]},
Ticks→Outer[Times,{5,5,2},{-1,0,1}],ImageSize→160],
{k,Range[Length[sollist]]},SaveDefinitions→True]
```

See Figure 6.2.

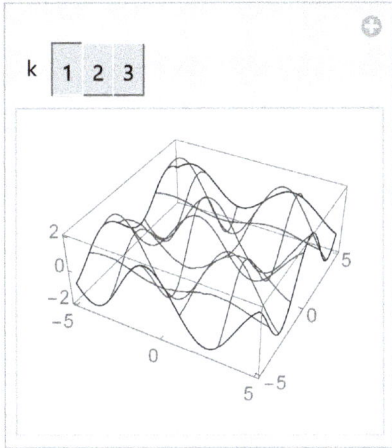

Figure 6.2: Dynamical plot of a particular solution for problem 6.1.

6.2. We introduce a Gaussian initial value problem for the one-dimensional hyperbolic equation,

$$\begin{cases} \partial_{t,t}u(t,x) = c^2\partial_{x,x}u(t,x), & t,x \in \mathbb{R}, \quad c > 0, \\ u(0,x) = \exp(-x^2), & \partial_t u(0,x) = 0. \end{cases}$$

6.2

```
Clear[c,m,t,x]
sol=Assuming[c>0,
eqn={∂_{t,t}u[t,x]==c²∂_{x,x}u[t,x],u[0,x]==Exp[-x²],
u^(1,0)[0,x]==0}; (* The problem *)
DSolve[eqn,u,{t,x}]//Flatten] (* Solve the problem *)
{u→Function[{t,x},½(e^{-(-c t+x)²}+e^{-(c t+x)²})]}
```

```
eqn/.sol//Simplify (* Check the solution *)
{True,True,True}
```

We plot a particular solution.

```
{Plot3D[u[t,x]/.sol/.c→1,{t,-5,5},{x,-6,6},
Ticks→Outer[Times,{5,6,1},{-1,0,1}],Mesh→3,
ImageSize→175,PlotStyle→{Black,Opacity[.15]},
PlotRange→All],
Manipulate[
Plot3D[Evaluate[u[t,x]/.sol/.c→m],##,ImageSize→160,
PlotRange→All,Mesh→3,PerformanceGoal→"Quality",
PlotStyle→{Black,Opacity[.15m]},Boxed→False,
Axes→None],&[{t,-5,5},{x,-6,6}],{m,Range[.1,1,.1]},
SaveDefinitions→True]}
```

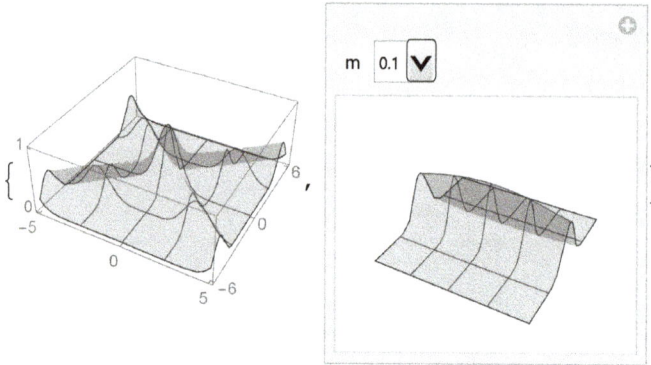

Figure 6.3: Plot of a particular solution for problem 6.2.

See Figure 6.3.

A continuous cross-section through the figure above is introduced.

```
Animate[
Plot[u[t,x]/.sol/.{c→1},{x,-5,5},PlotRange→All,
ImageSize→150,Ticks→{2Range[-2,2],.2Range[5]}
PlotStyle→Black],{{t,1},-6,6},
AnimationDirection→ForwardBackward,
DefaultDuration→15,SaveDefinitions→True,
AnimationRunning→False]
```

See Figure 6.4.

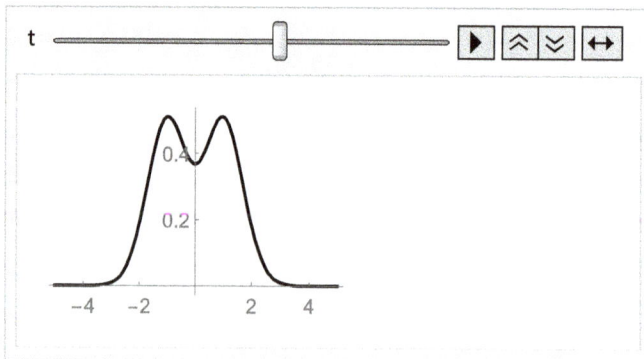

Figure 6.4: Continuous cross- section through the plot for problem 6.2.

6.3. We introduce an 1D linear hyperbolic equation with an initial condition so that its initial velocity is null [64, p. 117],

$$\begin{cases} \partial_{t,t}u(t,x) = a^2\partial_{x,x}u(t,x), & t,x \in \mathbb{R}, \quad a,h > 0, \quad 0 < c < l, \\ u(0,x) = \begin{cases} \frac{h}{c}x, & 0 \leq x \leq c, \\ \frac{h(l-x)}{l-c}, & c \leq x \leq l, \end{cases} & \partial_t u(0,x) = 0. \end{cases}$$

6.3 We plot the initial conditions.

```
Clear[t,x]
a=2.;l=2π;c=5.;h=1.;
f[x_]:=Piecewise[{{h/c x,0≤x≤c},{h(1-x)/(1-c),c≤x≤l}}]
g[x_]:=0.
Plot[{f[x],g[x]},{x,0,l},ImageSize→175,
PlotStyle→{{Black,Dashed},Black},
Ticks→{Range[6],{.5,1}},PlotLabel→Style["Initial
shape and speed",12,Bold,Black],
PlotLegends→"Expressions"]
```

See Figure 6.5.

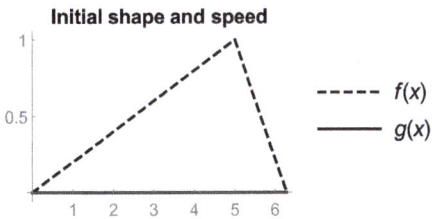

Figure 6.5: Plot of the initial conditions for problem 6.3.

```
eqncond={∂t,tu[t,x]==a²∂x,xu[t,x],u[0,x]==f[x],
u⁽¹,⁰⁾[0,x]==g[x]};(* The problem *)

sol=DSolve[eqncond,u,{t,x}]//Flatten (* Solve the
problem *)
```

$$\{u \to \text{Function}[\{t,x\},$$

$$0. + \frac{1}{2}\left(\left(\begin{bmatrix} 0.2(-2.t+x) & 0 \leq -2.t+x \leq 5. \\ 0.779311(2\pi+2.t-x) & 5. \leq -2.t+x \leq 2\pi \\ 0 & \text{True} \end{bmatrix}\right)\right.$$

$$+ \left(\begin{bmatrix} 0.2(2.t+x) & 0 \leq 2.t+x \leq 5. \\ 0.779311(2\pi-2.t-x) & 5. \leq 2.t+x \leq 2\pi \\ 0 & \text{True} \end{bmatrix}\right)\Big]\}$$

A continuous plot of the solution follows:

```
Animate[
{Plot3D[u[t,x]/.sol,{t,0.,τ},{x,0,1},ImageSize→160,
Mesh→3,PlotStyle→{Black,Opacity[.15]},
ViewPoint→{1.3,8,5},Ticks→{{0,5},3Range[0,2],{0,1}}],
SaveDefinitions→True,DefaultDuration→25,
AnimationRunning→False]
```

See Figure 6.6.

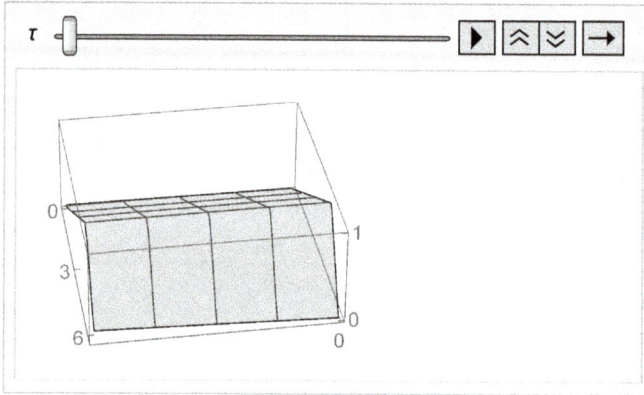

Figure 6.6: Continuous plot of the solution for problem 6.3.

6.4. We present an 1D hyperbolic equation

$$\partial_{t,t}u(t,x) = a^2\partial_{x,x}u(t,x), \quad t,x \in \mathbb{R}, \quad a > 0.$$

with a list of initial conditions having nonzero initial data.
1. $0 < b < l, h > 0,$

$$u(0,x) = \begin{cases} \frac{h}{l-b}(x+l), & -l \le x \le -b, \\ h, & -b \le x \le b, \\ \frac{h}{l-b}(l-x), & b \le x \le l, \end{cases} \quad \partial_t u(0,x) = .5.$$

2. $0 < b < l,$

$$u(0,x) = \begin{cases} \frac{h}{l-b}(x+l), & -l \le x \le -b, \\ h, & -b \le x \le b, \\ \frac{h}{l-b}(l-x), & b \le x \le l, \end{cases} \quad \partial_t u(0,x) = \exp(-x^2).$$

3. $b, c \in \mathbb{R}, m \in \mathbb{N}, u(0, x) = b \sin((m\pi/l)x), \partial_t u(0, x) = c.$
4. $0 \le x_1 \le x_2 \le l,$

$$u(0, x) = \frac{4h}{l^2} x(l - x), \quad \partial_t u(0, x) = \begin{cases} b, & x \in [x_1, x_2], \\ c, & x \notin [x_1, x_2]. \end{cases}$$

5. $h, b, c \in \mathbb{R}, 0 < c - \varepsilon < c + \varepsilon \le l$

$$\begin{cases} u(0, x) = \begin{cases} \frac{2h}{l} x, & 0 \le x \le l/2, \\ \frac{2h}{l}(l - x), & l/2 \le x \le l, \end{cases} \\ \partial_t u(0, x) = \begin{cases} b \cos(\pi(x - c)\varepsilon), & x \in [c - \varepsilon, c + \varepsilon], \\ c, & x \notin [c - \varepsilon, c + \varepsilon]. \end{cases} \end{cases}$$

6. $0 < l, u(0, x) = \frac{h}{l^2}(l^2 - x^2), \partial_t u(0, x) = \sin(\pi(x/l + 1)), -l \le x \le l.$
7. $0 < l, u(0, x) = \frac{h}{l^2}(l^2 - x^2), \partial_t u(0, x) = \cos(5\pi(x/l + 1)), -l \le x \le l.$

6.4
1.

```
Clear[t,x]
```

We plot the initial conditions.

```
a=2.;l=2π;b=5π/4;h=.25;
f[x_]:=Piecewise[{{(h/(1-b))(x+1),-1≤x≤-b},{h,-b≤x≤b},
{(h/(1-b))(1-x),b≤x≤1}}]
g[x_]:=.5
Plot[{f[x],g[x]},{x,-1,1},ImageSize→160,
Ticks→{3Range[-2,2],{.2}},PlotLegends→"Expressions",
PlotLabel→Style["Initial shape and speed",12,Bold,
Black],PlotStyle→{{Black,Dashed},Black}]
```

See Figure 6.7.

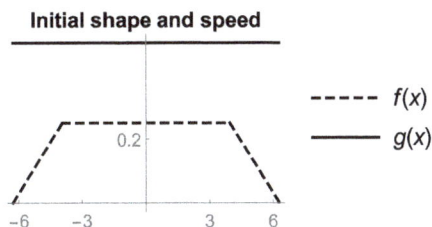

Figure 6.7: Plot of the initial conditions for problem 6.4 1.

```
eqncond={∂t,tu[t,x]==a²∂x,xu[t,x],u[0,x]==f[x],
u⁽¹,⁰⁾[0,x]==g[x]}; (* The problem *)

sol=DSolve[eqncond,u,{t,x}]//Flatten (* Solve the
problem *)
{u→Function[{t,x},
```

$$0.+\frac{1}{2}\left(\left(\left(\begin{bmatrix} 0.106103(2\pi-2.t+x) & -2\pi\le-2.t+x\le-\frac{5\pi}{4} \\ 0.25 & -\frac{5\pi}{4}\le-2.t+x\le\frac{5\pi}{4} \\ 0.106103(2\pi+2.t-x) & \frac{5\pi}{4}\le-2.t+x\le2\pi \\ 0 & True \end{bmatrix}\right)\right.\right.$$

$$+\left.\left.\begin{bmatrix} 0.106103(2\pi+2.t+x) & -2\pi\le2.t+x\le-\frac{5\pi}{4} \\ 0.25 & -\frac{5\pi}{4}\le2.t+x\le\frac{5\pi}{4} \\ 0.106103(2\pi-2.t-x) & \frac{5\pi}{4}\le2.t+x\le2\pi \\ 0 & True \end{bmatrix}\right)\right)]\}$$

```
Plot3D[u[t,x]/.sol,{t,0.,5.},{x,-1,1},ImageSize→160,
Mesh→3,PlotStyle→{Black,Opacity[.15]},
Ticks→{{0,5},3Range[-2,2],{0,.2}},ViewPoint→{2,2,2}]
```

See Figure 6.8.

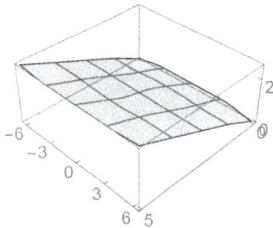

Figure 6.8: Plot of the solution for problem 6.4 1.

2.

```
Clear[t,x]
```

We plot the initial conditions.

```
a=2.;l=2π;c=5.;b=2;h=5.;
f[x_]:=Piecewise[{{(h/(1-b))(x+1),-1≤x≤-b},
{h,-b≤x≤b},{(h/(1-b))(1-x),b≤x≤1}}]
g[x_]:=Exp[-x²]
Plot[{f[x],g[x]},{x,-1,1},ImageSize→160,
Ticks→{3Range[-2,2],{1,5}},PlotLegends→
"Expressions",PlotLabel→Style["Initial shape and
speed",12,Bold,Black]
```

Initial shape and speed

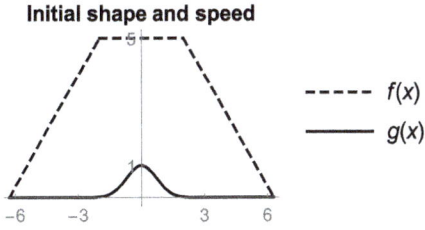

- - - - - $f(x)$

——— $g(x)$

Figure 6.9: Plot of the initial conditions for problem 6.4 2.

See Figure 6.9.

```
eqncond={∂t,tu[t,x]==a²∂x,xu[t,x],u[0,x]==f[x],
u⁽¹,⁰⁾[0,x]==g[x]}; (* The problem *)
```

```
sol=DSolve[eqncond,u,{t,x}]//Flatten
{u→Function[{t,x},
0.25(0.886227 Erf[2.t-1.x]+.886227 Erf[2.t+x])
```

$$+\frac{1}{2}\left(\left(\begin{cases} 1.16736(2\pi-2.t+x) & -2\pi \leq -2.t+x\leq -2 \\ 5. & -2\leq -2.t+x\leq 2 \\ 1.16736(2\pi+2.t-x) & 2\leq -2.t+x\leq 2\pi \\ 0 & \text{True} \end{cases}\right)\right.$$

$$+\left(\left(\begin{cases} 1.16736(2\pi+2.t+x) & -2\pi \leq 2.t+x\leq -2 \\ 5. & -2\leq 2.t+x\leq 2 \\ 1.16736(2\pi-2.t-x) & 2\leq 2.t+x\leq 2\pi \\ 0 & \text{True} \end{cases}\right)\right)]\}$$

```
Plot3D[u[t,x]/.sol,{t,0.,5.},{x,0,1},ImageSize→160,
Mesh→4,PlotStyle→{Black,Opacity[.15]},
Ticks→{{0,5},3Range[-2,2],{0,5}},ViewPoint→{2.6,4,3}]
```

See Figure 6.10.

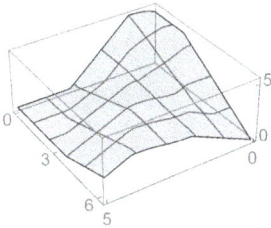

Figure 6.10: Plot of the solution for problem 6.4 2.

3.

```
Clear[t,x]
```

We plot the initial conditions.

```
a=2.;l=2π;m=4;c=5.;b=1.;
f[x_]:=b Sin[(mπ/l)x]
g[x_]:=c
Plot[{f[x],g[x]},{x,0.,l},ImageSize→160,
Ticks→{Range[6],{5}},PlotLegends→"Expressions",
PlotLabel→Style["Initial shape and speed",12,Bold,
Black],PlotStyle→{{Black,Dashed},Black}]
```

See Figure 6.11.

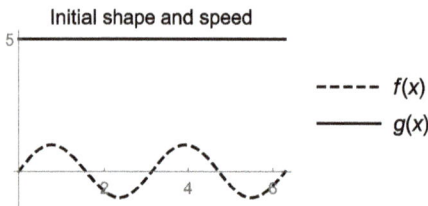

Figure 6.11: Plot of the initial conditions for problem 6.4 3.

```
eqncond={∂t,tu[t,x]==a²∂x,xu[t,x],u[0,x]==f[x],
u⁽¹,⁰⁾[0,x]==g[x]}; (* The problem *)
```

```
sol=DSolve[eqncond,u,{t,x}]//Flatten
{u→Function[{t,x},
5.t+1/2(1.Sin[2(-2.t+x)]+1.Sin[2(2.t+x)])]]
```

```
Plot3D[u[t,x]/ sol,{t,0.,5.},{x,0.,l},ImageSize→160,
Mesh→3,PlotStyle→{Black,Opacity[.15]},
Ticks→Outer[Times,{2.5,3,12},{0,1,2}]]
```

See Figure 6.12.

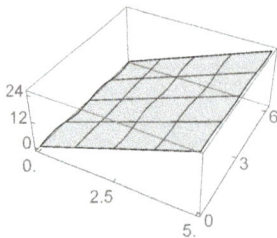

Figure 6.12: Plot of the solution for problem 6.4 3.

4.

```
<< Notation`
```

```
Symbolize[x_]
```

```
Clear[t,x]
```

We plot the initial conditions.

```
a=2.;l=2π;c=5.;b=2;h=5.;x₁=1.;x₂=4.;
f[x_]:=(4h/l²)x(l-x);
g[x_]:=Piecewise[{{c,0.≤x≤x₁},{b,x₁ ≤x≤x₂},{c,x₂ ≤x≤l}}]
Plot[{f[x],g[x]},{x,0.,l},ImageSize→160,
AxesOrigin→{0,0},Ticks→{Range[6],Range[5]},
PlotLabel→Style["Initial shape and speed",12,Bold,
Black],PlotLegends→"Expressions"]
```

See Figure 6.13.

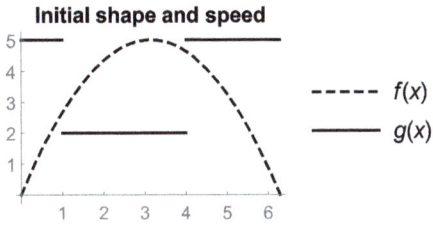

Figure 6.13: Plot of the initial conditions for problem 6.4 4.

```
eqncond={∂_{t,t}u[t,x]==a²∂_{x,x}u[t,x],u[0,x]==f[x],
u^(1,0)[0,x]==g[x]}; (* The problem *)
```

```
sol=DSolve[eqncond,u,{t,x}]//Flatten
{u→Function[{t,x},½
(.506606(2π+2.t-x)(-2.t+x)+.506606(2π-2.t-x)(2.t+x))
```

$$
+.25\,\text{Integrate}\left[\begin{cases} 5. & 0.\le K[1]\le1. \\ 2 & 1.\le K[1]\le4. \\ 5. & 4.\le K[1]\le2\pi \\ 0 & \text{True} \end{cases}\right.,
$$

```
{K[1],-2.t+x,2.t+x},Assumptions→True]]}
```

```
Plot3D[u[t, x]/.sol,{t,0.,5.},{x,0.,1},ImageSize→160,
Mesh→3,PlotStyle→{Black,Opacity[.15]},
Ticks→{2{0,1},3Range[0,2],20{-2,-1,0}}]
```

See Figure 6.14.

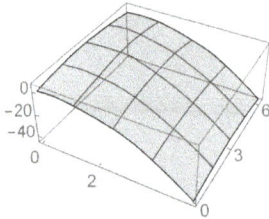

Figure 6.14: Plot of the solution for problem 6.4 4.

5.

```
Clear[t,x]
```

We plot the initial conditions.

```
a=2.;l=2π;c=3.;b=2.;h=2.;ε=.5;
f[x_]:=Piecewise[{{(2h/l)x,0.≤x≤1/2},
{(2h/l)(1-x),1/2≤x≤1}}]
g[x_]:=Piecewise[{{0.,0≤x≤c-ε},{b Cos[π(x-c)/(2ε)],
c-ε≤x≤c+ε},{0.,c+ε ≤x≤1}}]
Plot[{f[x],g[x]},{x,0.,1},ImageSize→160,
AxesOrigin→{0,0},Ticks→{Range[6],{1,2}},
PlotLabel→Style["Initial shape and speed",12,Bold,
Black],PlotStyle→{{Black,Dashed},Black},
PlotLegends→"Expressions"]
```

See Figure 6.15.

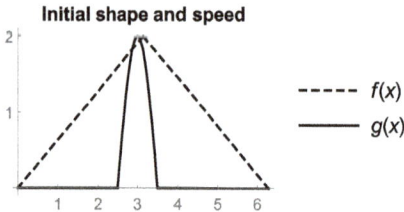

Figure 6.15: Plot of the initial conditions for problem 6.4 5.

```
eqncond={∂t,tu[t,x]==a²∂x,xu[t,x],u[0,x]==f[x],
u^(1,0)[0,x]==g[x]}; (* The problem *)
```

```
sol=DSolve[eqncond,u,{t,x}]//Flatten
{u→Function[{t,x},
```

$$0.25\text{Integrate}\left[\begin{cases} 0. & 0{\le}K[1]{\le}2.5 \\ 2.\text{Cos}[3.14159(-3.K[1])] & 2.5{\le}K[1]{\le}3.5 \\ 0. & 3.5{\le}K[1]{\le}2\pi \\ 0 & \text{True} \end{cases}\right.,$$

```
{K[1],-2.t+x,2.t+x},Assumptions→True]
```

$$+\frac{1}{2}\left(\left(\begin{cases} 0.63662(-2.t+x) & 0.{\le}-2.t+x {\le} \pi \\ 0.63662(2\pi+2.t-x) & \pi {\le} -2.t+x{\le}2\pi \\ 0 & \text{True} \end{cases}\right.\right.$$

$$+\left(\begin{cases} 0.63662(2.t+x) & 0.{\le}2.t+x {\le} \pi \\ 0.63662(2\pi-2.t-x) & \pi {\le} 2.t+x{\le}2\pi \\ 0 & \text{True} \end{cases}\right.\left.\left.\right)\right]\}$$

```
Plot3D[u[t,x]/.sol,{t,0.,5.},{x,0.,1},ImageSize→160,
PlotStyle→{Black,Opacity[.15]},
Ticks→{{0,5},3Range[0,2],Range[-1,2]}]
```

See Figure 6.16.

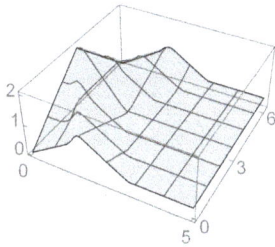

Figure 6.16: Plot of the solution for problem 6.4 5.

6.

```
Clear[t,x]
```

We plot the initial conditions.

```
a=2.;l=2π;b=5π/4;h=.25;
f[x_]:=(h/l²)(l²-x²)
g[x_]:=Sin[π(x/l+1)]
Plot[{f[x],g[x]},{x,-1,1},ImageSize→160,
Ticks→{3Range[-2,2],{-1,.2,1}},
PlotLabel→Style["Initial shape and speed",12,Bold,
Black],PlotStyle→{{Black,Dashed},Black},
PlotLegends→"Expressions"]
```

Initial shape and speed

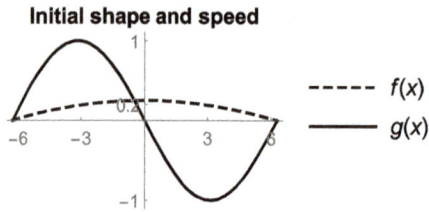

Figure 6.17: Plot of the initial conditions for problem 6.4 6.

See Figure 6.17.

```
eqncond={∂t,tu[t,x]==a²∂x,xu[t,x],u[0,x]==f[x],
u⁽¹,⁰⁾[0,x]==g[x]}; (* The problem *)
```

```
sol=DSolve[eqncond,u,{t,x}]//Chop//Flatten
{u→Function[{t,x},1/2(0.00633257(4π²-(-2.t+x)²)
+0.00633257(4π²-(2.t+x)²))-1.Sin[1.t]Sin[.5x]]
```

```
Plot3D[u[t,x]/.sol,{t,0.,5.},{x,-1,1},ImageSize→160,
Mesh→5,PlotStyle→{Black,Opacity[.15]},
Ticks→{{0,5},3Range[-2,2,2],{-1,0,1}},
ViewPoint→{2.6,8,2}]
```

See Figure 6.18.

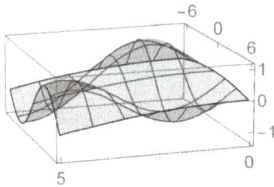

Figure 6.18: Plot of the solution for problem 6.4 6.

7.

```
Clear[t,x]
```

We plot the initial conditions.

```
a=2.;l=2π;b=5π/4;h=25;
f[x_]:=(h/l²)(l²-x²)
g[x_]:=(l²-x²)Cos[5π(x/l+1)]
Plot[{f[x],g[x]},{x,-1,1},ImageSize→160,
AxesOrigin→{0,0},Ticks→{Range[-6,6,3],20{-1,1}},
PlotLabel→Style["Initial shape and speed",12,Bold,
Black],PlotStyle→{{Black,Dashed},Black},
PlotLegends→"Expressions"]
```

Initial shape and speed

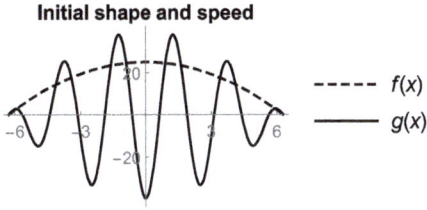

- - - - - $f(x)$
——— $g(x)$

See Figure 6.19.

```
eqncond={∂t,tu[t,x]==a²∂x,xu[t,x],u[0,x]==f[x],
u⁽¹,⁰⁾[0,x]==g[x]};  (* The problem *)
```

```
sol=DSolve[eqncond,u,{t,x}]//Chop//Flatten
{u→Function[{t,x},1/2((25(4π²-(-2.t+x)²))/(4π²)
+(25(4π²-(2.t+x)²))/(4π²))
+0.25((0.64 t-.32 x)Cos[5.t]Cos[2.5 x]
+Sin[5.t]((-31.8387+3.2t²+.8x²)Cos[2.5x]
-.64x Sin[2.5x])
+Cos[5.t]((.64 t+.32 x)Cos[2.5 x]+3.2t x Sin[2.5x]))]}
```

```
Plot3D[u[t,x]/.sol,{t,0.,5.},{x,-1,1},ImageSize→160,
Mesh→5,PlotStyle→{Black,Opacity[.15]},
Ticks→{{0,5},6{-1,1},10{-6,2}}]
```

See Figure 6.20.

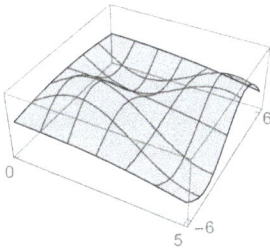

6.5. We study the next hyperbolic equation with certain initial conditions,

$$\begin{cases} \partial_{t,t}u(t,x) + 2\partial_{t,x}u(t,x) - 3\partial_{x,x}u(t,x) = 0, & t, x \in \mathbb{R}, \\ u(0,x) = x^2, & \partial_t u(0,x) = 0. \end{cases}$$

6.5 The solution to this initial value problem is an elliptic paraboloid whose horizontal section is an ellipse. Several kinds of meshes are exposed.

```
Clear[a,t,x]
```

```
sol=DSolve[{-3 u^(0,2)[t,x]+2 u^(1,1)[t,x]+u^(2,0)[t,x]==0,
u[0,x]==x²,u^(1,0)[0,x]==0},u,{t,x}]//First
{u→Function[{t,x},3t²+x²]}
```

Now we plot the solution having different meshes.

```
Plot3D[u[t,x]/.sol,{t,-6,6},{x,-5,5},ImageSize→160,
Mesh→5,PlotStyle→{Black,Opacity[.15]},MeshStyle→
{{Black,Dashed},{Black,Dotted}},Ticks→Outer[Times,
{6,5,60},Range[-1,2,1]],#]&/@{MeshFunctions→{#1&,#2&},
MeshFunctions→{Re[Sqrt[(#1+I#2)³]]&,
Im[Sqrt[(#1+I#2)³]]&},MeshFunctions→{#2&},
MeshFunctions→{#3&}}
```

See Figure 6.21.

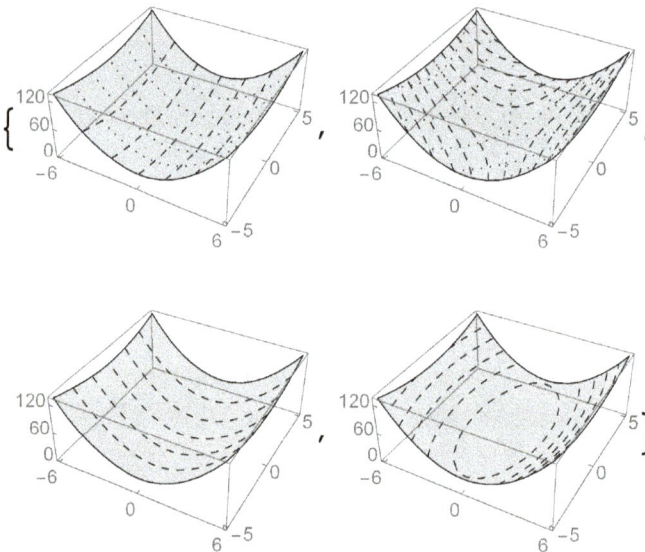

Figure 6.21: Plot of the meshes for problem 6.5.

We look for the genesis of the level curves supposing that the domain of definition is a circle of varying radius.

```
{Plot3D[u[t,x]/.sol,##,ImageSize→150,
RegionFunction→Function[{t,x,z},t²+x² ≤4],Mesh→7,
PlotStyle→{Black,Opacity[.15]},MeshStyle→Blanck,
MeshFunctions→{#3&},
Ticks→Outer[Times,2{1,1,3},Range[-1,2,1]]],
Manipulate[Plot3D[u[t,x]/.sol,##,ImageSize→160,
RegionFunction→Function[{t,x,z},t²+x² ≤a],Mesh→7,
PlotStyle→{Black,Opacity[.1]},MeshStyle→Black,
MeshFunctions→{IntegerPart[#3/2]&},
Ticks→{2{-1,0,1},2{-1,0,1},Automatic}],{a,.001,4},
SaveDefinitions→True]}&[{t,-2,2},{x,-2,2}]
```

See Figure 6.22.

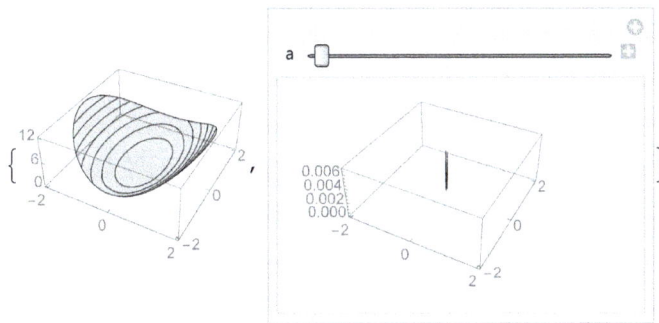

Figure 6.22: Plot of the solution for problem 6.5.

6.6. Next, we study an initial value problem, i. e.,

$$\begin{cases} \partial_{t,t}u(t,x) - \partial_{t,x}u(t,x) - 6\partial_{x,x}u(t,x) = 0, & t,x \in \mathbb{R}, \\ u(t,0) = \cos(3t) - \sin(2t), & \partial_x u(t,0) = -\cos(2t) - \sin(3t). \end{cases}$$

6.6

```
Clear[a,t,x,y]

sol=DSolve[{u^(2,0)[t,x]-u^(1,1)[t,x]-6u^(0,2)[t,x]==0,
u[t,0]==Cos[3t]-Sin[2t],u^(0,1)[t,0]==-Cos[2t]-Sin[3t]},
u,{t,x}]//Flatten (* Solve the problem *)
{u→Function[{t,x},
1/5(5Cos[3(t+x/3)]+Sin[2(t-x/2)]-6 Sin[2(t+x/3)])]}
```

We plot the solution and the initial conditions.

```
Show[
{Plot3D[u[t,x]/.sol,{t,-π,π},{x,-π,π},Mesh→3,
Ticks→Outer[Times,{π,π,2},{-1,0,1}],ImageSize→160,
PlotStyle→{Black,Opacity[.15]}],
ParametricPlot3D[{t,0,Cos[3t]-Sin[2t]},{t,-π,π},
PlotStyle→{Black,Thick}]}]
```

See Figure 6.23.

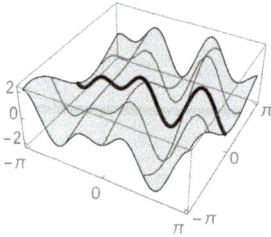

Figure 6.23: Plot of the solution for problem 6.6.

We color the surface by the `Rainbow` option and project the colors on the horizontal plane $y = -10$.

```
Show[
Plot3D[u[t,x]/.sol,{t,-π,π},{x,-π,π},Mesh→None,
ImageSize→160,ColorFunction→"Rainbow",
PlotRange→{{-π,π},{-π,π},{-10,3}},
PlotStyle→Directive[Opacity[.1],
Specularity[White,50]]],
SliceContourPlot3D[u[t,x]/.sol,y==-10,{t,-π,π},
{x,-π,π},{y,-10,3},ColorFunction→"Rainbow",
ContourStyle→Opacity[.45],Boxed→False]]
```

See Figure 6.24.

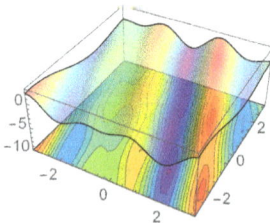

Figure 6.24: Different colors of the solution for problem 6.6.

```
Animate[(* Animate the previous figure *)
Show[Plot3D[u[t+a,x-a]/.sol,{t,-π,π},{x,-π,π},
Mesh→None,ImageSize→160,Boxed→False,Ticks→None,
PlotStyle→Directive[Opacity[.1],
Specularity[White,50]],ColorFunction→"Rainbow",
PlotRange→{{-π,π},{-π,π},{-10,3}}],
SliceContourPlot3D[u[t+a,x-a]/.sol,y==-10,{t,-π,π},
{x,-π,π},{y,-10,3},ColorFunction→"Rainbow",
ContourStyle→Opacity[.1],Boxed→False]],
{a,0,2π,π/10},SaveDefinitions→True,
DefaultDuration→75,AnimationRunning→False,
AnimationDirection→ForwardBackward]
```

See Figure 6.25.

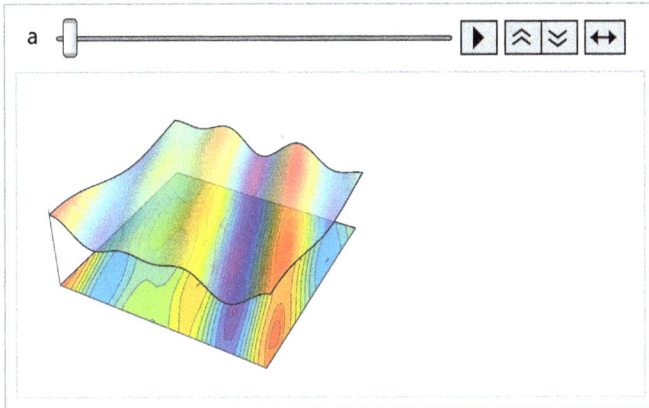

Figure 6.25: Animate figure of the solution for problem 6.6.

6.7. We find the solution to the following problem of a finite vibrating string:

$$\begin{cases} \partial_{t,t}u(t,x) = 4^2\partial_{x,x}u(t,x) = 0, & t \in [0,10], \quad x \in [0,5], \\ u(0,x) = \sin(\frac{x\pi}{5}), \quad \partial_t u(0,x) = x(5-x), & \text{(Initial conditions)} \\ u(t,0) = u(t,5) = 0, & \text{(Boundary conditions)}. \end{cases}$$

6.7

```
Clear[t,x]
```

```
eqn={∂t,tu[t,x]==4²∂x,xu[t,x],u[t,0]==u[t,5]==0,
u[0,x]==Sin[xπ/5],u⁽¹,⁰⁾[0,x]==x(5-x)}; (* The
problem *)
sol=DSolve[%,u,{t,x}]/.{K[1]→n,∞ →100}//Flatten
(* Solve the equation, change the index of summation,
and keep the partial sum of the first 100 terms *)
{u→Function[{t,x},
```
$$\sum_{n=1}^{100} \frac{125(-1+(-1)^n)\sin[(4\pi t n)/5]\sin[(\pi x n)/5]}{\pi^4 n^4}]\}$$

We define the function

```
asol[t_,x_]:=Activate[u[t,x]/.sol]
```

and plot it. We also project the resulted surface on planes xOy and yOz.

```
Show[Plot3D[asol[t,x],{t,-2,2},{x,-2,3},Mesh→None,
ImageSize→175,PlotStyle→Directive[Opacity[.45],
Specularity[White,50]],ColorFunction→"Rainbow",
PlotTheme→"Detailed",FaceGrids→None,
PlotRange→{{-2,2},{-2,3},{-8,4}}],
SliceContourPlot3D[asol[t,x],{t==-2,y==-8},{t,-2,2},
{x,-2,3},{y,-8,4},ColorFunction→"Rainbow",
Boxed→False]]
```

See Figure 6.26.

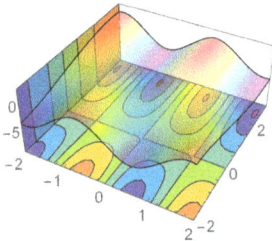

Figure 6.26: Plot of the solution for problem 6.7.

6.8. We look for a solution to the following Dirichlet and Neumann problem of a hyperbolic equation defined on a square:

$$\begin{cases} \partial_{x,x}u(x,y) = 2\partial_{y,y}u(x,y), & \Omega = \{(x,y) \mid 0 \le x \le 1 \wedge 0 \le y \le 1\} \subset \mathbb{R}^2, \\ u(x,y) = \begin{cases} 1, & x > 0, \quad y = 0, \\ 5y, & x = 0, \quad y > 0, \end{cases} & \text{(Dirichlet conditions)} \\ \partial_n u(x,y) = 1, & x > 0, \quad \text{(Neumann condition).} \end{cases}$$

6.8

```
Clear[t,x]
```

We set the domain

```
bound={x,1-x,y,1-y};
Ω=ImplicitRegion[And@@(#≥0&/@bound),{x,y}];
```

and plot it.

```
Show[RegionPlot[Ω,PlotStyle→{Black,Opacity[.1]}],
ContourPlot[Evaluate[Thread[bound==0]],{x,0,1},{y,0,1},
ContourStyle→Black],ImageSize→8]
```

See Figure 6.27.

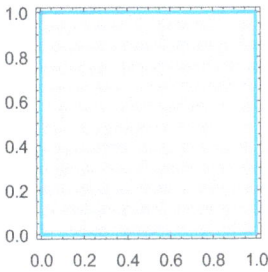

Figure 6.27: Plot of the domain of integration for problem 6.8.

We define the problem

```
sol=NDSolveValue[{∂_{x,x}u[x,y]==2∂_{y,y}u[x,y]
+NeumannValue[1,x>0],
DirichletCondition[u[x,y]==1,x>0&&y==0],
DirichletCondition[u[x,y]==5 y,x==0&&y>0]},u,{x,0,1},
{y,0,1}];
```

and plot its solution.

```
Plot3D[sol[x,y],{x,0,1},{y,0,1},PlotRange→All,
PlotStyle→{Black,Opacity[.15]},PlotPoints→50,
ImageSize→200,Mesh→3]
```

See Figure 6.28.

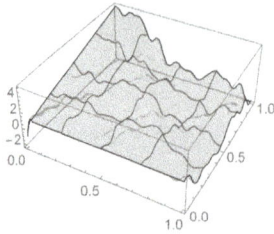

Figure 6.28: Plot of the numerical solution for problem 6.8.

6.9. We study the inhomogeneous hyperbolic equation with zero initial and boundary conditions [64, p. 118],

$$\begin{cases} \partial_{t,t}u(t,x) - c^2\partial_{x,x}u(t,x) = t + x^2, & t \in [0,10], \quad x \in [0,5], \quad c > 0, \\ u(0,x) = \partial_t u(0,x) = 0, & \text{(Initial conditions)} \\ u(t,0) = u(t,5) = 0, & \text{(Boundary conditions).} \end{cases}$$

6.9

```
Clear[t,x] (* Clean values and definitions *)

eqncond={∂_t,t u[t,x]-c²∂_x,x u[t,x]==t+x²,u[0,x]==0,
u^(1,0)[0,x]==0,u[t,0]==u[t,5]==0}; (* The problem *)

Assuming[c>0,DSolve[eqncond,u,{t,x}]]; (* No
closed-form solution is returned *)
```

We follow a numerical way and plot the solution.

```
c=2;

sol=NDSolve[eqncond,u,{t,0,10},{x,0,5},
PrecisionGoal→2]; (* Numerical solution *)

Plot3D[u[t,x]/.sol,{t,0,10},{x,0,5},ImageSize→175,
MeshStyle→Black,PlotStyle→{Black,Opacity[.15]},
Ticks→{5{0,1,2},{0,5},15{0,1}},Mesh→3]
```

See Figure 6.29.

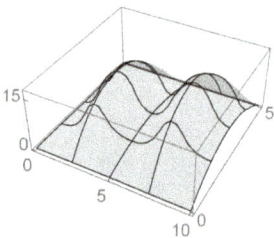

Figure 6.29: Plot of the numerical solution for problem 6.9.

A dynamical approach is presented as follows:

```
τ=.;
Animate[
Plot3D[u[t+τ,x]/.sol,{t,0.,5.},{x,0.,5.},Mesh→3,
ImageSize→160,MeshStyle→Black,
Ticks→{{0,5},{0,5},Automatic},
PlotStyle→{Black,Opacity[.15]}],
{τ,0.,5.},AnimationDirection→ForwardBackward,
DefaultDuration→35,AnimationRunning→False]
```

See Figure 6.30.

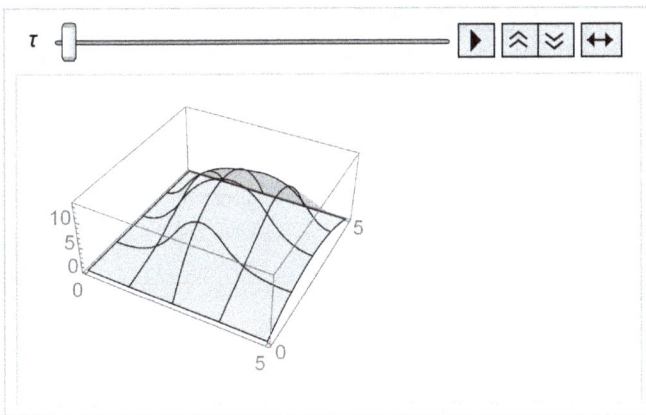

Figure 6.30: Plot of the dynamical solution for problem 6.9.

6.2 Hyperbolic equations with variable coefficients

6.10. We study a hyperbolic problem with initial and boundary conditions. The equation has variable coefficients,

$$\begin{cases} t\,x\partial_{t,t}u(t,x) - t^2\partial_{t,x}u(t,x) - x\,\partial_t u(t,x) = 0, & t > 0, \quad x > 0, \\ u(0,x) = x^2(1 - 2\sin(x)), & \text{(Initial condition)} \\ u(t,0) = t^2, \quad u(t,2\pi) = 4\pi^2 + t^2, & \text{(Boundary conditions).} \end{cases}$$

6.10

```
Clear[t,x]
```

Let us see the initial condition and the boundary conditions

```
{Plot[x²(1-2Sin[x]),{x,0,2π},ImageSize→140,
PlotLabel→Style["Initial condition",12,Bold,Black]],
Plot[{t²,4π²+t²},{t,0,5},ImageSize→140,
PlotStyle→{Black,{Black,Dashed}},
PlotLabel→Style["Boundary conditions",12,Bold,
Black]]}
```

See Figure 6.31.

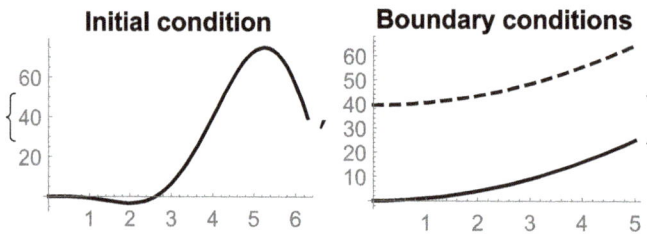

Figure 6.31: Plot of the initial condition and the boundary conditions for problem 6.10.

First, we try solving the equation.

```
eqn=t x ∂t,tu[t,x]-t²∂t,xu[t,x]-x∂tu[t,x]==0;  (* The
equation *)
Assuming[t>0&& x>0,
DSolve[%,u,{t,x}]]//First (* Solve the equation *)
%%/.%//Simplify (* Check the general solution *)
{u→Function[{t,x},C[1][½(t²+x²)]+C[2][x]]}
True
```

Thus, the equation has a general solution. Now we write the whole problem

```
eqncond={eqn,u[0,x]==x²(1-2Sin[x]),u[t,0]==t²,
u[t,2π]==4π²+t²};
```

We try finding a closed-form solution to this problem

```
DSolve[eqncond,u,{t,x}];
```

and remark that no closed-form solution is returned.

   A tentative leads us to the following particular solution, which checks the equation, initial condition and boundary conditions

```
v[t_,x_]:=t²+x²-2x²Sin[x]
Assuming[t>0&& x>0,
{tx∂t,tv[t,x]-t²∂t,xv[t,x]-x∂tv[t,x]==0,v[t,0]==t²,
v[t,2π]==4π²+t²,v[0,x]==x²(1-2Sin[x])}//Simplify]
{True,True,True,True}
```

Thus, function v satisfies the problem and we plot it.

```
Plot3D[v[t,x],{t,0,5},{x,0,2π},ImageSize→160,Mesh→3,
MeshStyle→Black,PlotStyle→{Black,Opacity[.15]},
Ticks→Outer[Times,{2.5,π,50},{0,1,2}]]
```

See Figure 6.32.

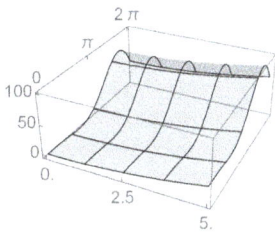

Figure 6.32: Plot of the analytical solution for problem 6.10.

We also study the problem numerically.

```
Clear[u]
sol=NDSolve[eqncond,u,{t,0,5},{x,0,2π}];
```

```
Plot3D[u[t,x]/.sol,{t,0,5},{x,0,2π},ImageSize→175,
Ticks→{{0,5},π{0,1,2},100{-2,0,3}},Mesh→2,
PlotStyle→{Gray,Opacity[.1]}]
```

See Figure 6.33.

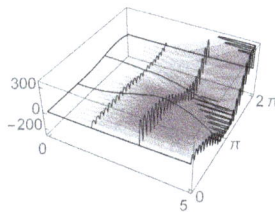

Figure 6.33: Numerical solution for problem 6.10.

By visual inspection, we observe that the last two surfaces differ. Indeed, a numerical confirmation immediately comes

```
FindMaximum[Abs[v[t,x]-u[t,x]/.sol],{{t,3},{x,3}}]
{131.651,{t→2.85738,x→2.7057}}
```

6.11. We introduce an initial and boundary value problem for an inhomogeneous hyperbolic equation,

$$\begin{cases} t\,\partial_{t,t}u(t,x) - x\,\partial_{x,x}u(t,x) = t\,x, & t \in [0,5], \quad x \in [0,6], \\ u(0,x) = 6\cos(\pi\,x/3), & \text{(Initial condition)} \\ u(t,0) = t\exp(-t^2), \quad u(t,6) = 0, & \text{(Boundary conditions).} \end{cases}$$

6.11

```
Clear[a,t,x]
```

```
eqn=t ∂t,t u[t,x]-x ∂x,x u[t,x]==t x; (* The equation *)
```

```
Assuming[t>0&&x>0,
DSolve[eqn,u,{t,x}]];  (* No closed-form solution is
returned *)
```

```
Assuming[t>0&&x>0,
DSolve[{eqn,u[t,0]==t e^{-t^2},u[t,6]==0,
u[0,x]==6Cos[πx/3]},u,{t,x}]];  (* No closed-form
solution is returned to this problem *)
```

We try finding a numerical solution

```
sol=NDSolve[{eqn,u[t,0]==t e^{-t^2},u[t,6]==0,
u[0,x]==6Cos[πx/3]},u,{t,0,5},{x,0,6}];
```

and plot it.

```
{Plot3D[u[t,x]/.sol,{t,0,5},{x,0,6},Mesh→2,
MeshStyle→Black,PlotStyle→{Black,Opacity[.15]},
Ticks→{2.5{0,1,2},3{0,1,2},400{-1,0,1}},
ImageSize→175],
Manipulate[
Plot3D[u[t,x]/.sol,{t,0,a},{x,0,6},ImageSize→160,
MeshStyle→Black,PlotStyle→{Black,Opacity[.15]},
Ticks→{{0,2.5,5},{0,3,6},Automatic},Mesh→2],
{a,.01,5},SaveDefinitions→True]}
```

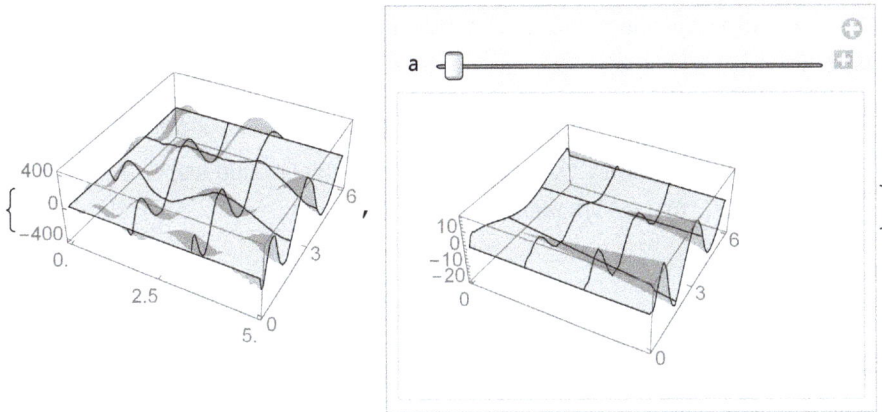

Figure 6.34: Static and dynamical representation of a particular solution for problem 6.11.

See Figure 6.34.

We want to know whether the boundary conditions are satisfied, namely how exact the numerical solution on boundaries is.

```
{Plot[t e^{-t^2}-u[t,x]/.sol/.x→0,{t,0,5},##],
Plot[u[t,x]/.sol/.x→6,{t,0,5},##],
Plot[6 Cos[π x/3]-u[t,x]/.sol/.t→0,{x,0,6},##]}&[
PlotStyle→Black,ImageSize→175]
```

See Figure 6.35.

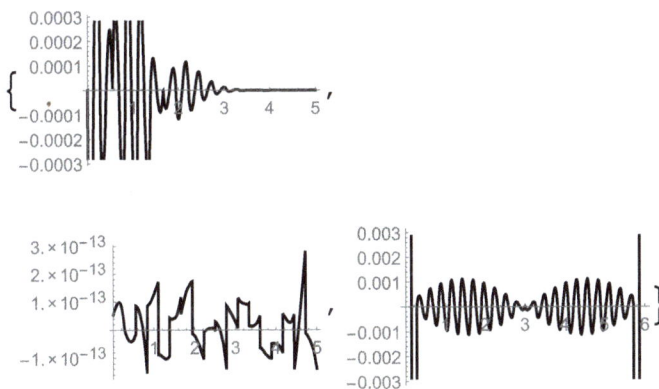

Figure 6.35: Numerical solution on boundaries for problem 6.11.

6.12. It follows a Gaussian initial value problem with a boundary condition for an in-homogeneous hyperbolic equation,

$$
\begin{cases}
\partial_{t,t}u(t,x) = a^2\partial_{x,x}u(t,x) + \sin(2t\pi/5), & a > 0, \quad t,x \in \mathbb{R}, \\
u(0,x) = \exp(x^{-2}), \quad \partial_t u(0,x) = 0, & \text{(Initial conditions)} \\
u(t,-6) = 0, \quad \text{(Boundary condition)}.
\end{cases}
$$

6.12

```
<< Notation`
Symbolize[p_]

Clear[a,t,τ,x,ξ]

eqn=∂t,tu[t,x]==a²∂x,xu[t,x]+Sin[2t π/5];  (* The
equation *)

Assuming[a>0,
DSolve[eqn,u,{t,x}]]//First (* General solution of
the equation *)
{u→Function[{t,x},-((25Sin[(2πt)/5])/(4π²))+C[1][a t+x]
+C[2][-a t+x]]}

eqn/.%//Simplify (* Check the solution *)
True
```

Thus, the equation has a general solution. Now we try finding the particular solution.

```
eqncond={eqn,u[t,-6]==0,u⁽¹,⁰⁾[0,x]==0,u[0,x]==e⁻ˣ²};
```

```
Assuming[a>0,
DSolve[%,u,{t,x}]]; (* No closed-form solution is
returned *)
```

We try finding a numerical solution

```
a=1;
sol=NDSolve[eqncond,u,{t,-5,5},{x,-6,6}]//Flatten;
```

and plot it.

```
{Show[Plot3D[u[t,x]/.sol,{t,-5,5},{x,-6,6},Mesh→None,
PlotStyle→{Black,Opacity[.15]},ImageSize→160,
PlotRange→All,Ticks→Outer[Times,{5,6,4},{-1,0,1}]],
Table[ParametricPlot3D[{t,px,u[t,px]/.sol},{t,-5,5},
PlotStyle→Black],{px,-6,6,3}],
Table[ParametricPlot3D[{pt,x,u[pt,x]/.sol},{x,-6,6},
PlotStyle→{Black,Dashed}],{pt,-5,5,2.5}]],
Manipulate[
Show[Plot3D[u[t,x]/.sol,{t,-5,5},{x,-6,6},Mesh→None,
PlotStyle→{Black,Opacity[.15]},ImageSize→160,
PlotRange→All,Ticks→Outer[Times,{5,6,4},{-1,0,1}]],
ParametricPlot3D[{τ,x,u[τ,x]/.sol},{x,-6,6},##],
ParametricPlot3D[{t,ξ,u[t,ξ]/.sol},{t,-5,5},##]]&[
PlotStyle→Black],{τ,-5,5},{ξ,-6,6,2},
SaveDefinitions→True]}
```

See Figure 6.36.

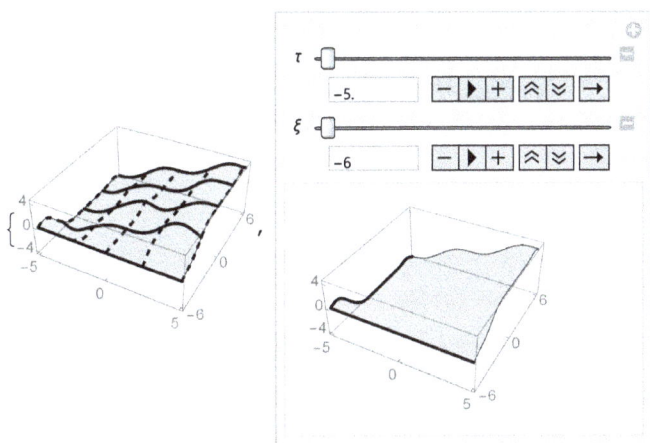

Figure 6.36: Numerical solution and dynamical plot for problem 6.12.

6.13. An inhomogeneous initial and boundary value problem with nonsmooth data is introduced,

$$\begin{cases} \partial_{t,t}u(t,x) = c^2\partial_{x,x}u(t,x) + tx, & c = \pi, \quad t \in [0,6], \quad x \in [0,5], \\ u(0,x) = 10(1 - \frac{2}{5}|x - \frac{5}{2}|), \quad \partial_t u(0,x) = 0, \quad \text{(Initial conditions)} \\ u(t,0) = \cos(ct) - 1, \quad u(t,5) = 0, \quad \text{(Boundary conditions)}. \end{cases}$$

6.13

```
Clear[t,x]
eqn=∂_{t,t}u[t,x]==c²∂_{x,x}u[t,x]+t x

Assuming[t≥0&&x≥0,
DSolve[eqn,u,{t,x}]]; (* No closed-form solution is
returned *)

Assuming[t≥0&&x≥0,
DSolve[{eqn,u[t,0]==Cos[c t]-1,u[t,5]==0,
u[0,x]==10(1-2/5 Abs[x-5/2]),u^(1,0)[0,x]==0},
u,{t,x}]]; (* No closed-form solution is returned *)
```

We try finding a numerical solution,

```
With[{c=π},
sol=NDSolve[{∂_{t,t}u[t,x]==c²∂_{x,x}u[t,x]+t x,
u[0,x]==10(1-2/5 Abs[x-5/2]),u^(1,0)[0,x]==0,
u[t,0]==Cos[c t]-1,u[t,5]==0},u,{t,0,6},{x,0,5},
Method→{"MethodOfLines","SpatialDiscretization"→
{"TensorProductGrid",MaxPoints→500}},
PrecisionGoal→2]]//Flatten;
```

and plot it with a cross-section.

```
{Plot3D[u[t,x]/.sol,{t,0,6},{x,0,5},ImageSize→175,
Mesh→{3,2},PlotRange→All,MeshStyle→Black,
Ticks→Outer[Times,{3,5,10},Range[-1,2]],
PlotStyle→Opacity[.2]],
Animate[
Plot[u[t,x]/.sol,{x,0,5},ImageSize→150,PlotRange→All,
PlotStyle→Black],{t,0,6},SaveDefinitions→True,
DefaultDuration→25,AnimationRunning→False,
AnimationDirection→ForwardBackward]}
```

See Figure 6.37.

6.14. Below we study the *Euler–Darboux equation* with mixed conditions,

$$\begin{cases} \partial_{t,x}u(t,x) - \frac{1}{t-x}\partial_t u(t,x) + \frac{1}{t-x}\partial_x u(t,x) = 0, & t \in [0,\pi/2], \quad x \in [\pi, 2\pi], \\ u(0,x) = \sin(x), \quad u(t,2\pi) = -\sin(t), & \text{(Mixed conditions).} \end{cases}$$

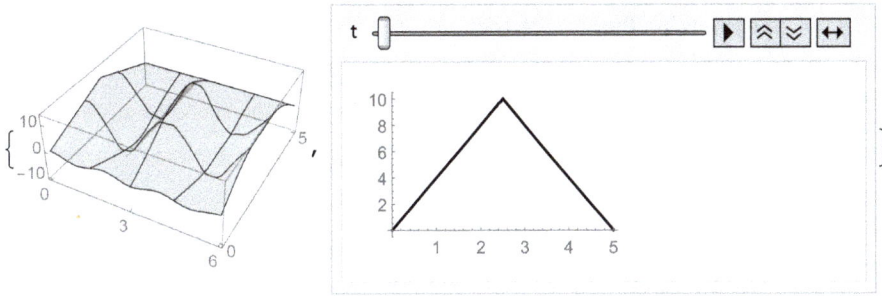

Figure 6.37: Numerical solution and a cross-section for problem 6.13.

6.14

```
Clear[t,τ,x,ξ]
```

$$eqncond=\{\partial_{t,x}u[t,x]-\frac{1}{t-x}\partial_t u[t,x]+\frac{1}{t-x}\partial_x u[t,x]==0,$$
```
u[0,x]==Sin[x],u[t,2π]==-Sin[t]};
DSolve[eqncond,u,{t,x}];(* No closed-form solution is
returned *)
```

We try a numerical approach

```
sol=NDSolve[eqncond,u,{t,0,π/2},{x,π,2π}//Flatten;
```

and plot this solution.

```
Plot3D[u[t,x]/.sol,{t,0,π/2},{x,π,2π},ImageSize→175,
Mesh→2,PlotStyle→{Black,Opacity[.15]},
MeshStyle→Black,Ticks→{{0,π/2},{π,2π},{-3,0}}]
```

See Figure 6.38.

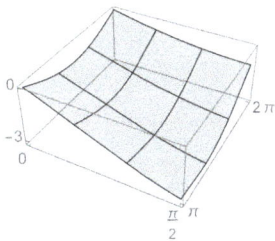

Figure 6.38: Plot of the numerical solution for problem 6.14.

If two lines move on the surface, then

```
Animate[
Show[
Plot3D[u[t,x]/.sol,{t,0,π/2},{x,π,2π},Mesh→None,
ImageSize→175,PlotStyle→{Black,Opacity[.15]},
Ticks→{π/2{0,1},π{1,2},{-3,0}}],
ParametricPlot3D[{τ,x,u[τ,x]/.sol},{x,π,2π},
PlotStyle→{Black,Thick}],
ParametricPlot3D[{t,ξ,u[t,ξ]/.sol},{t,0,π/2},
PlotStyle→{Black,Thick}],{τ,0,π/2},{τ,π,2π},
AnimationRunning→False,DefaultDuration→20,
AnimationDirection→ForwardBackward,
SaveDefinitions→True]
```

See Figure 6.39.

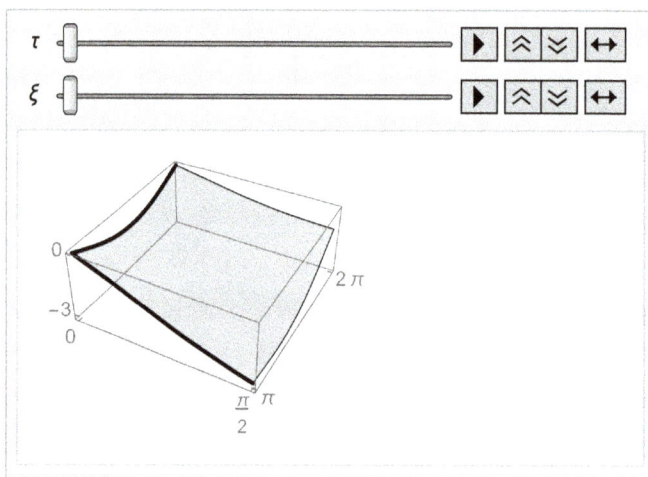

Figure 6.39: Dynamic plot of the numerical solution for problem 6.14.

6.3 Hyperbolic equations on curvilinear domains

6.15. We introduce a boundary value problem defined on an annulus sector whose boundaries are given by the ImplicitRegion built-in function. The equation considered here is

$$\partial_{t,t}u(t,x) = c^2\partial_{x,x}u(t,x), \quad t,x \in \mathbb{R}, \quad c > 0.$$

6.15

```
Clear[t,x]
```

First, we show the equations of the boundary,

```
bound={t²+x²-1,4-t²-x²,x-√3/3 t, √3t-x};
```

specify the region,

```
Ω=ImplicitRegion[And@@(#≥0&/@bound),{t,x}];
```

and plot the domain of integration.

```
Show[RegionPlot[Ω],
ContourPlot[Evaluate[Thread[bound==0]],{t,-.01,2},
{x,0,2},ContourStyle→{Black,{Black,Dashed},
{Black,Dotted},{Black,DotDashed}}],PlotRange→{{0,2},
{0,2}},AspectRatio→Automatic,ImageSize→140]
```

See Figure 6.40.

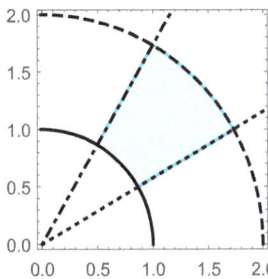

Figure 6.40: Plot of the domain of integration for problem 6.15.

Then we solve the hyperbolic equation on this domain considering some boundary conditions.

```
c=5;
nsolv=NDSolveValue[{∂_{t,t}u[t,x]==c²∂_{x,x}u[t,x],
DirichletCondition[u[t,x]==t²-x²,bound[[1]]==0],
DirichletCondition[u[t,x]==x²-t²,bound[[2]]==0],
DirichletCondition[u[t,x]==-1,bound[[3]]==0],
DirichletCondition[u[t,x]==1,bound[[4]]==0]},u,
{t,x}∈ Ω,Method→{"FiniteElement","MeshOptions"→
{"BoundaryMeshGenerator"→"Continuation"}}];
```

Below is the corresponding figure.

```
Plot3D[nsolv[t,x],{t,x}∈ Ω,PlotPoints→50,Mesh→3,
Ticks→{{.5,1,1.6},{.5,1,1.6},{-2,1,4}},ImageSize→200,
PlotStyle→{Black,Opacity[.15]}]
```

See Figure 6.41.

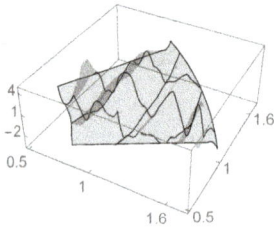

Figure 6.41: Plot of the numerical solution for problem 6.15.

We study a similar problem with two boundaries changed. The resulted domain is nonconvex.

```
Clear[t,x,u]
```

The new boundaries follow:

$$bound=\{t^2+x^2-1/4,t^2+x^2-5t-5x+17/2,x-\tfrac{\sqrt{3}}{3}t,\sqrt{3}t-x\};$$
$$\Omega=\text{ImplicitRegion}[\text{And@@}(\#≥0\&/@bound),\{t,x\}];$$

Here is the associated plot:

```
Show[RegionPlot[Ω],
ContourPlot[Evaluate[Thread[bound==0]],{t,-.1,1.51},
{x,0,1.51},ContourStyle→{Black,{Black,Dashed},
{Black,Dotted},{Black,DotDashed}}],ImageSize→140,
PlotRange→{{0,1.5},{0,1.5}},AspectRatio→Automatic]
```

See Figure 6.42.

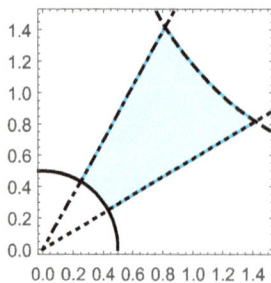

Figure 6.42: Plot of the other domain of integration for problem 6.15.

We consider other boundary values:

```
c=5;
nsolv=NDSolveValue[{∂t,tu[t,x]==c²∂x,xu[t,x],
DirichletCondition[u[t,x]==-x²+t²,bound[[1]]==0],
DirichletCondition[u[t,x]==x²-t²,bound[[2]]==0],
DirichletCondition[u[t,x]==-1,bound[[3]]==0],
DirichletCondition[u[t,x]==1,bound[[4]]==0]},u,
{t,x}∈ Ω,Method→{"FiniteElement","MeshOptions"→
{"BoundaryMeshGenerator"→"Continuation"}}];
```

and plot the figure.

```
{{Plot3D[nsolv[t,x],{t,x}∈ Ω,PlotPoints→50,Mesh→5,
PlotStyle→{Black,Opacity[.15]},ImageSize→180,
Ticks→{{.5,1,1.4},.45{1,2,3},Automatic}],
Plot[nsolv[t,t],{t,1/(2Sqrt[2]),1.09},AxesOrigin→
{1/(2Sqrt[2]),0},PlotStyle→Black,ImageSize→175]}
```

See Figure 6.43.

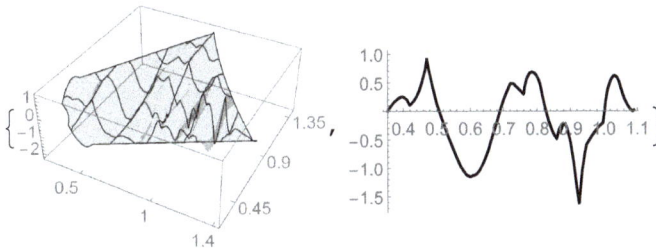

Figure 6.43: Plot of the numerical solution on the second domain of integration for problem 6.15 and a cross-section through it.

We again change a boundary.

```
Clear[t,f,x,u]
```

First, we show the equations of the boundary,

```
f[t_]:=Which[t<1,0,t≥1,t-1]
```

```
bound={t²+x²-1,4-t²-x²,x-(-f[t]+1)Sin[4πf[t]],
√3 t-x t-x};
```

specify the region,

```
Ω=ImplicitRegion[And@@(#≥0&/@bound),{t,x}];
```

and the domain of integration.

```
Show[RegionPlot[Ω],
ContourPlot[Evaluate[Thread[bound==0]],{t,-.01,2},
{x,0,2},ContourStyle→{Black,{Black,Dashed},
{Black,Dotted},{Black,DotDashed}}],ImageSize→140,
PlotRange→{{0,2},{0,2}},AspectRatio→Automatic]
```

See Figure 6.44.

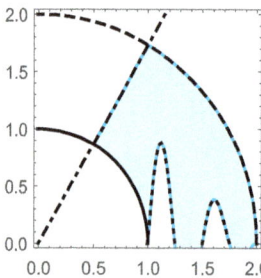

Figure 6.44: Plot of the third domain of integration for problem 6.15.

Then we solve the hyperbolic equation on this domain

```
c=5;
nsolv=NDSolveValue[{∂_{t,t}u[t,x]==c²∂_{x,x}u[t,x],
DirichletCondition[u[t,x]==t²-x²,bound[[1]]==0],
DirichletCondition[u[t,x]==x²-t²,bound[[2]]==0],
DirichletCondition[u[t,x]==-1,bound[[3]]==0],
DirichletCondition[u[t,x]==1,bound[[4]]==0]},u,
{t,x}∈ Ω,Method→{"FiniteElement","MeshOptions"→
{"BoundaryMeshGenerator"→"Continuation"}}];
```

and plot the numerical solution.

```
Plot3D[nsolv[t,x],{t,x}∈ Ω,PlotPoints→50,Mesh→3,
MeshStyle→Black,ImageSize→200,PlotRange→All,
PlotStyle→{Black,Opacity[.15]},
Ticks→{{.5,2},{-.5,1.5},{0,15}}]
```

See Figure 6.45.

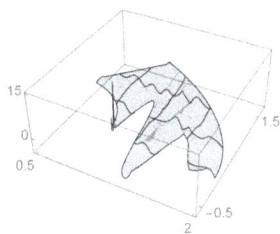

Figure 6.45: Plot of the numerical solution on the third domain of integration for problem 6.15.

6.16. Here, we introduce another boundary value problem whose boundaries are given by the DirichletCondition and the NeumannValue built-in functions. The problem considered is of the form

$$\begin{cases} \partial_{t,t}u(t,x) = c^2\partial_{x,x}u(t,x), & t, x \in \mathbb{R}, \quad c > 0, \\ t^4 + t^2x^2 + x^4 - t(t^2 + x^2) \leq 0, & \text{(Bean-shaped domain).} \end{cases}$$

6.16

```
Clear[t,x]
```

First, we show the equation of the boundary:

```
boundbean=t⁴+t²x²+x⁴-t(t²+x²);
Ω=ImplicitRegion[boundbean≤0,{t,x}];
```

and plot the domain of integration.

```
Show[RegionPlot[Ω],
ContourPlot[Evaluate[Thread[boundbean==0]],{t,0,1},
{x,-.7,.7},ContourStyle→Black],ImageSize→135]
```

See Figure 6.46.

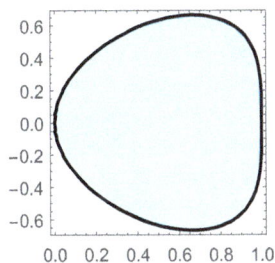

Figure 6.46: Plot of the domain of integration for problem 6.16.

```
c=5;
```

First, we consider a Dirichlet condition and look for a numerical solution,

```
nsolv1=NDSolveValue[{∂_{t,t}u[t,x]==c²∂_{x,x}u[t,x],
DirichletCondition[u[t,x]==t x,boundbean==0]},u,
{t,x}∈ Ω,Method→{"FiniteElement","MeshOptions"→
{"BoundaryMeshGenerator"→"Continuation"}}];
```

We pass to an inhomogeneous equation with a Dirichlet condition and follow the same path as above.

```
nsolv2=NDSolveValue[{∂_{t,t}u[t,x]-c²∂_{x,x}u[t,x]==
√(t+Abs[x]),DirichletCondition[u[t,x]==t x,boundbean==0]
},u,{t,x}∈ Ω,Method→{"FiniteElement","MeshOptions"→
{"BoundaryMeshGenerator"→"Continuation"}}]
```

For the same inhomogeneous equation, we set Neumann and Dirichlet boundary conditions.

```
nsolv3=NDSolveValue[{∂_{t,t}u[t,x]-c²∂_{x,x}u[t,x]
-√(t+Abs[x])==NeumannValue[If[x≥0,1,-1],t≥.7],
DirichletCondition[u[t,x]==t x,t<.7]},u,{t,x}∈ Ω,
Method→{"FiniteElement","MeshOptions"→
{"BoundaryMeshGenerator"→"Continuation"}}];
```

Now we plot the three solutions.

```
{Plot3D[nsolv1[t,x],##],
Plot3D[nsolv2[t,x],##],
Plot3D[nsolv3[t,x],##]}&[{t,x}∈ Ω,Mesh→{2,5},
PlotPoints→50,PlotStyle→{Black,Opacity[.15]},
Ticks→{{0,.5,1},.6{-1,0,1},.5{-1,0,1}},
ImageSize→175,]
```

See Figure 6.47.

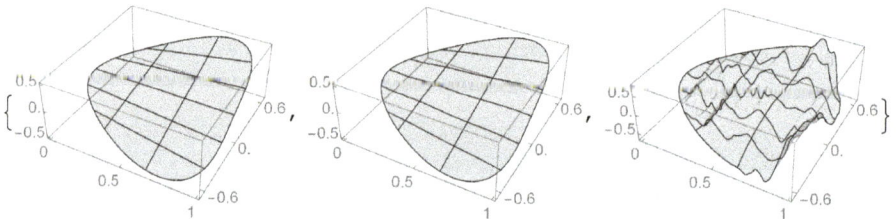

Figure 6.47: Plot of the numerical solutions for Problem 6.16.

We change the Neumann boundary condition

```
nsolv4=NDSolveValue[{∂t,tu[t,x]-c²∂x,xu[t,x]
-√t+Abs[x]==NeumannValue[If[x≥0,-1,1],t≥.7],
DirichletCondition[u[t,x]==t x,t<.7]},u,{t,x}∈ Ω,
Method→{"FiniteElement","MeshOptions"→
{"BoundaryMeshGenerator"→"Continuation"}}];
```

and

```
nsolv5=NDSolveValue[{∂t,tu[t,x]-c²∂x,xu[t,x]
-√t+Abs[x]==NeumannValue[If[x≥0,1,-1],.4≤t≤.6]
+NeumannValue[If[x≥0,-1,0],.8≤t≤1],
DirichletCondition[u[t,x]==t x,t<.4]},u,{t,x}∈ Ω,
Method→{"FiniteElement","MeshOptions"→
{"BoundaryMeshGenerator"→"Continuation"}}];
```

We plot these two pictures:

```
{Plot3D[nsolv4[t,x],##],
Plot3D[nsolv5[t,x],##]}&[{t,x}∈ Ω,Mesh→{2,5},
PlotPoints→50,PlotStyle→{Black,Opacity[.15]},
Ticks→{{0,.5,1},.6{-1,0,1},.5{-1,0,1}},
ImageSize→175,]
```

See Figure 6.48.

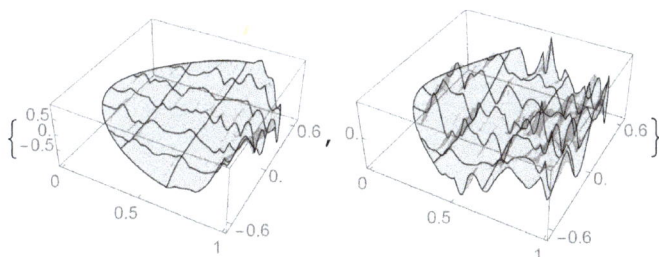

Figure 6.48: Plot of the other numerical solutions for problem 6.16.

6.17. We introduce a boundary value problem that uses the ParametricRegion built-in function. The problem considered here is

$$\begin{cases} \partial_{t,t}u(t,x) = c^2\partial_{x,x}u(t,x), & t,x \in \mathbb{R}, \quad c > 0, \\ \text{on the interior of the finite domain defined as} \\ \{t,(1+x)t^2 - x\}, \quad -1 \le t \le 1, \quad 0 \le x \le 1. \end{cases}$$

6.17

```
Clear[t,x]
```

First we introduce the equations of the boundary.

```
bound={t,(1+x)t²-x};
Ω=ImplicitRegion[
RegionMember[
ParametricRegion[{{t,(1+x)t²-x},-1≤t≤1&&0≤x≤1},
{t,x}],
{t,x}],
{t,x}];
```

This is the curvilinear domain.

```
RegionPlot[Ω,PlotStyle→LightGray,ImageSize→140,
BoundaryStyle→Black]
```

See Figure 6.49.

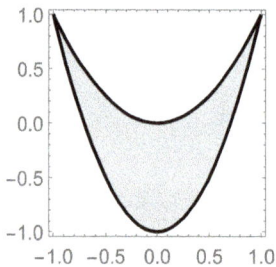

Figure 6.49: Plot of the domain of integration for problem 6.17.

We set the problem:

```
c=5;
nsolv=NDSolveValue[{∂t,tu[t,x]==c²∂x,xu[t,x],
DirichletCondition[u[t,x]==2t x,First[bound]≥0],
DirichletCondition[u[t,x]==-2Sin[t x],First[bound]<0]},
u,{t,x}∈ Ω,Method→{"FiniteElement","MeshOptions"→
{"BoundaryMeshGenerator"→"Continuation"}}];
```

and plot its solution.

```
Plot3D[nsolv[t,x],{t,x}∈ Ω,PlotPoints→50,Mesh→{3,5},
PlotRange→All,ImageSize→175,PlotStyle→{Opacity[.15],
Black},Ticks→{{-1,0,1},{-1,0,1},{0,1,2}}]
```

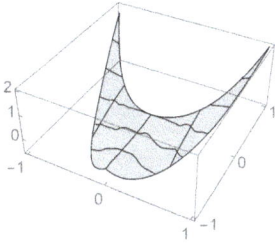

Figure 6.50: Plot of the numerical solution for problem 6.17.

See Figure 6.50.

6.18. We look for a solution to the following Dirichlet and Neumann problem of a hyperbolic equation defined on a *lemniscate*, namely

$$\begin{cases} \partial_{x,x}u(x,y) = 2\partial_{y,y}u(x,y) \text{ on } \Omega = \{(x,y) \mid (x^2+y^2)^2 - (x^2-y^2) \leq 0\} \subset \mathbb{R}^2, \\ u(x,y) = \begin{cases} \sin(5\pi\, x\, y), & x < 0, \quad y < 0, \\ \cos(x+y), & x < 0, \quad y > 0, \end{cases} \quad \text{(Dirichlet conditions)} \\ \partial_n u(x,y) = \begin{cases} 1, & x > 0, \quad y > 0, \\ -1, & x > 0, \quad y < 0, \end{cases} \quad \text{(Neumann conditions).} \end{cases}$$

6.18

```
Clear[t,x]
```

We introduce the equations of the boundary:

```
boundlemniscate=(x²+y²)²-(x²-y²);
Ω=ImplicitRegion[boundlemniscate≤0,{x,y}];
```

and plot the domain.

```
Show[RegionPlot[Ω,PlotStyle→{Gray,Opacity[.15]},
ContourPlot[Evaluate[Thread[boundlemniscate==0]],
{x,-1,1},{y,-.4,.4},ContourStyle→Black],
AspectRatio→3/4,ImageSize→160]
```

See Figure 6.51.

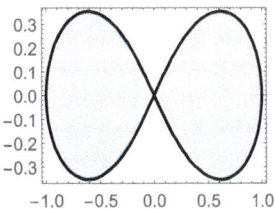

Figure 6.51: Plot of the domain of integration for problem 6.18.

We set the equation with Dirichlet and Neumann boundary conditions:

```
solv=NDSolveValue[{∂x,xu[x,y]==2∂y,yu[x,y]
+NeumannValue[1,x>0&&y>0]+NeumannValue[-1,x>0&&y<0],
DirichletCondition[u[x,y]==Sin[5πx y],x<0&&y<0],
DirichletCondition[u[x,y]==Cos[x+y],x<0&&y>0]},u,
{x,y}∈ Ω];
```

and plot its solution.

```
Plot3D[solv[x,y],{x,y}∈ Ω,ImageSize→175,Mesh→{5,3},
MeshStyle→Black,PlotPoints→50,
Ticks→Outer[Times,{1,.3,4},{-1,0,1}],
PlotStyle→{Black,Opacity[.15]}]
```

See Figure 6.52.

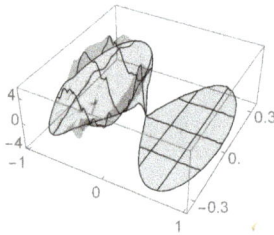

Figure 6.52: Plot of the numerical solution for problem 6.18.

6.19. Find a solution to the following Dirichlet problem of a hyperbolic equation defined on an *astroid* in the first quadrant:

$$\begin{cases} \partial_{x,x}u(x,y) = 2\partial_{y,y}u(x,y), & \Omega = \{(x,y) \mid x^{2/3} + y^{2/3} \le 1\} \subset \mathbb{R}^2, \\ 0, & x > 0, \quad y = 0, \\ u(x,y) = 5y, & x = 0, \quad y > 0, \\ x^2 y^2, & x^{2/3} + y^{2/3} = 1 \text{ and } x, y > 0, \end{cases} \quad \text{(Dirichlet conditions).}$$

6.19

```
Clear[x,y]
```

We specify the boundary,

```
boundastro={-x,-y,x^(2/3)+y^(2/3)-1};
```

define the domain,

```
Ω=ImplicitRegion[And@@(#≤0&/@boundastro),{x,y}];
```

and plot it.

```
Show[RegionPlot[Ω],
ContourPlot[Evaluate[Thread[boundastro==0]],{x,0,1},
{y,0,1},ContourStyle→Black],ImageSize→130,
PlotRange→{{-.01,1},{-.01,1}}]
```

See Figure 6.53.

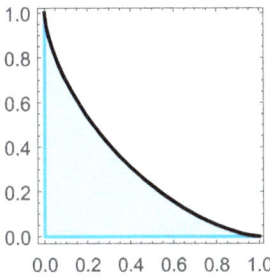

Figure 6.53: Plot of the domain of integration for problem 6.19.

Then we solve the hyperbolic equation with certain Dirichlet conditions on this domain:

```
c=√2;
nsolv=NDSolveValue[{∂t,tu[t,x]==c²∂x,xu[t,x],
DirichletCondition[u[x,y]==x²y²,boundastro[[3]]==0],
DirichletCondition[u[x,y]==0,boundastro[[1]]==0],
DirichletCondition[u[x,y]==0,boundastro[[2]]==0]},u,
{x,y}∈ Ω,Method→{"FiniteElement","MeshOptions"→
{"BoundaryMeshGenerator"→"Continuation"}}];
```

and plot its solution

```
Plot3D[nsolv[x,y],{x,y}∈ Ω,Mesh→3,ImageSize→175,
MeshStyle→Black,PlotStyle→{Black,Opacity[.15]},
Ticks→{.5{0,1,2},.5{0,1,2},None},PlotRange→All]
```

See Figure 6.54.

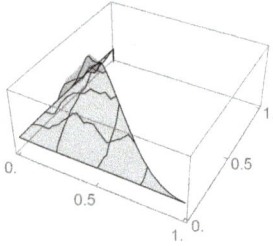

Figure 6.54: Plot the numerical solution for problem 6.19.

6.4 Hyperbolic equations in solid space

6.20. We introduce a 2D hyperbolic equation with initial conditions [64, p. 112],

$$\begin{cases} \partial_{t,t} u(t,x,y) = \nabla^2_{x,y} u(t,x,y), & t,x,y,\alpha,\beta \in \mathbb{R}, \quad a,b > 0, \\ \text{dom} = \{(x,y) \mid |x - \frac{a}{2}| \le \frac{a}{2}, \quad |y - \frac{b}{2}| \le \frac{b}{2}\} \subset \mathbb{R}^2, \\ u(0,x,y) = \alpha(x-1), \quad \partial_t u(0,x,y) = \beta(y+1), \quad (x,y) \in \text{dom}. \end{cases}$$

6.20

```
Clear[t,x,y]
a=1;α=5;b=2;β=4;c=8; (* Initial data *)
```

Here is the domain of the problem.

```
dom=ImplicitRegion[Abs[x-a/2]≤ a/2&&Abs[y-b/2]≤ b/2,{x,y}];
RegionPlot[dom,AspectRatio→Automatic,ImageSize→75,
BoundaryStyle→Black,PlotStyle→LightGray];
```

We set the initial value problem.

```
eqncond={∂t,t u[t,x,y]==∇²{x,y}u[t,x,y],
DirichletCondition[u[t,x,y]==α(x-1),t==0],
u⁽¹,⁰,⁰⁾[0,x,y]==β(y+1)};
```

```
DSolveValue[eqncond,u,{t,x,y}]; (* No closed-form
solution is returned *)
```

We look for a numerical solution

```
sol=NDSolveValue[eqncond,u,{t,0,c},{x,y}∈dom]//Chop;
```

and show a three-dimensional contour plot.

```
ContourPlot3D[sol[t,x,y],{t,0,c},{x,0,a},{y,0,b},
PlotRange→All,Mesh→None,BoundaryStyle→Black,
ImageSize→160,Ticks→Outer[Times,{4,1,1},{0,1,2}],
ContourStyle→Directive[Black,Opacity[.15]],
ImageSize→160]
```

See Figure 6.55.

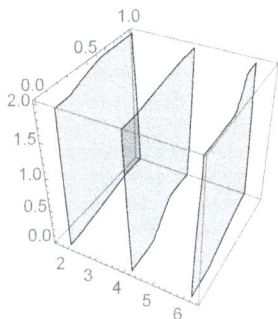

Figure 6.55: Contour plot for problem 6.20.

6.21. We introduce a 2D hyperbolic equation with initial conditions [64, p. 112],

$$\begin{cases} \partial_{t,t}u(t,x,y) = \nabla^2_{x,y}\, u(t,x,y), \quad t,x,y,\alpha,\beta \in \mathbb{R}, \quad a,b > 0, \\ \text{dom} = \{(x,y) \mid |x - \frac{a}{2}| \le \frac{a}{2},\ |y - \frac{b}{2}| \le \frac{b}{2}\} \subset \mathbb{R}^2, \\ u(0,x,y) = \alpha(x+1), \quad \partial_t u(0,x,y) = \beta(y-1), \quad (x,y) \in \text{dom}. \end{cases}$$

6.21

```
Clear[t,x,y]
a=1;α=5;b=2;β=4;c=8; (* Initial data *)
```

Here is the domain of the problem.

```
dom=ImplicitRegion[Abs[x-a/2]≤a/2&&Abs[y-b/2]≤b/2,{x,y}];
RegionPlot[dom,AspectRatio→4/5,ImageSize→75,
BoundaryStyle→Black,PlotStyle→LightGray];
```

We set the initial value problem.

```
eqncond={∂t,tu[t,x,y]==∇²{x,y}u[t,x,y],
DirichletCondition[u[t,x,y]==α(x+1),t==0],
u^(1,0,0)[0,x,y]==β(y-1)};
```

We look for an analytical solution

```
DSolveValue[eqncond,u,{t,x,y}]; (* No closed-form
solution is returned *)
```

We look for a numerical solution

```
sol=NDSolveValue[eqncond,u,{t,0,c},{x,y}∈dom]//Chop;
```

and show a three-dimensional contour plot and a cross-section through it.

```
{ContourPlot3D[sol[t,x,y],{t,0,c},{x,0.,a},{y,0.,b},
PlotRange→All,Mesh→None,Contours→1,ImageSize→160,
ContourStyle→Directive[LightGray,Opacity[.2]],
Ticks→Outer[Times,{4,1,1},{0,1,2}]],
Manipulate[
Show[{Plot3D[sol[t,x,y],{x,y}∈dom,PlotRange→All,
Mesh→None,Ticks→None,ImageSize→160,
PlotStyle→Directive[Black,Opacity[.1]]],
Plot3D[{t,x,y},{x,y}∈dom,PlotRange→All,
PlotStyle→Directive[Black,Opacity[.12]],Mesh→None]}],
{t,0,c},SaveDefinitions→True]}
```

See Figure 6.56.

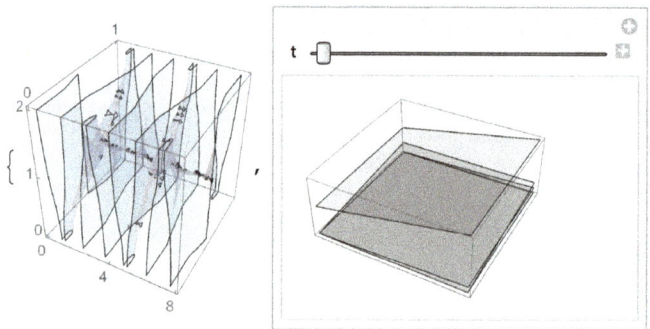

Figure 6.56: Contour plot and cross-section for problem 6.21.

6.22. Below we introduce another 2D hyperbolic equation with initial conditions, [64, p. 124],

$$\begin{cases} \partial_{t,t}u(t,x,y) = \nabla^2_{x,y}\,u(t,x,y), & t,x,y,a \in \mathbb{R}, \quad a,b > 0, \\ \text{dom} = \{(x,y) \mid 0 \le x \le a,\ 0 \le y \le b\} \subset \mathbb{R}^2, \\ u(0,x,y) = 0, \quad \partial_t u(0,x,y) = a\,x(a-x)(b-y), \quad (x,y) \in \text{dom}. \end{cases}$$

6.22

```
Clear[t,x,y,u]
a=1;a=5;b=2;c=8;  (* Initial data *)
```

First, we show the equations of the boundary

```
bound={t,x,a-x,y,b-y};
Ω=ImplicitRegion[And@@(#≥0&/@bound),{t,x,y}];
```

and the domain of integration.

```
{RegionPlot3D[Ω,PlotPoints→35,PlotRange→All,
PlotStyle→Directive[Black,Opacity[.15]],Mesh→None,
ImageSize→125],
ContourPlot3D[Evaluate[Thread[bound==0]],{t,0,c},
{x,0,a},{y,0,b},ImageSize→125,Mesh→None,
ContourStyle→Directive[Black,Opacity[.15]]]};
```

Then we solve the initial value problem on this domain.

```
eqncond={∂t,tu[t,x,y]==∇²{x,y}u[t,x,y],
DirichletCondition[u[t,x,y]==0,t==0],
u^(1,0,0)[0,x,y]==a x(a-x)(b-y)};
```

```
DSolveValue[eqncond,u,{t,x,y}];  (* No closed-form
solution is returned *)
```

We look for a numerical solution.

```
sol=NDSolveValue[eqncond,u,{t,0,c},{x,0,a},{y,0,b}];
```

```
ContourPlot3D[sol[t,x,y],{t,0,c},{x,0.,a},{y,0.,b},
ContourStyle→Directive[Black,Opacity[.2]],
Contours→2,Mesh→None,PlotRange→All,ImageSize→160]
```

See Figure 6.57.

6.23. We introduce a 2D hyperbolic equation with initial conditions, [64, p. 125],

$$\begin{cases} \partial_{t,t}u(t,x,y) = \nabla^2_{x,y} u(t,x,y), & 0 \le t \le c, \quad -a \le x \le a, \quad -b \le y \le b, \\ u(0,x,y) = (x^2-y^2)(a^2-x^2-y^2), & \partial_t u(0,x,y) = x(a-x)(b-y). \end{cases}$$

6.23

```
Clear[t,x,y]
a=1;b=2;c=8;  (* Set the parameters *)
```

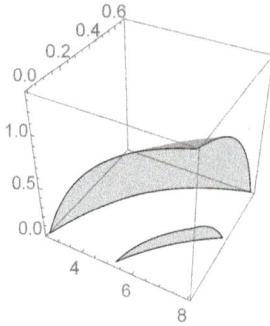

First, we show the equations of the boundary

```
bound={t,c-t,a+x,a-x,b+y,b-y};
Ω=ImplicitRegion[And(#≥0&/bound),{t,x,y}];
```

and then the domain of integration.

```
{RegionPlot3D[Ω,PlotRange→All,
PlotStyle→Directive[Black,Opacity[.15]],##],
ContourPlot3D[Evaluate[Thread[bound==0]],{t,0,c},
{x,-a,a},{y,-b,b},ContourStyle→Directive[Black,
Opacity[.15]],Ticks→{{0,c},{-a,a},{-b,b}},##]}
&[Mesh→None,ImageSize→125]
```

We solve the hyperbolic equation on this domain.

```
eqncond={∂t,tu[t,x,y]==∇²{x,y}u[t,x,y],
DirichletCondition[u[t,x,y]==(x²-y²)(a²-x²-y²),t==0],
u⁽¹,⁰,⁰⁾u[0,x,y]==x(a-x)(b-y)};

DSolveValue[eqncond,u,{t,x,y}]; (* No closed-form
solution is returned *)
```

We look for a numerical solution.

```
sol=NDSolveValue[eqncond,u,{t,0,c},{x,-a,a},{y,-b,b}];

ContourPlot3D[sol[t,x,y],{t,0,c},{x,-a,a},{y,-a,a},
ContourStyle→Directive[Black,Opacity[.15]],
Contours→2,Mesh→None,ImageSize→160,
Ticks→{{0,c/2,c},{-a,0,a},{-b,0,b}}]
```

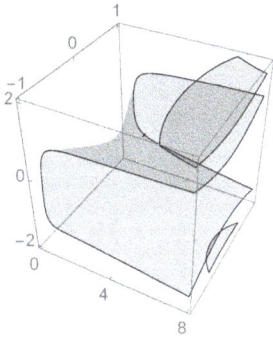

Figure 6.58: Contour plots for problem 6.23.

See Figure 6.58.

6.24. We introduce another 2D hyperbolic equation with Dirichlet and Neumann conditions,

$$\begin{cases} \partial_{t,t}u(t,x,y) = 5^2\nabla^2_{x,y}u(t,x,y), & t \in [0,20], \quad x \in [0,2], \quad y \in [0,3], \\ u(0,x,y) = 0, \quad \text{(Initial condition)}, \\ \partial_n u(0,x,y) = 2\pi(xy-3), \quad \text{(Neumann condition)}, \\ u(t,0,y) = u(t,2,y) = 0, \quad u(t,x,0) = u(t,x,3) = 2, \quad \text{(Dirichlet conditions)}. \end{cases}$$

6.24 Let us try finding a closed-form solution.

```
Clear[t,τ,x,y]

eqncond={∂t,tu[t,x,y]-5²∇²{x,y}u[t,x,y]
==NeumannValue[2π(x y-3),t==0],u[0,x,y]==0,
u[t,0,y]==u[t,2,y]==u[t,x,0]==u[t,x,3]==0};

DSolve[%,u,{t,x,y}]; (* No closed-form solution is
returned *)
```

We try a numerical approach

```
sol=NDSolve[eqncond,u,{t,0,20},{x,0,2},{y,0,3},
PrecisionGoal→2]//Flatten;
```

and represent the plot and the contour plot.

```
{Plot3D[u[t,x,y]/.sol/.t→c/2,{x,0,a},{y,0,b},
PlotStyle→Directive[Black,Opacity[.15]],Mesh→{5,3},
Ticks→Outer[Times,{a,b,2},{0,1}],MeshStyle→Black],
ContourPlot3D[u[t,x,y]/.sol,{t,0.,c/2},{x,0,a},{y,0,b},
Contours→1,PlotRange→All,ImageSize→150,
Ticks→Outer[Times,{c/2,a,b},{0,1}],Mesh→None,
ContourPlot→[Black,Opacity[.05]],
ContourStyle→Opacity[.1]]}
```

See Figure 6.59.

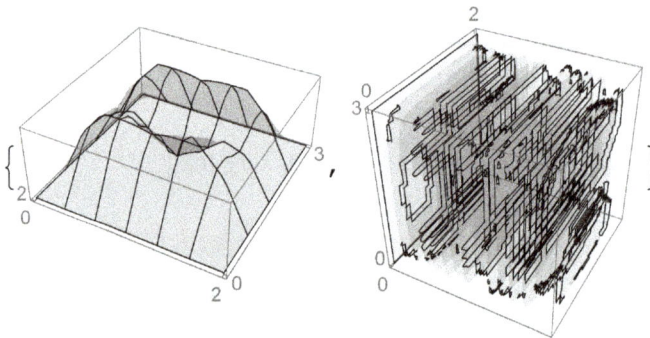

6.25. Next, we introduce a 2D inhomogeneous hyperbolic equation with initial conditions [64, p. 124],

$$\begin{cases} \partial_{t,t}u(t,x,y) = \nabla^2_{x,y}\,u(t,x,y) + \beta \sin(\alpha\,t)H(-(x-x_0))H(-(y-y_0)), \\ a,\beta \in \mathbb{R}, \quad 0 \le x,\,x_0 \le a, \quad 0 \le y,\,y_0 \le b, \\ u(0,x,y) = 0, \quad \partial_t u(0,x,y) = 0, \end{cases}$$

where H is the unit step function; see Section 1.2.4.

6.25

```
Clear[t,x]
a=1;b=2;c=8;x₀=.5;y₀=1;α = π/e;β=4;  (* Initial data *)
```

First, we show the equations of the boundary

```
bound={t,x,a-x,y,b-y};
Ω=ImplicitRegion[And@@(#≥0&/@bound),{t,x,y}];
```

Then we solve the problem on this domain.

```
eqncond={∂t,tu[t,x,y]==∇²{x,y}u[t,x,y]
+βSin[αt]UnitStep[-(x-x₀)]UnitStep[-(y-y0)],
DirichletCondition[u[t,x,y]==0,t==0],
u^(1,0,0)[0,x,y]==0};
```

```
DSolveValue[eqncond,u,{t,x,y}];  (* No closed-form
solution is returned *)
```

We look for a numerical solution.

```
sol=NDSolveValue[eqncond,u,{t,0,c},{x,0,a},{y,0,b}];
```

```
ContourPlot3D[sol[t,x,y],{t,0,c},{x,0,a},{y,0,b},
PlotRange→All,Contours→2,Mesh→None,ImageSize→160,
ContourStyle→Directive[Black,Opacity[.15]]]
```

See Figure 6.60.

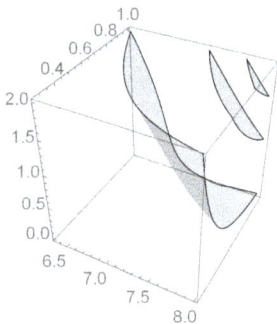

Figure 6.60: Contours for the numerical solution to problem 6.25.

6.26. We introduce a 2D inhomogeneous hyperbolic equation with initial conditions [64, p. 116],

$$\begin{cases} \partial_{t,t}u(t,x,y) = \nabla^2_{x,y}\,u(t,x,y) + t\,x\,y, & 0 \le x \le a, \quad 0 \le y \le b, \quad 0 \le t \le c, \\ u(0,x,y) = x\,y^2, & \partial_t u(0,x,y) = x + y. \end{cases}$$

6.26

```
Clear[t,x,y,u]
a=1;b=2;c=8;  (* Initial data *)
```

First, we show the equations of the boundary

```
bound={t,x,a-x,y,b-y};
Ω=ImplicitRegion[And@@(#≥0&/@bound),{t,x,y}]
```

Here, we solve the hyperbolic equation on this domain.

```
eqncond={∂_{t,t}u[t,x,y]==∇²_{x,y}u[t,x,y]+t x y,
DirichletCondition[u[t,x,y]==x y²,t==0],
u^(1,0,0)u[0,x,y]==x+y};
```

```
DSolveValue[eqncond,u,{t,x,y}]; (* No closed-form
solution is returned *)
```

We look for a numerical solution.

```
sol=NDSolveValue[eqncond,u,{t,0.,c},{x,0.,a},{y,0.,b}];
```

```
ContourPlot3D[sol[t,x,y],{t,0.,c},{x,0.,a},{y,0.,b},
PlotRange→All,Mesh→None,Contours→2,
ContourStyle→Directive[Black,Opacity[.15]],
Ticks→{{0,c/2,c},{0,a},{0,b}},ImageSize→160]
```

See Figure 6.61.

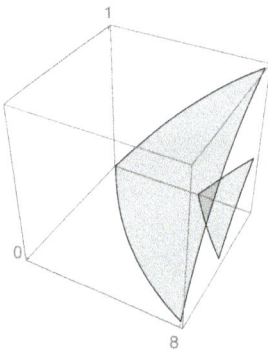

Figure 6.61: Contours for the numerical solution to problem 6.26.

6.5 Klein–Gordon equation

The scalar *Klein–Gordon equation* is of the form

$$\partial_{t,t}u(t,x) - a^2\partial_{x,x}u(t,x) + b\,u(t,x) = 0, \quad a,b,t,x \in \mathbb{R}, \quad a,b \neq 0.$$

6.27. Some particular solutions of the scalar Klein–Gordon equation are listed below,
http://eqworld.ipmnet.ru/en/solutions/lpde/lpde203.pdf:

1.
$$u(t,x) = \cos(\lambda x)(a\cos(\mu t) + \beta\sin(\mu t)), \quad b = -a^2\lambda^2 + \mu^2,$$

2.
$$u(t,x) = \sin(\lambda x)(a\cos(\mu t) + \beta\sin(\mu t)), \quad b = -a^2\lambda^2 + \mu^2,$$

3.
$$u(t,x) = \exp(\pm\mu t)(a\cos(\lambda x) + \beta\sin(\lambda x)), \quad b = -a^2\lambda^2 - \mu^2,$$

4.
$$u(t,x) = \exp(\pm\lambda x)(a\cos(\mu t) + \beta\sin(\mu t)), \quad b = a^2\lambda^2 + \mu^2,$$

5.
$$u(t,x) = \exp(\pm\lambda x)(a\exp(\mu t) + \beta\exp(-\mu t)), \quad b = a^2\lambda^2 - \mu^2,$$

6.
$$u(x,t) = aJ_0(\xi) + \beta Y_0(\xi), \quad \text{where } \xi = \frac{\sqrt{b}}{a}\sqrt{a^2(t+c_1)^2 - (x+c_2)^2}, \quad b > 0,$$

7.
$$u(t,x) = aI_0(\xi) + \beta K_0(\xi), \quad \text{where } \xi = \frac{\sqrt{-b}}{a}\sqrt{a^2(t+c_1)^2 - (x+c_2)^2}, \quad b < 0,$$

where a, β, c_1, and c_2 are arbitrary real constants, $J_0(\xi)$ and $Y_0(\xi)$ are the Bessel functions, and $I_0(\xi)$ and $K_0(\xi)$ are the modified Bessel functions.

6.27 We check that the functions in the list are indeed solutions of the Klein–Gordon equation.

```
Clear[t,x,a,α,β,λ,μ]
eqn=∂t,tu[t,x]-a²∂x,xu[t,x]+b u[t,x];

1.
b=-a²λ²+μ²;
u[t_,x_]:=Cos[λ x](α Cos[μ t]+β Sin[μ t])
eqn//Simplify
0

2.
b=-a²λ²+μ²;
u[t_,x_]:=Sin[λ x](α Cos[μ t]+β Sin[μ t])
eqn//Simplify
0

3.
b=-a²λ²-μ²;
u[t_,x_]:=Exp[±μ t](α Cos[λ x]+β Sin[λ x])
eqn//Simplify
```
$e^{t(\pm\mu)}(-\mu^2 + (\pm\mu)^2)(\alpha Cos[x\lambda]+\beta Sin[x\lambda])$ (* Null *)

4.
```
b=a²λ²+μ²;
u[t_,x_]:=Exp[±λx](α Cos[μt]+β Sin[μt])
eqn//Simplify
```
$-a^2((\pm\lambda)^2 - \lambda^2)e^{x(\pm\lambda)}(\alpha Cos[\mu t]+\beta Sin[\mu t])$ (* Null *)

5.
```
b=a²λ²-μ²;
u[t_,x_]:=Exp[±λx](α Exp[μt]+β Exp[-μt])
eqn//Simplify
```
$a^2 e^{-t\mu+x(\pm\lambda)}(e^{2t\mu}\alpha + \beta)(\lambda^2 - (\pm\lambda)^2)$ (* Null *)

6.
```
Clear[b]
Assuming[b>0,{ξ[t_,x_] := √b/a √(a²(t+c₁)² - (x-c₂)²;
u[t_,x_]:=α BesselJ[0,ξ[t,x]]+β BesselY[0,ξ[t,x]];
eqn//FullSimplify}]
```
{0}

7.
```
Assuming[b<0,{ξ[t_,x_]:=√-b/a √(a²(t+c₁)² - (x+c₂)²;
u[t_,x_]:=α BesselI[0,ξ[t,x]]+β BesselK[0,ξ[t,x]];
eqn//FullSimplify}]
```
{0}

Remark. The solutions in cases 6 and 7 contain Bessel functions; see Section 1.6. ☐

6.28. We introduce a scalar Klein–Gordon equation with initial and boundary conditions,

$$\begin{cases} \partial_{t,t}u(t,x) - \partial_{x,x}u(t,x) + u(t,x) = 0, & t \in [0,50], \quad x \in [0,20], \\ u(0,x) = 0, \quad \partial_t u(0,x) = -u_0 \sqrt{1+k^2} \sin(k x), & \text{(Initial conditions)}, \\ u(t,0) = 0, \quad \text{(Boundary condition)}. \end{cases}$$

6.28 Our aim is to emphasize how much the figures differ when the PlotRange option is changed from Automatic to All.

```
Clear[k,t,x,u]
```

```
eqncond={∂_{t,t}u[t,x]-∂_{x,x}u[t,x]+u[t,x]==0,u[0,x]==0,
∂_t u[0,x]==-u₀ √(1+k²)Sin[k x],u[t,0]==0}; (* The
problem *)
```

```
u₀=4;k=π; (* Initial data *)
```

```
DSolve[eqncond,u,{t,x}]; (* No closed-form solution is
returned *)
```

We look for a numerical solution,

```
sol=NDSolve[eqncond,u,{t,0,50},{x,0,20}];
```

and plot the solution by two plot ranges: `Automatic` and `All`.

```
Plot3D[u[t,x]/.sol,{t,0,50},{x,0,20},Mesh→2,
MeshStyle→Black,PlotRange→#,
Ticks→{25Range[0,2],10{0,1,2},Automatic},
PlotLabel→Style[#,12,Bold,Black],
PlotStyle→Directive[Black,Opacity[.15],
ImageSize→175]&/@{Automatic,All}
```

See Figure 6.62.

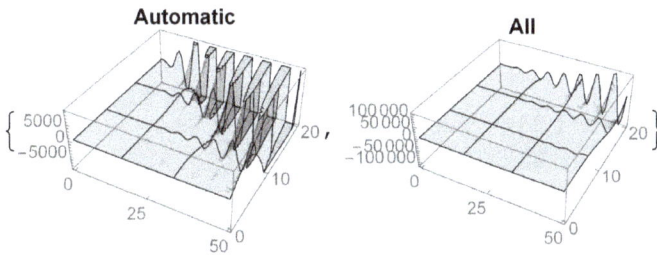

Figure 6.62: Plot of the numerical solution for problem 6.28.

6.29. One introduces a scalar Klein–Gordon equation with initial and boundary conditions,

$$
\begin{cases}
\partial_{t,t}u(t,x) - \partial_{x,x}u(t,x) + u(t,x) = 0, \quad l > 0, \quad t \in [0,l], \quad x \in [-l,l], \\
u(0,x) = u(l,x) = \begin{cases}
-5, & -l < x \le -l/2, \\
.5\,x, & -l/2 \le x \le 0, \\
0, & 0 \le x \le l/4, \quad \text{(Initial and terminal conditions)}, \\
x - 5, & l/4 \le x \le l/2, \\
5, & l/2 \le x < l,
\end{cases} \\
u(t,-l) = .5, \quad u(t,l) = .5, \quad \text{(Boundary conditions)}.
\end{cases}
$$

6.29

```
Clear[a,t,x,u]
```

```
l=20.; (* Initial datum *)
```

The initial condition is written now

```
cond={{-5,-1<x≤-1/2},{.5x,-1/2≤x≤0},{0,0≤x≤1/4},
{x-5,1/4≤x≤1/2},{5,1/2≤x<1}}; (* Initial condition *)
```

and its shape is given below.

```
Plot[Piecewise[cond],{x,-1,1},ImageSize→160,
PlotStyle→Black,Ticks→{10Range[-2,2],Range[-5,5,2]}]
```

See Figure 6.63.

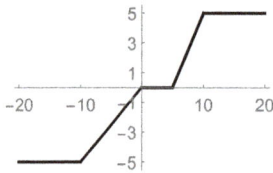

Figure 6.63: The initial conditions for problem 6.29.

```
eqn=∂t,tu[t,x]-∂x,xu[t,x]+u[t,x]==0 (* The equation *)
```

```
DSolve[{eqn,u[t,-1]==-.5,u[t,1]==.5,u[0,x]
==Which[-1<x≤-1/2,-5,-1/2≤x≤0,.5x,0≤x≤1/4,0,
1/4≤x≤1/2,x-5,1/2≤x<1,5],u[1,x]==Piecewise[cond]},
u,{t,x}]//Flatten; (* No closed-form solution is
returned *)
```

We look for a numerical solution,

```
sol=NDSolve[{eqn,u[t,-1]==-.5,u[t,1]==.5,u[0,x]
==Which[-1<x≤-1/2,-5,-1/2≤x≤0,.5x,0≤x≤1/4,0,
1/4≤x≤1/2,x-5,1/2≤x<1,5],u[1,x]==Piecewise[cond]},
u,{t,0,1},{x,-1,1}]//Flatten;
```

and plot the solution.

```
Block[{k=1/2},
{Show[{
Plot3D[u[t,x]/.sol,{t,0,k},{x,-k,k},ImageSize→160,
Mesh→5,MeshStyle→Black,PlotPoints→50,
PlotStyle→Directive[Black,Opacity[.15]],
Ticks→{{0,k},{-k,k},Automatic}],
{ParametricPlot3D[{0,x,u[0,x]/.sol},{x,-k,k},##],
ParametricPlot3D[{k,x,u[k,x]/.sol},{x,-k,k},##]}&[
ImageSize→180,PlotStyle→Black]}]}]
```

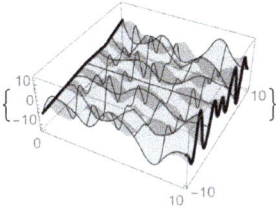

Figure 6.64: Plot of the solution for problem 6.29.

See Figure 6.64.

6.30. One introduces a scalar Klein–Gordon equation with initial and boundary conditions,

$$\begin{cases} \partial_{t,t}u(t,x) = \partial_{x,x}u(t,x) + u(t,x), & l > 0, \quad t \in [0,1], \quad x \in [-l,l], \\ u(0,x) = u(1,x) = x\,|\sin(x)|, & \text{(Initial and terminal conditions)} \\ u(t,-l) = -l/4, \quad u(t,l) = l/4, & \text{(Boundary conditions).} \end{cases}$$

6.30

```
Clear[t,x,u]

l=4; (Initial datum)
```

We look for a numerical solution,

```
sol=NDSolve[{∂_{t,t}u[t,x]==∂_{x,x}u[t,x]+u[t,x],
u[t,-l]==-1/4,u[t,l]==1/4,u[0,x]==x Abs[Sin[x]],
u[1,x]==x Abs[Sin[x]]},u,{t,0,1},{x,-1,1}]//Flatten;
```

and plot it.

```
Plot3D[u[t,x]/.sol,{t,0,1},{x,-1,1},Mesh→4,
MeshStyle→Black,PlotPoints→50,ImageSize→180,
PlotStyle→Directive[Black,Opacity[.15]],
Ticks→{1/2{0,1,2},l{-1,0,1},4{-1,0,1}}]
```

See Figure 6.65.

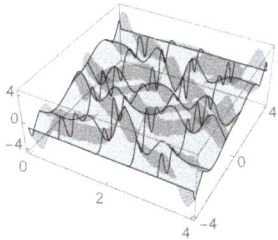

Figure 6.65: Plot of the solution for problem 6.30.

7 Nonlinear and higher-order hyperbolic equations

This chapter is dedicated to certain nonlinear and higher-order hyperbolic equations. It focuses on the sine-Gordon equation, modified Klein–Gordon equation, double sine-Gordon equation, nonlinear hyperbolic equations, and nonlinear hyperbolic equations in higher dimensions.

A *solitary wave* is a wave that propagates without any temporal evolution in shape or size when viewed in the reference frame moving with the group velocity of the wave.

A *soliton* is a solitary wave with the property where other solitons can pass through it without changing its shape and speed, [65, p. 324] and [16].

7.1 Sine-Gordon equation

The scalar *sine-Gordon equation* is of the form [71, §7.3.3],

$$\partial_{t,t}u(t,x) = a\partial_{x,x}u(t,x) + b\sin(\lambda u(t,x)), \quad t,x,b,\lambda \in \mathbb{R}, \ b,\lambda \neq 0, \ a > 0.$$

The 2D sine-Gordon equation is of the form,

$$\partial_{t,t}u(t,x,y) = a\nabla^2_{x,y}u(t,x,y) + b\sin(\lambda u(t,x,y)), \quad t,x,y,b,\lambda \in \mathbb{R}, \ b,\lambda \neq 0, \ a > 0.$$

7.1. We check that functions u introduced below satisfy the sine-Gordon equation and plot them for certain values of the parameters [69, /npde/npde2106.pdf].

1.

$$u_1(t,x) = \frac{4}{\lambda}\arctan\left(\exp\left(\frac{b\lambda(kx + \mu t + a)}{\sqrt{b\lambda(\mu^2 - ak^2)}}\right)\right), \quad b\lambda(\mu^2 - ak^2) > 0,$$

$$u_2(t,x) = \frac{4}{\lambda}\arctan\left(\exp\left(-\frac{b\lambda(kx + \mu t + a)}{\sqrt{b\lambda(\mu^2 - ak^2)}}\right)\right), \quad b\lambda(\mu^2 - ak^2) > 0;$$

2.

$$u_3(t,x) = -\frac{\pi}{\lambda} + \frac{4}{\lambda}\arctan\left(\exp\left(\frac{b\lambda(kx + \mu t + a)}{\sqrt{b\lambda(ak^2 - \mu^2)}}\right)\right), \quad b\lambda(ak^2 - \mu^2) > 0,$$

$$u_4(t,x) = -\frac{\pi}{\lambda} + \frac{4}{\lambda}\arctan\left(\exp\left(-\frac{b\lambda(kx + \mu t + a)}{\sqrt{b\lambda(ak^2 - \mu^2)}}\right)\right), \quad b\lambda(ak^2 - \mu^2) > 0;$$

3.

$$u_5(t,x) = \frac{4}{\lambda}\arctan\left(\frac{\mu\sinh(kx + a)}{k\sqrt{a}\cosh(\mu t + b)}\right), \quad \mu = \sqrt{ak^2 + b\lambda},$$

$$u_6(t,x) = \frac{4}{\lambda}\arctan\left(-\frac{\mu\sinh(kx + a)}{k\sqrt{a}\cosh(\mu t + b)}\right), \quad \mu = \sqrt{ak^2 + b\lambda};$$

https://doi.org/10.1515/9783111411392-009

4.
$$u_7(t,x) = \frac{4}{\lambda} \arctan\left(\frac{\mu \sin(k\,x + a)}{k\,\sqrt{a}\,\cosh(\mu\,t + b)}\right), \quad \mu = \sqrt{b\lambda - a\,k^2},$$

$$u_8(t,x) = \frac{4}{\lambda} \arctan\left(-\frac{\mu \sin(k\,x + a)}{k\,\sqrt{a}\,\cosh(\mu\,t + b)}\right), \quad \mu = \sqrt{b\lambda - a\,k^2};$$

5.
$$u_9(t,x) = 4\,\arctan\left(\exp\left(\frac{x - x_0 - v\,t}{\sqrt{1 - v^2}}\right)\right), \quad 0 < v < 1,$$

$$u_{10}(t,x) = 4\,\arctan\left(\exp\left(-\frac{x - x_0 - v\,t}{\sqrt{1 - v^2}}\right)\right), \quad 0 < v < 1.$$

7.1
1.

```
Clear[a,b,λ,θ,k,μ,t,x]
u₁[t_,x_]:=4/λ ArcTan[Exp[bλ(kx+μt+a)/√(bλ(μ²-ak²))]] (* For positive
bλ(μ² - ak²) *)
u₂[t_,x_]:=4/λ ArcTan[Exp[-bλ(kx+μt+a)/√(bλ(μ²-ak²))]](* For positive
bλ(μ² - ak²) *)
```

Functions u_1 and u_2 satisfy the sine-Gordon equation,

```
Assuming[a>0&&b λ(μ²-ak²)>0,
∂t,t#-a∂x,x#==b Sin[λ#]&/@{u₁[t,x],u₂[t,x]}//
FullSimplify]
{True,True}
```

We consider some particular values of the constants $a = b = 1$, $\lambda = \pi$, $a = -1$, $k = 2$, and $\mu = 5\pi/3$. The assumption for these particular values is satisfied because the denominator is positive,

```
bλ(μ²-ak²)//N
73.5622
```

The corresponding animate plot follows.

```
Animate[
Plot[{u₁[t,x],u₂[t,x]},{x,-27,27},PlotRange→All,
Ticks→{9Range[-3,3,1],{1,2}},ImageSize→150,
PlotStyle→{Black,{Black,Dashed}}],{t,-10,10},
SaveDefinitions→True,DefaultDuration→20,
AnimationDirection→ForwardBackward,
AnimationRunning→False]
```

See Figure 7.1.

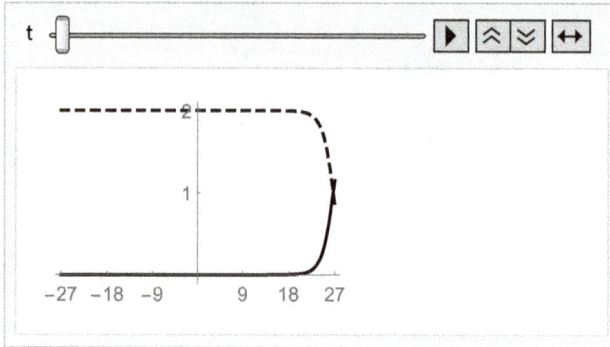

Figure 7.1: Animate plot of the solutions for problem 7.1 1.

2.

```
Clear[a,b,λ,α,k,μ,t,x]
```

$u_3[t\_,x\_]:=-\frac{\pi}{\lambda}+\frac{4}{\lambda}$ArcTan$[Exp[\frac{b\lambda(k x+\mu t+a)}{\sqrt{b\lambda(a k^2-\mu^2)}}]]$ (* For

positive $b\lambda(a k^2 - \mu^2)$ *)

$u_4[t\_,x\_]:=-\frac{\pi}{\lambda}+\frac{4}{\lambda}$ArcTan$[Exp[-\frac{b\lambda(k x+\mu t+a)}{\sqrt{b\lambda(a k^2-\mu^2)}}]]$ (* For

positive $b\lambda(a k^2 - \mu^2)$ *)

Both functions satisfy the sine-Gordon equation.

```
Assuming[a>0&&bλ(a k²-μ²)>0,
∂t,t#-a∂x,x#==b Sin[λ#]&/@{u₃[t,x],u₄[t,x]}//
FullSimplify]
{True,True}
```

We consider some particular values of the constants $a = b = 2, \lambda = \pi, \alpha = -1, k = 2$, and $\mu = 2\pi/3$.
The assumption is satisfied for these values since the denominator is positive,

```
bλ(a k²-μ²)//N
22.7043
```

The corresponding plot follows.

```
Animate[
Plot[{u₃[t,x],u₄[t,x]},{x,-28,27},PlotStyle→
{Black,{Black,Dashed}},PlotRange→All,ImageSize→160,
Ticks→{9Range[-3,3,1],{-1,1}}},{t,-22,23},
SaveDefinitions→True,DefaultDuration→20,
AnimationDirection→ForwardBackward,
AnimationRunning→False]
```

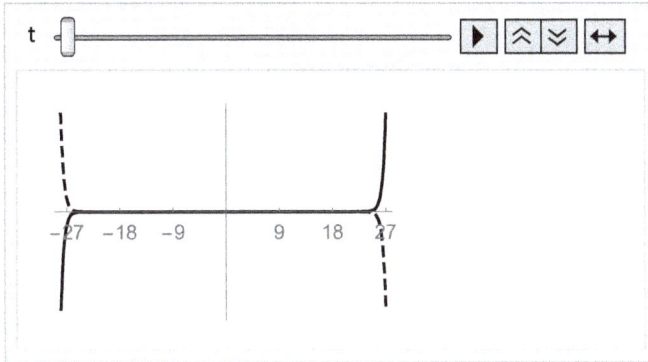

Animate plot of the solution for problem 7.1 2.

See Figure 7.2.

3.

```
Clear[a,b,λ,k,t,x]
```

$$\mu=\text{Sqrt}[a\,k^2+b\,\lambda];$$

```
u₅[t_,x_]:=4/λ ArcTan[μSinh[k x+a]/(k Sqrt[a] Cosh[μt+b])]
u₆[t_,x_]:=4/λ ArcTan[-μSinh[k x+a]/(k Sqrt[a] Cosh[μt+b])]
```

Functions u_5 and u_6 satisfy the sine-Gordon equation.

```
∂t,t#-a∂x,x#==b Sin[λ#]&/@{u₅[t,x],u₆[t,x]}//
FullSimplify
{True,True}
```

We consider some particular values of the constants. $a = b = 1$, $\lambda = \pi$, and $k = 2$. The corresponding plot follows.

```
Manipulate[
Plot[{u₅[t,x],u₆[t,x]},{x,-22,21},ImageSize→160,
PlotStyle→{Black,{Black,Dashed}},PlotRange→All,
Ticks→{10Range[-2,2],Range[-2,2]}],{t,-15,15},
SaveDefinitions→True]
```

See Figure 7.3.

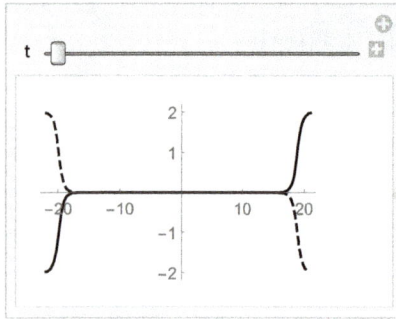

Figure 7.3: Animate plot of the solution for problem 7.1 3.

```
Animate[
RevolutionPlot3D[u₅[t,x],{x,-7,6},Mesh→2,
MeshStyle→{Black,{Black,Dashed}},PlotStyle→
Directive[Black,Opacity[.15]],PlotRange→All,
ImageSize→175,RevolutionAxis→"X",Ticks→
Outer[Times,{6,2,2},Range[-1,1]]],{t,-4,4},
SaveDefinitions→True,DefaultDuration→25,
AnimationDirection→ForwardBackward,
AnimationRunning→False]
```

See Figure 7.4.

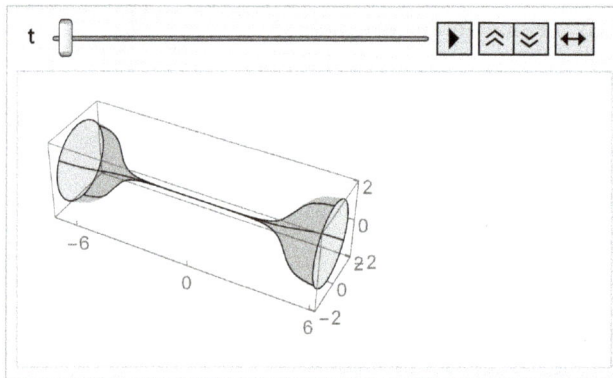

Figure 7.4: Rotated plot of the solution for problem 7.1 3.

4.

```
Clear[a,b,λ,k,μ,t,x]

μ=Sqrt[b λ-a k²];
```

$$u_7[t\_,x\_]:=\frac{4}{\lambda}\text{ArcTan}[\frac{\mu\text{Sin}[k\,x+a]}{k\,\text{Sqrt}[a]\text{Cosh}[\mu\,t+b]}]$$
$$u_8[t\_,x\_]:=\frac{4}{\lambda}\text{ArcTan}[-\frac{\mu\text{Sin}[k\,x+a]}{k\,\text{Sqrt}[a]\text{Cosh}[\mu\,t+b]}]$$

Functions u_7 and u_8 satisfy the sine-Gordon equation.

```
∂t,t#-a∂x,x#==b Sin[λ#]&/@{u₇[t,x],u₈[t,x]}//
FullSimplify
{True,True}
```

We consider some particular values of the constants $a = 1$, $b = 10$, $\lambda = \pi$, and $k = 2$. Here is the corresponding plot.

```
Animate[
Plot[{u₇[t,x],u₈[t,x]},{x,-7,6},ImageSize→160,
PlotStyle→{Black,{Black,Dashed}},PlotRange→All,
Ticks→{3Range[-2,2],Automatic}],{t,-4,3},
SaveDefinitions→True,DefaultDuration→25,
AnimationDirection→ForwardBackward,
AnimationRunning→False]
```

See Figure 7.5.

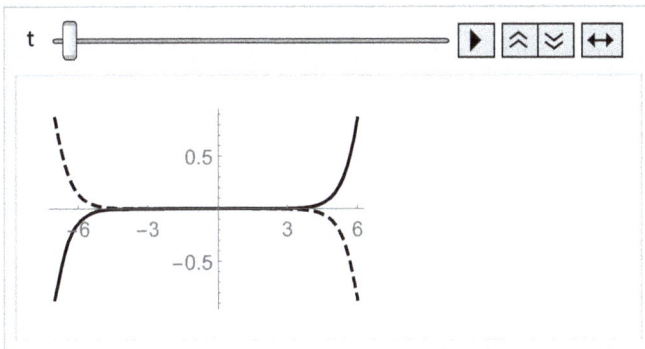

Figure 7.5: Animate plot of the solution for problem 7.1 4.

5.

```
Clear[t,x,x₀,v]
```

$$u_9[t\_,x\_]:=4ArcTan[Exp[\tfrac{x-x_0-v\,t}{Sqrt[1-v^2]}]]$$
$$u_{10}[t\_,x\_]:=4ArcTan[Exp[-\tfrac{x-x_0-v\,t}{Sqrt[1-v^2]}]]$$

Functions u_9 and u_{10} satisfy the sine-Gordon equation.

```
Assuming[-1<v<1,∂t,t#-a∂x,x#==b Sin[λ#]&/@
{u₉[t,x],u₁₀[t,x]}//FullSimplify
{True,True}
```

We set the values of the constants $v = .75$ and $x_0 = 0$ and plot the corresponding figures.

```
Animate[
Show[Plot[{u₉[t,x],u₁₀[t,x]},{x,-10,10},PlotStyle→
{Black,{Black,Dashed}},ImageSize→160,AxesLabel→
{x,u},Ticks→Outer[Times,{5,10,6},{-1,0,1}]]],
{t,-9,9},SaveDefinitions→True,DefaultDuration→25,
AnimationDirection→ForwardBackward,
AnimationRunning→False]
```

See Figure 7.6.

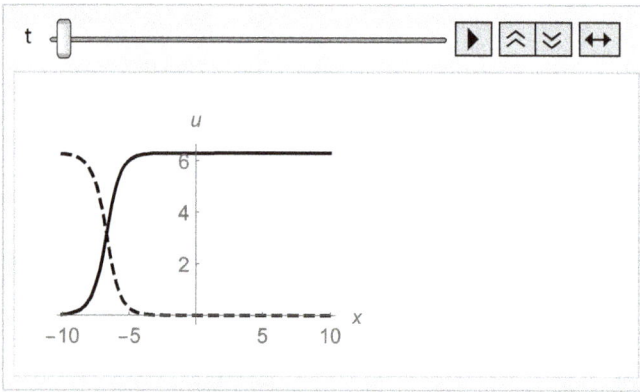

Figure 7.6: Animate plot of the solution for problem 7.1 5.

and

```
{Plot3D[u₉[t,x],##],Plot3D[u₁₀[t,x],##]}&[{t,-5,5},
{x,-10,10},PlotStyle→Directive[Black,Opacity[.15]],
Mesh→3,ImageSize→175,AxesLabel→{t,x,u},
Ticks→{5{-1,0,1},10{-1,0,1},3Range[0,2]}]
```

See Figure 7.7.

7.2. We look for the solution of the scalar sine-Gordon equation with a Gaussian initial condition and boundary conditions [92, p. 221],

$$\begin{cases} \partial_{t,t}u(t,x) = \partial_{x,x}u(t,x) + \sin(u(t,x)), & l > 0, \quad 0 \le t, \quad 0 \le x \le l, \\ u(0,x) = \exp(-x^2), \quad \partial_t u(0,x) = 0, & \text{(Initial conditions)} \\ u(t,-l) = u(t,l), & \text{(Boundary conditions)}. \end{cases}$$

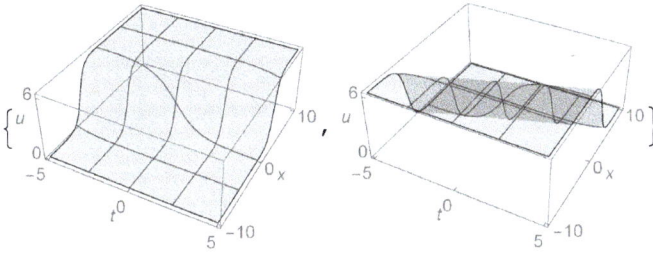

Figure 7.7: Spacial representation of the solution for problem 7.1 5.

7.2

```
Clear[l,t,x]
```

```
eqn=∂t,tu[t,x]==∂x,xu[t,x]+Sin[u[t,x]];
ic={u[0,x]==Exp[-x²],u^(1,0)[0,x]==0}; (* Initial
conditions *)
bc={u[t,-l]==u[t,l]}; (* Boundary conditions *)
```

```
DSolve[{eqn,ic,bc},u,{t,x}]; (* No closed-form
solution is returned *)
```

We look for a numerical solution

```
l=16;
sol=NDSolve[{eqn,ic,bc},u,{t,0,l},{x,0,l}]//Flatten;
```

and plot it.

```
Plot3D[u[t,x]/.sol,{t,0,l},{x,0,l},PlotRange→All,
ImageSize→175,Mesh→{3,5},PlotStyle→{Black,
Opacity[.15]},Ticks→Outer[Times,{2,2,1}/4,{0,1,2}]]
```

See Figure 7.8.

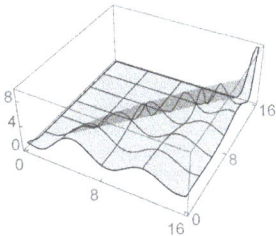

Figure 7.8: Numerical solution for problem 7.2.

7.3. We try finding the solution u of the initial value problem for the sine-Gordon equation and plot it together with a sine transform,

$$\begin{cases} \varepsilon^2 \partial_{t,t} u(t,x) - \varepsilon^2 \partial_{x,x} u(t,x) + \sin(u(t,x)) = 0, \\ l_x > 0, \quad -l_x \le x \le l_x, \quad 0 < t < l_t, \\ u(0,x) = 0, \quad \varepsilon \partial_t u(0,x) = -3 \operatorname{sech}(x), \quad \varepsilon \in \{.5, 2.2\}. \end{cases}$$

7.3

```
Clear[t,x,u]
```

```
l_t=5;l_x=2.5; (* Initial data *)
f[x_]:=0;g[x_]:=-3Sech[x]
```

For ε, we consider the values in the list {.5, 2.2}.

```
{eqncond={#²∂_t,tu[t,x]-#²∂_x,xu[t,x]+Sin[u[t,x]]==0,
u[0,x]==f[x],#u^(1,0)[0,x]==g[x]}; (* The problem *)
DSolve[{#²∂_t,tu[t,x]-#²∂_x,xu[t,x]+Sin[u[t,x]]==0,
u[0,x]==f[x],#u^(1,0)[0,x]==g[x]},u,{t,x}]&/@{.5,2.2};
(* No solution is returned *)}
```

We look for the numerical solution, contour plot of it, sine transform of it, cross- section, and plot them for $\varepsilon \in \{.5, 2.2\}$, successively.

```
{sol=NDSolve[{#²∂_t,tu[t,x]-#²∂_x,xu[t,x]+Sin[u[t,x]]==0,
u[0,x]==f[x],#u^(1,0)[0,x]==g[x]},u,{t,0,l_t},{x,-l_x,l_x},
PrecisionGoal→3];
{Plot3D[u[t,x]/.sol,{t,0,l_t},{x,-l_x,l_x},PlotPoints→25,
PlotStyle→Directive[Black,Opacity[.15]],Mesh→3,
ImageSize→175,PlotRange→All,
Ticks→{l_t/2{0,1,2},l_x{-1,0,1},Automatic},
PlotLabel→Style["u[t,x]",12,Black]],
Plot3D[Sin[u[t,x]/.sol],{t,0,l_t},{x,-l_x,l_x},
PlotPoints→25,ImageSize→175,Mesh→3,
PlotStyle→Directive[Black,Opacity[.15]],
Ticks→{l_t{0,1,2}/2,l_x{-1,0,1},Automatic},
PlotLabel→Style["Sin[u[t,x]]",12,Black]]}&/@{.5,2.2}
//Flatten
```

See Figure 7.9.

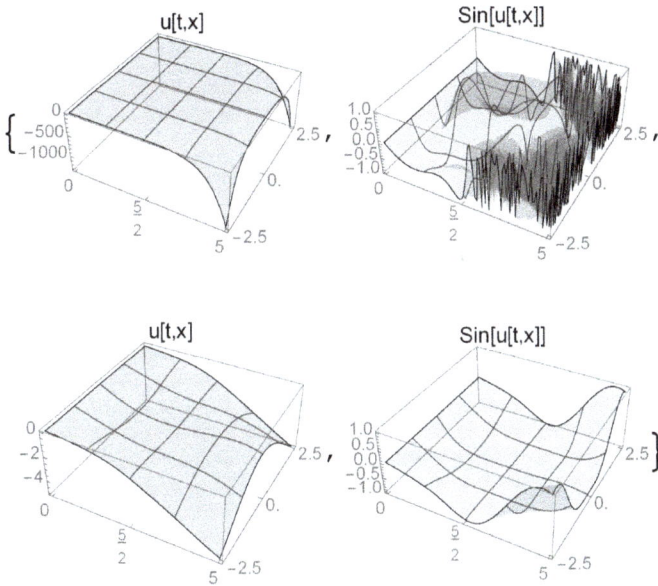

Figure 7.9: Solutions to sine-Gordon equation with ε.

7.4. We introduce an initial and boundary values problem for the sine-Gordon equation resulting a *kink* solution [16, p. 92],

$$
\begin{cases}
\partial_{t,t}u(t,x) - \partial_{x,x}u(t,x) + \sin(u(t,x)) = 0, & l_x > 0, \; -l_x \le x \le l_x, \; 0 \le t \le l_t, \\
\varepsilon = .2, \quad l_t = 5, \quad l_x = 20, \\
u(0,x) = f(x), \quad \partial_t u(0,x) = g(x), \quad \text{(Initial conditions)} \\
\partial_x u(t,-l_x) = \partial_x u(t,l_x) = 0, \quad \text{(Boundary conditions)} \\
f(x) = 4\arctan\left(\exp\left(\dfrac{x}{\sqrt{1-\varepsilon^2}}\right)\right), \quad g(x) = -3\dfrac{\varepsilon}{\sqrt{1-\varepsilon^2}}\,\mathrm{sech}(\dfrac{x}{\sqrt{1-\varepsilon^2}}).
\end{cases}
$$

7.4

```
Clear[t,x,u]

ε=.2;lₜ=5;lₓ=20; (* Initial data *)

f[x_]:=4 ArcTan[Exp[x/Sqrt[1-ε²]]] (* Initial shape *)
g[x_]:=-3 ε/Sqrt[1-ε²] Sech[x/Sqrt[1-ε²]] (* Initial speed *)
```

We look for a numerical solution,

```
sol=NDSolve[{∂t,tu[t,x]-∂x,xu[t,x]+Sin[u[t,x]]==0,
u[0,x]==f[x],u(1,0)[0,x]==g[x],u(0,1)[t,-lx]==u(0,1)[t,lx]
==0},u,{t,0,lt},{x,-lx,lx},MaxStepSize→.01]//
Simplify;
```

and plot it.

```
Plot3D[u[t,x]/.sol,{t,0,lt},{x,-lx,lx},ImageSize→160,
PlotStyle→Directive[Black,Opacity[.15]],Mesh→{3,7},
PlotPoints→50,Ticks→{{0,lt/2,lt},{-lx,0,lx},{0,6}}]
```

See Figure 7.10.

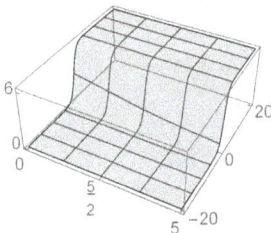

Figure 7.10: Numerical solution for problem 7.4.

7.5. We introduce an initial and boundary values problem for the sine-Gordon equation resulting an *antikink* solution [16, p. 92],

$$
\begin{cases}
\partial_{t,t}u(t,x) - \partial_{x,x}u(t,x) + \sin(u(t,x)) = 0, & -l_x \le x \le l_x, \quad 0 \le t \le l_t, \\
\varepsilon = .2, \quad l_t = 5, \quad l_x = 20, \\
u(0,x) = f(x), \quad \partial_t u(0,x) = g(x), \quad \text{(Initial conditions)} \\
\partial_x u(t,-l_x) = \partial_x u(t,l_x) = 0, \quad \text{(Boundary conditions)} \\
f(x) = 4\arctan\left(\exp\left(\dfrac{-x}{\sqrt{1-\varepsilon^2}}\right)\right), \quad g(x) = -3\dfrac{\varepsilon}{\sqrt{1-\varepsilon^2}}\,\text{sech}(\dfrac{x}{\sqrt{1-\varepsilon^2}}).
\end{cases}
$$

7.5

```
Clear[t,x]
```

```
ε=.2;lt=5;lx=20; (* Initial data *)
f[x_]:=4 ArcTan[Exp[ -x/Sqrt[1-ε²] ]] (* Initial shape *)
g[x_]:=-3 ε/Sqrt[1-ε²] Sech[ x/Sqrt[1-ε²] ] (* Initial speed *)
```

We look for a numerical solution

```
sol=NDSolve[{∂t,tu[t,x]-∂x,xu[t,x]+Sin[u[t,x]]==0,
u[0,x]==f[x],u(1,0)[0,x]==g[x],
u(0,1)[t,-lx]==u(0,1)[t,lx]==0},u,
{t,0,lt},{x,-lx,lx},MaxStepSize→.01];
```

and plot it.

```
Plot3D[u[t,x]/.sol,{t,0,lt},{x,-lx,lx},ImageSize→160,
PlotStyle→Directive[Black,Opacity[.15]],Mesh→{3,7},
Ticks→{lt/2{0,1,2},lx{-1,0,1},Automatic},
PlotPoints→50]
```

See Figure 7.11.

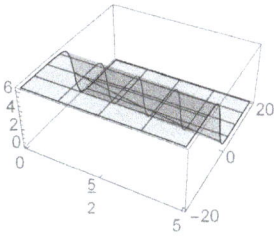

Figure 7.11: Numerical solution for problem 7.5.

7.6. We introduce an initial and boundary values problem for the sine-Gordon equation with kink and antikink solutions.

$$
\begin{cases}
\partial_{t,t}u(t,x) - \partial_{x,x}u(t,x) + \sin(u(t,x)) = 0, \quad -l_x \le x \le l_x, \quad 0 \le t \le l_t, \\
\varepsilon = .2, \quad l_t = 5, \quad l_x = 20, \\
u(0,x) = f(x), \quad \partial_t u(0,x) = g(x), \quad \text{(Initial conditions)} \\
\partial_x u(t,-l_x) = \partial_x u(t,l_x) = 0, \quad \text{(Boundary conditions)} \\
f_1(x) = 4\arctan\left(\exp\left(\frac{x+l_x/2}{\sqrt{1-\varepsilon^2}}\right)\right) + 4\arctan\left(\exp\left(\frac{x-l_x/2}{\sqrt{1-\varepsilon^2}}\right)\right), \\
g_1(x) = -2\frac{\varepsilon}{\sqrt{1-\varepsilon^2}}\operatorname{sech}\left(\frac{x+l_x/2}{\sqrt{1-\varepsilon^2}}\right) + 2\frac{\varepsilon}{\sqrt{1-\varepsilon^2}}\operatorname{sech}\left(\frac{x-l_x/2}{\sqrt{1-\varepsilon^2}}\right), \\
f_2(x) = 4\arctan\left(\exp\left(\frac{x+l_x/2}{\sqrt{1-\varepsilon^2}}\right)\right) + 4\arctan\left(\exp\left(-\frac{x-l_x/2}{\sqrt{1-\varepsilon^2}}\right)\right), \\
g_2(x) = -2\frac{\varepsilon}{\sqrt{1-\varepsilon^2}}\operatorname{sech}\left(\frac{x+l_x/2}{\sqrt{1-\varepsilon^2}}\right) - 2\frac{\varepsilon}{\sqrt{1-\varepsilon^2}}\operatorname{sech}\left(\frac{x-l_x/2}{\sqrt{1-\varepsilon^2}}\right),
\end{cases}
$$

7.6

```
Clear[t,x]
```

```
ε=.2;lt=5;lx=20; (* Initial data *)
```

$$f_1[x\_]:=4\,\mathsf{ArcTan}[\mathsf{Exp}[\tfrac{x+1_x/2}{\mathsf{Sqrt}[1-\varepsilon^2]}]]+4\,\mathsf{ArcTan}[\mathsf{Exp}[\tfrac{x-1_x/2}{\mathsf{Sqrt}[1-\varepsilon^2]}]]$$

$$g_1[x\_]:=-2\tfrac{\varepsilon}{\mathsf{Sqrt}[1-\varepsilon^2]}\,\mathsf{Sech}[\tfrac{x+1_x/2}{\mathsf{Sqrt}[1-\varepsilon^2]}]+2\tfrac{\varepsilon}{\mathsf{Sqrt}[1-\varepsilon^2]}\,\mathsf{Sech}[\tfrac{x-1_x/2}{\mathsf{Sqrt}[1-\varepsilon^2]}]$$

$$f_2[x\_]:=4\,\mathsf{ArcTan}[\mathsf{Exp}[\tfrac{x+1_x/2}{\mathsf{Sqrt}[1-\varepsilon^2]}]]+4\,\mathsf{ArcTan}[\mathsf{Exp}[-\tfrac{x-1_x/2}{\mathsf{Sqrt}[1-\varepsilon^2]}]]$$

$$g_2[x\_]:=-2\tfrac{\varepsilon}{\mathsf{Sqrt}[1-\varepsilon^2]}\,\mathsf{Sech}[\tfrac{x+1_x/2}{\mathsf{Sqrt}[1-\varepsilon^2]}]-2\tfrac{\varepsilon}{\mathsf{Sqrt}[1-\varepsilon^2]}\,\mathsf{Sech}[\tfrac{x-1_x/2}{\mathsf{Sqrt}[1-\varepsilon^2]}]$$

We look for the numerical solutions of the problems

```
{sol₁,sol₂}
=NDSolve[{∂t,tu[t,x]-∂x,xu[t,x]+Sin[u[t,x]]==0,
u[0,x]==First[#],u⁽¹,⁰⁾[0,x]==Last[#],u⁽⁰,¹⁾[t,-lₓ]
==u⁽⁰,¹⁾[t,lₓ]==0},u,{t,0,lₜ},{x,-lₓ,lₓ},
MaxStepSize→.01]&/@{{f₁[x],g₁[x]},{f₂[x],g₂[x]}};
```

and plot them.

```
{Plot3D[u[t,x]/.sol₁,##]
Plot3D[u[t,x]/.sol₂,##]}&[{t,0,lₜ},{x,-lₓ,lₓ},
PlotPoints→25,ImageSize→175,Mesh→{3,7},
PlotStyle→Directive[Black,Opacity[.15]],
Ticks→{lₜ/2{0,1,2},lₓ{-1,0,1},Automatic}]
```

See Figure 7.12.

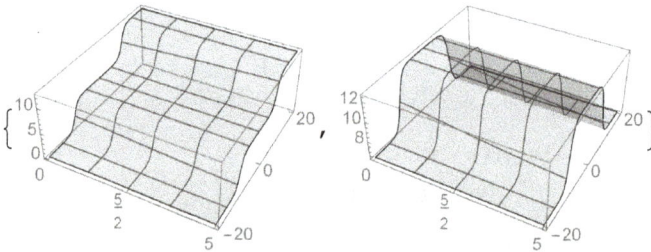

Figure 7.12: Numerical solutions for problem 7.6.

We plot together the two solutions.

```
Plot3D[u[t,x]/.{sol₁,sol₂},{t,0,lₜ},{x,-lₓ,lₓ},
ColorFunction→Function[{x,y,z},Hue[z/10]],
ColorFunctionScaling→False,PlotStyle→Opacity[.3],
PlotPoints→25,ImageSize→175,Mesh→1,
MeshStyle→{Blue,Red},Ticks→{lₜ/2{0,1,2},lₓ{-1,0,1},
Automatic}]
```

See Figure 7.13.

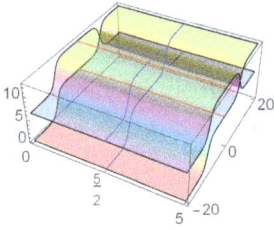

Figure 7.13: Numerical solutions for problem 7.6.

We introduce a dynamical picture of the kink and antikink solutions.

```
p=.;
```

```
Animate[
Show[
{Plot3D[u[t,x]/.sol₁,##,PlotStyle→Directive[
Black,Opacity[.2]]],
Plot3D[u[t,x]/.sol₂,##,PlotStyle→Directive[
Blue,Opacity[.05]]]}&[{t,0,lₜ},{x,-lₓ,plₓ},
PlotPoints→25,ImageSize→160,Mesh→1,
Ticks→{lₜ/2{0,1,2},lₓ{-1,0,1},Automatic},
PlotRange→All]],{p,-.99,.999},SaveDefinitions→True,
DefaultDuration→75,AnimationRunning→False,
AnimationDirection→ForwardBackward]
```

See Fig 7.14.

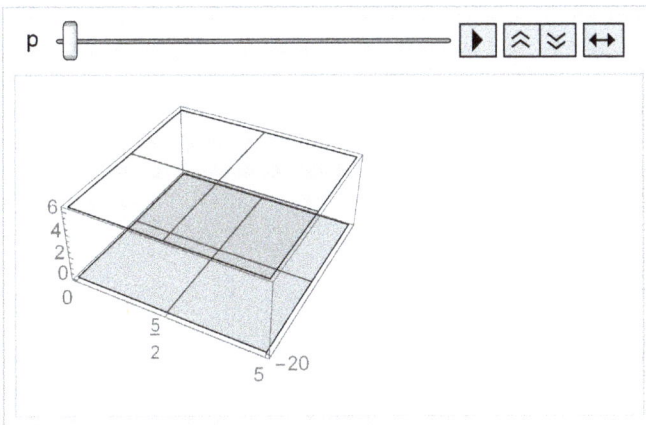

Figure 7.14: Dynamical solutions for problem 7.6.

7.2 Modified Klein–Gordon equation

7.7. We study a problem for a nonlinear and inhomogeneous hyperbolic equation which connected to the sine-Gordon equation,

$$\partial_{t,t} u(t,x) - \partial_{x,x} u(t,x) = u^3(t,x) - u(t,x).$$

7.7

```
Clear[t,x]
```

```
sol=DSolve[∂_{t,t}u[t,x]-∂_{x,x}u[t,x]==u[t,x]³-u[t,x],
u[t,x],{t,x}]//Simplify (* Solve the equation *)
{{u[t, x]→-Tanh[x C[2]-t Sqrt[1/2+C[2]²]+C[3]]},
{u[t,x]→Tanh[x C[2]-t Sqrt[1/2+C[2]²]+C[3]]},
{u[t,x]→-Tanh[x C[2]+t Sqrt[1/2+C[2]²]+C[3]]},
{u[t,x]→Tanh[x C[2]+t Sqrt[1/2+C[2]²]+C[3]]}}
```

Thus we got four solutions and plot them for some particular values of the constants of integrations.

```
Plot3D[u[t,x]/.sol/.{C[2]→2,C[3]→1},{t,-4,4},{x,0,1},
Mesh→None,ImageSize→175,ColorFunction→"Temperature",
PlotStyle→Opacity[.3],PlotLegends→Automatic,
Ticks→Outer[Times,{4,1,1},{-1,0,1}]]
```

See Figure 7.15.

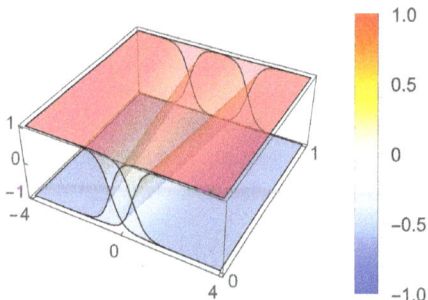

Figure 7.15: Numerical solution for problem 7.7.

We show a dynamical variant of the previous picture.

```
p=0.;
```

```
Animate[
Plot3D[u[t,x]/.sol/.{C[2]→2,C[3]→1},{t,-4p,4p},
{x,0,1},Mesh→None,ImageSize→175,
ColorFunction→"Temperature",PlotStyle→Opacity[.3],
Ticks→Outer[Times,{IntegerPart[4p*100]/100,1,1},
{-1,0,1}]],{p,.1,1},SaveDefinitions→True,
DefaultDuration→75,AnimationRunning→False,
AnimationDirection→ForwardBackward]
```

See Figure 7.16.

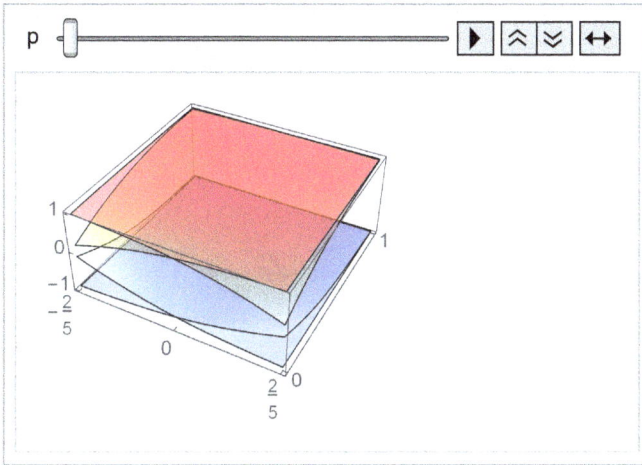

Figure 7.16: Dynamical numerical solution for problem 7.7.

7.3 Double sine-Gordon equation

7.8. We study the *double sine-Gordon equation* with constant coefficients [78],

$$\partial_{t,x}u(t,x) = \lambda \sin(u(t,x)) + \mu \sin(2u(t,x)).$$

7.8

```
Clear[t,x,λ,μ]
```

```
eqn=∂t,xu[t,x]==λSin[u[t,x]]+μSin[2u[t,x]]; (* The
equation *)
```

```
DSolve[eqn,u[t,x],{t,x}]; (* No closed-form
solution is returned *)
```

We try a numerical approach:

```
Block[{λ=#[[1]],μ=#[[2]]},
{nsol=NDSolve[{eqn,u[0,x]==Sin[x],u[t,0]==0},u,
{t,0,10},{x,0,10}];
Plot3D[u[t,x]/.nsol,{t,0,10},{x,0,10},Mesh→3,
ImageSize→175,PlotPoints→25,PlotRange→All,
Ticks→{{0,10},{0,10},Automatic},
PlotStyle→{Black,Opacity[.1]}]
&/@{{1,1},{.1,.1},{.1,1},{1,.1}}//Flatten}]
```

See Figure 7.17.

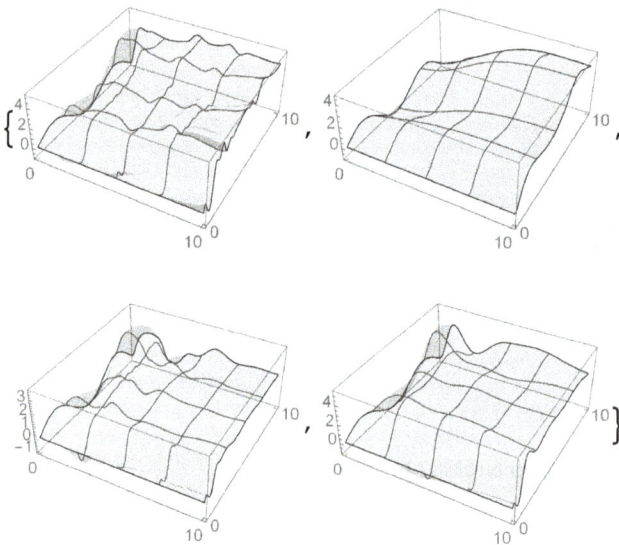

Figure 7.17: Numerical discrete solutions for problem 7.8.

We introduce a continuous representation of the numerical solution to the double sine-Gordon equation.

```
doubleSineGordon[λ_,μ_]:=
Module[{t,x,u},
eqn=∂_{t,x}u[t,x]==λSin[u[t,x]]+μSin[2u[t,x]];
nsol=NDSolve[{eqn,u[0,x]==Sin[x],u[t,0]==0},u,
{t,0,10},{x,0,10}];
Plot3D[u[t,x]/.nsol,{t,0,10},{x,0,10},Mesh→3,
ImageSize→175,PlotPoints→25,PlotRange→All,
Ticks→{{0,10},{0,10},Automatic},
PlotStyle→{Black,Opacity[.1]}]]
```

```
Manipulate[
Quiet@doubleSineGordon[λ,μ],{λ,.1,1},{μ,0,1},
SynchronousUpdating→False,SaveDefinitions→True]
```

See Figure 7.18.

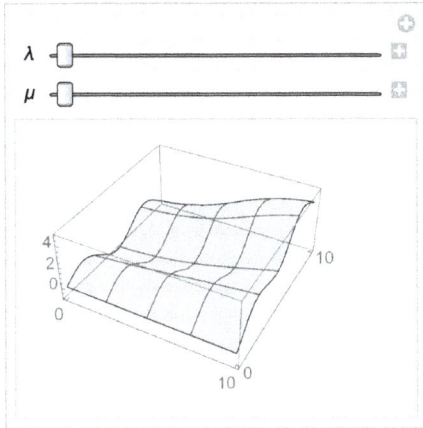

Figure 7.18: Continuous dependence of the solution for problem 7.8.

7.4 Nonlinear hyperbolic equations

7.9. Here, we study an initial and boundary value problem for a nonlinear inhomogeneous hyperbolic equation,

$$\begin{cases} u(t,x)\,\partial_{t,t}u(t,x) = \partial_{x,x}u(t,x) + 2t^2\sin(x) + t^2\sin^2(x), \\ u(t,0) = 0, \quad u_x(t,0) = t^2, \quad u_t(0,x) = 0, \quad 0 \le t \le 1, \quad x \in [0,\pi/2]. \end{cases}$$

7.9

```
Clear[t,τ,x,ξ]

eqn={u[t,x]∂t,tu[t,x]==∂x,xu[t,x]+2t²Sin[x]+t²Sin[x]²,
u[t,0]==0,u⁽⁰,¹⁾[t,0]==t²,u⁽¹,⁰⁾[0,x]==0};

DSolve[eqn,u,{t,x}]; (* No closed-form solution is
returned *)
```

Therefore, we are looking for a numerical solution, if any,

```
sol=NDSolve[eqn,u,{t,0,1},{x,0,π/2}]//Flatten;
```

and plot the solution.

```
{Plot3D[u[t,x]/.sol,{t,0,1},{x,0,π/2},ImageSize→175,
PlotStyle→{Black,Opacity[.15]},Mesh→2,
Ticks→Outer[Times,{1,1,1},{0,1}]],
Manipulate[
Plot3D[u[t,x]/.sol,{t,0,τ},{x,0,ξ},ImageSize→220,
PlotStyle→{Black,Opacity[.15]},Mesh→2,
Ticks→{{0,τ},{0,ξ},Automatic}],{τ,.01,1},{ξ,.01,π/2},
SaveDefinitions→True]}
```

See Figure 7.19.

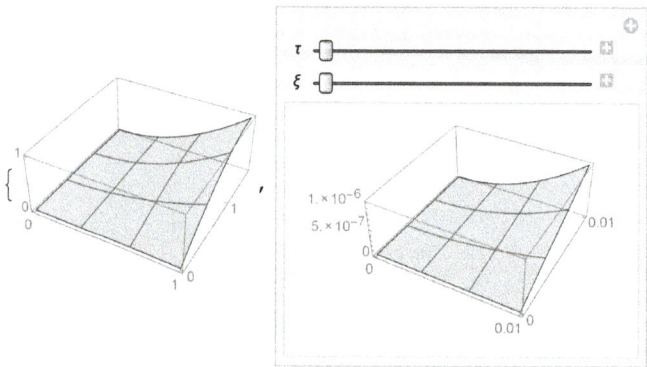

Figure 7.19: Numerical solutions for problem 7.9.

Remark. The analytical solution to this problem is $u(t, x) = t^2 \sin(x)$. ☐

7.10. We study an initial and boundary value problem for a nonlinear hyperbolic equation,

$$\begin{cases} u(t,x)\partial_{t,t}u(t,x) = \partial_{x,x}u(t,x) + \frac{t^2+x^2}{2} - 1, \\ \partial_x u(t,0) = 0, \quad \partial_t u(0,x) = 0, \quad u(t,0) = \frac{t^2}{2}, \quad t \in [0,1], \quad x \in [0,\pi/2]. \end{cases}$$

7.10

```
Clear[t,τ,x,ξ]
```

```
eqn={u[t,x]∂t,tu[t,x]==∂x,xu[t,x]+t²+x²/2-1,u[t,0]==t²/2,
u⁽⁰,¹⁾[t,0]==0,u⁽¹,⁰⁾[0,x]==0};
```

```
DSolve[eqn,u,{t,x}]; (* No closed-form solution is
returned *)
```

Therefore, we are looking for a numerical approach:

```
sol=NDSolve[eqn,u,{t,0,1},{x,0,π/2}]//Flatten;
```

and plot the solution.

```
{Plot3D[u[t,x]/.sol,{t,0,1},{x,0,π/2},ImageSize→175,
PlotStyle→{Black,Opacity[.15]},Mesh→{3,1},
Ticks→Outer[Times,{1,π/2,1.5},{0,1/2,1}]],
Manipulate[
Plot3D[Evaluate[u[t,x]/.sol],{t,0,τ},{x,0,ξ},
PlotStyle→{Black,Opacity[.15]},ImageSize→220,
Mesh→{3,1},Ticks→{{0,τ},{0,ξ},Automatic}],{τ,.001,1},
{ξ,.001,π/2},SaveDefinitions→True]}
```

See Figure 7.20.

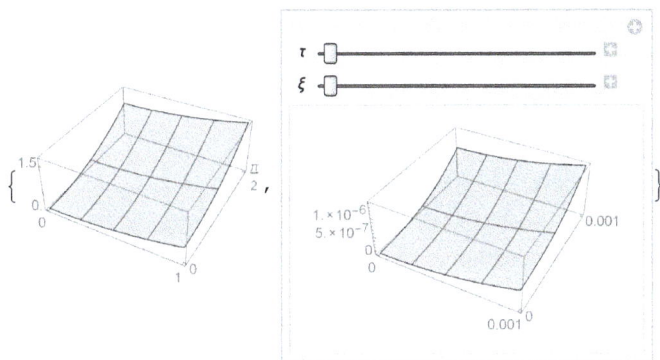

Figure 7.20: Numerical solutions for problem 7.10.

Remark. The analytical solution to this problem is $u(t, x) = (t^2 + x^2)/2$. ☐

7.11. The next problem of a nonlinear second-order partial differential equation may be found at https://reference.wolfram.com/language/howto/PlotTheResultsOfNDSolve.html,

$$\begin{cases} \partial_{t,t} u(t, x) = \partial_{x,x} u(t, x) + (1 - u^2(t, x))(1 + 2u(t, x)), \\ 0 \le t \le 10, \quad -10 \le x \le 10, \\ u(0, x) = \exp(-x^2), \quad \partial_t u(0, x) = 0, \quad \text{(Initial conditions)} \\ u(t, -10) = u(t, 10), \quad \text{(Boundary conditions)}. \end{cases}$$

7.11

```
Clear[t,x]
```

We solve the problem

```
sol=NDSolve[{∂_t,t u[t,x]==∂_x,x u[t,x]
+(1-u[t,x]²)(1+2u[t,x]),u[0,x]==e^{-x²},u^{(1,0)}[0,x]==0,
u[t,-10]==u[t,10]},u,{t,0,10},{x,-10,10}];
```

and plot its solution.

```
Plot3D[u[t,x]/.sol,{t,0,10},{x,-10,10},Mesh→2,
PlotStyle→Directive[Black,Opacity[.15]],
ImageSize→175,Ticks→{5{0,1,2},10{-1,0,1},{-1,0,1}}]
```

See Figure 7.21.

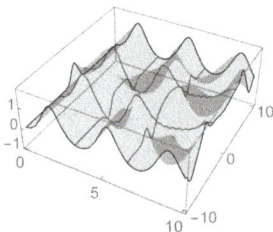

Figure 7.21: Numerical solutions for problem 7.9.

Here, we present a dynamical cross-section through the above figure.

```
plots=Table[Plot[u[t,x]/.sol,{x,-6,6},PlotStyle→Black,
Ticks→{3Range[-2,2],Range[-1,2]},PlotRange→{-1,2}],
{t,0,10,.25}];
ListAnimate[plots,ImageSize→175,DefaultDuration→30,
SaveDefinitions→True,AnimationRunning→False,
AnimationDirection→ForwardBackward]
```

See Figure 7.22.

7.12. We study the system of inhomogeneous hyperbolic equations with initial and boundary conditions,

$$
\begin{cases}
\partial_{x,x} u(t,x) - v(t,x)\partial_{t,t} u(t,x) - u(t,x)\partial_{t,t} v(t,x) = 2 - 2t^2 - 2x^2, \\
\partial_{x,x} v(t,x) - v(t,x)\partial_{t,t} v(t,x) + u(t,x)\partial_{t,t} u(t,x) = 1 + 3t^2/2 + 3x^2/2, \\
u(t,0) = t^2, \quad \partial_x u(t,0) = 0, \quad \partial_t u(0,x) = 0, \\
v(t,0) = t^2/2, \quad \partial_x v(t,0) = 0, \quad \partial_t v(0,x) = 0, \\
t \in [0,1], \quad x \in [0,1].
\end{cases}
$$

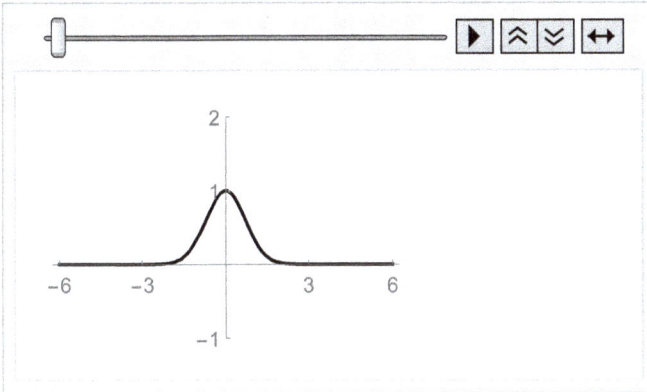

Figure 7.22: Dynamical cross- section through the numerical solution for problem 7.9

7.12

```
Clear[t,τ,x,ξ]
```

We introduce the system of differential equations.

```
eqncond={∂x,xu(t,x)-v(t,x)∂t,tu(t,x)-u(t,x)∂t,tv(t,x)==
2-2t²-2x²,
∂x,xv(t,x)-v(t,x)∂t,tv(t,x)-u(t,x)∂t,tu(t,x)==
2+3t²/2+3x²/2,
u[t,0]==t²,u⁽¹,⁰⁾[0,x]==0,u⁽⁰,¹⁾[t,0]==0,
v[t,0]==t²/2,v⁽¹,⁰⁾[0,x]==0,v⁽⁰,¹⁾[t,0]==0};
```

We look for analytical solutions.

```
DSolve[eqncond,{u,v},{t,x}]; (* No closed-form solution
is returned *)
```

Therefore, we are looking for a numerical approach.

```
sol=NDSolve[eqncond,{u,v},{t,0,1},{x,0,1},
PrecisionGoal→2]//Flatten;
{Plot3D[u[t,x]/.sol,{t,0,1},{x,0,1},ImageSize→160,
PlotStyle→{Black,Opacity[.15]},Mesh→2,
Ticks→{{0,1},{0,1},{0,1,2}}],
Manipulate[
Plot3D[u[t,x]/.sol,##,ImageSize→160,
PlotStyle→{Black,Opacity[.15]},Mesh→2,
Ticks→{{0,τ},{0,ξ},None}]&[{t,0,τ},{x,0,ξ}],
{τ,.01,1},{ξ,.01,1},SaveDefinitions→True]}
```

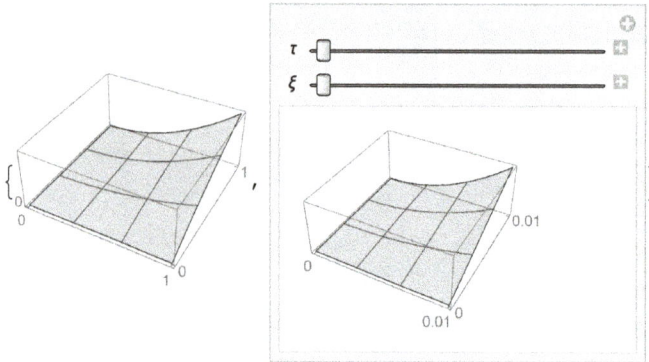

Figure 7.23: The first solution for problem 7.12.

See Figure 7.23.

We plot the second solution.

```
{Plot3D[v[t,x]/.sol,{t,0,1},{x,0,1},Mesh→2,
ImageSize→160,PlotStyle→{Black,Opacity[.15]},
Ticks→Outer[Times,{1,1,1},{0,1}]],
Manipulate[
Plot3D[v[t,x]/.sol,{t,0,τ},{x,0,ξ},Mesh→2,
PlotStyle→{Black,Opacity[.15]},ImageSize→160,
Ticks→{{0,τ},{0,ξ},None}],{τ,.01,1},
{ξ,.01,1},SaveDefinitions→True]}
```

See Figure 7.24.

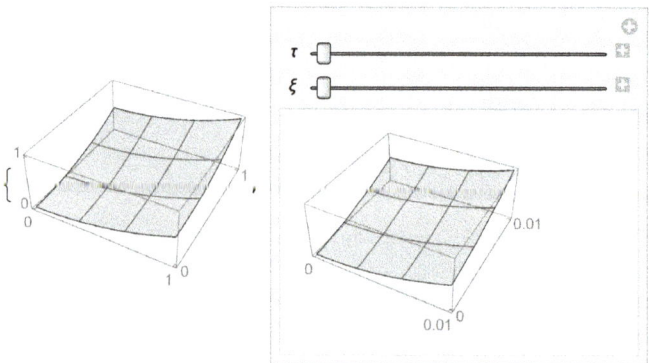

Figure 7.24: The second solution for problem 7.12.

Remark. The analytical solutions to the present problem are $u(t,x) = t^2 + x^2$ and $v(t,x) = (t^2 + x^2)/2$. □

7.5 Nonlinear hyperbolic equations in higher dimension

7.13. One introduces a 2D sine-Gordon equation with Gaussian initial conditions and periodic boundary conditions [92, p. 8],

$$\begin{cases} \partial_{t,t} u(t,x,y) = \nabla^2_{x,y} u(t,x,y) + \sin(u(t,x,y)), & l > 0, \quad t \in [0, \frac{l}{2}], \quad x,y \in [-l, l], \\ u(0,x,y) = \exp(-(x^2 + y^2)), \quad \partial_t u(0,x,y) = 0, & \text{(Initial conditions)} \\ u(t,-l,y) = u(t,l,y), \quad u(t,x,-l) = u(t,x,l), & \text{(Boundary conditions)}. \end{cases}$$

7.13

```
Clear[l,t,τ,x,y]
```

We introduce the problem,

```
eqn=∂t,tu[t,x,y]==∇²{x,y}u[t,x,y]+Sin[u[t,x,y]];
ic={u[0,x,y]==e^(-(x²+y²)),u^(1,0,0)==0};
bc={u[t,-l,y]==u[t,l,y],u[t,x,-l]==u[t,x,l]};
```

Let us try finding a closed-form solution.

```
DSolve[{eqn,ic,bc},u,{t,x,y}]//Flatten;
```

No closed-form solution is returned. We look for a numerical solution and set

```
l=4; (* Initial datum *)
```

```
sol=NDSolve[{eqn,ic,bc},u,{t,0,l/2},{x,-1,1},{y,-1,1}]
//Flatten;
```

and plot it.

```
{Plot3D[u[l/2,x,y]/.sol,{x,-1,1},{y,-1,1},Mesh→5,
Ticks→{4{-1,0,1},4{-1,0,1},.2{0,1,2}},ImageSize→175,
PlotStyle→{Black,Opacity[.15]},PlotPoints→50]
Manipulate[
Plot3D[u[t,x,y]/.sol,##,ImageSize→175,Mesh→3,
PlotStyle→{Black,Opacity[.15]},PlotRange→All,
Axes→False]&[{x,-1,1},{y,-1,1}],{t,0,l/2},
SaveDefinitions→True]}
```

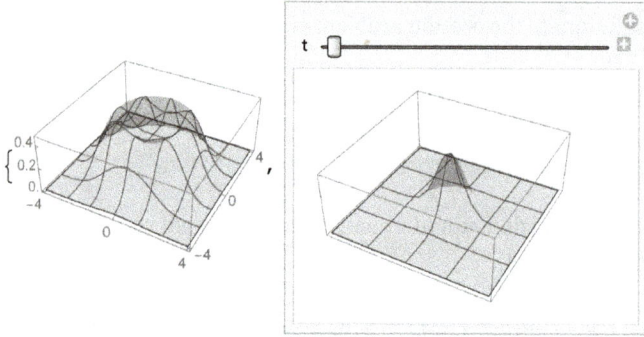

Figure 7.25: Numerical solutions for problem 7.13.

See Figure 7.25.

8 Elliptic partial differential equations

This chapter deals with harmonic functions, Laplace equations on rectangles, Laplace equations on arbitrary domains, Poisson equations on rectangles, Poisson equations on arbitrary domains, Laplace equations in higher dimensions, and Poisson equations in higher dimensions.

We recall from Section 1.2.6 that a real valued function u of two real variables, which is twice differentiable on an nonempty, open, and connected domain $D \subset \mathbb{R}^2$ is said to satisfy the Laplace equation on D if it fulfills the following equation:

$$\nabla_{x,y}^2 u(x,y) = \frac{\partial^2 u(x,y)}{\partial x^2} + \frac{\partial^2 u(x,y)}{\partial y^2} = 0, \quad \forall\, (x,y) \in D.$$

A real valued function u of n variables, which is twice differentiable function on a nonempty, open, and connected domain D in $\mathbb{R}^n$, $n \in \mathbb{N}, n \geq 2$ is said to satisfy the Laplace equation on D if it fulfills the following equation:

$$\nabla_{x_1,x_2,\ldots,x_n}^2 u(x_1, x_2, \ldots, x_n) = \sum_{k=1}^{n} \frac{\partial^2 u(x_1, x_2, \ldots, x_n)}{\partial_{x_k}^2} = 0 \quad \text{on } D.$$

8.1 Harmonic functions

8.1. We study the simplest Laplace equation,

$$\nabla_{x,y}^2 u(x,y) = 0, \quad \forall\, x, y \in \mathbb{R}.$$

8.1 We use the power of *Mathematica*.

```
Clear[x,y]
```

```
∇²_{x,y}u[x,y]==0;  (* The equation *)
sol=DSolve[%,u,{x,y}]//First  (* The general solution
depends on two twice differentiable functions *)
%%/.%  (* Laplace equation is satisfied *)
{u→Function[{x,y},C[1][I x+y]+C[2][-I x+y]]}
True
```

Let us consider a particular solution,

```
fn=u[x,y]/.sol/.{C[1][t_]→Log[Re[t]²+Im[t]²],
C[2][t_]→ArcTan[Im[t]/Re[t]]};
```

and plot it

https://doi.org/10.1515/9783111411392-010

```
Plot3D[fn,{x,-7,7},{y,-7,7},ImageSize→175,Mesh→3,
Ticks→Outer[Times,{7,7,6},{-1,0,1}],PlotRange→{-6,6},
PlotStyle→{Black,Opacity[.15]}]
```

See Figure 8.1.

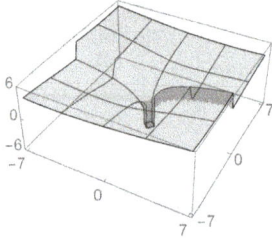

Figure 8.1: Particular solution for problem 8.1.

8.2. We show that function u below componentwise satisfies the Laplace equation,

$$u(x,y) = \frac{(x,y)}{x^2 + y^2}, \quad \forall\, x,y \in \mathbb{R}, \text{ such that } x^2 + y^2 \neq 0.$$

8.2

```
Clear[x,y]
```

```
u[x_,y_]:=\frac{{x,y}}{x^2+y^2}  (* The function *)
∇²_{x,y}u[x,y]//Simplify (* The components of
function u satisfy Laplace equation *)
{0,0}
```

Now we plot this harmonic function and its components.

```
{Plot3D[u[x, y],{x,-2,2},{y,-2,2},PlotRange→{-1,1},
PlotStyle→{Black,Opacity[.15]},PlotPoints→60,
ViewPoint→{-1.3,.1,1.5},ImageSize→175,Mesh→{5,3},
Ticks→Outer[Times,{2,2,1},{-1,0,1}],
BoxRatios→{1,1,1},PlotLabel→"u[x,y]"],
Plot3D[First[u[x,y]],{x,-2,2},{y,-2,2},
PlotLabel→"x/(x²+y²)",##],
Plot3D[Last[u[x,y]],{x,-2,2},{y,-2,2},
PlotLabel→"y/(x²+y²)",##]}&[PlotPoints→40,Mesh→{5,3},
Ticks→Outer[Times,{1,1,1},2Range[-1,1]],
ImageSize→175,PlotStyle→{Black,Opacity[.15]}]
```

See Figure 8.2.

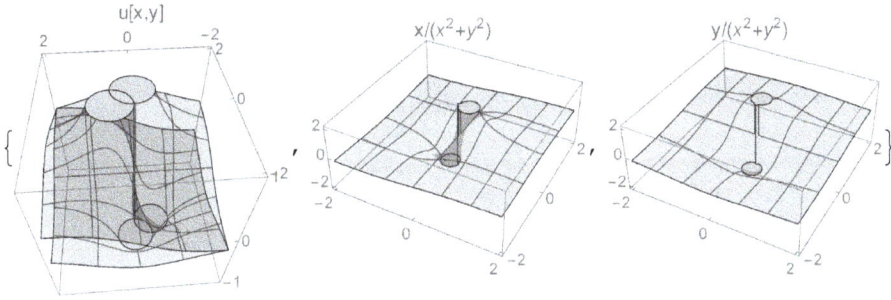

Figure 8.2: Solution for problem 8.2.

8.3. We check that the following functions u satisfy the Laplace equation:

$$u(x,y) = \frac{f(x+iy) + f(x-iy)}{2}, \quad \text{with } f(t) = \begin{cases} \ln(1 + |t|^2), \\ t\exp(t), \\ 1 - 1/t, \\ 1/(1 + t^2). \end{cases}$$

8.3

```
Clear[x,y,t,f,n]
```

```
$Assumptions={x,y}∈Reals&&n∈Integers&&n>0;
```

```
u[x_,y_]:=(f[x+I y]+f[x-I y])/2
```

Function u satisfies the Laplace equation for any twice differentiable function f.

```
∇²_{x,y}u[x,y]//Simplify
0
```

We set some particular functions f.

```
functions=u[x,y]/.f[t_]→{Log[1+Abs[t]²],t Exp[t],
1-1/t,1/(1+t²)};
```

```
icons=Table[function→Plot3D[function,
{x,-2,2},{y,-2,2},PlotStyle→{Gray,Opacity[.15]},
PlotPoints→60,ViewPoint→{-1.3,.1,1.5},
ImageSize→70,Mesh→3,Ticks→None,BoxRatios→{1,1,1}],
{function,functions}];
```

We plot the functions just defined.

```
Manipulate[
Plot3D[function,{x,-2,2},{y,-2,2},PlotPoints→60,
PlotStyle→{Black,Opacity[.15]},BoxRatios→{1,1,1},
ViewPoint→{-1.3,.1,1.5},ImageSize→180,Mesh→{5,3},
Ticks→Outer[Times,{2,2,1},{-1,0,1}],PlotRange→{-1,1}],
PlotLabel→Style[function,Black,10],
{{function,icons[[1,1]]},icons,Setter}]
```

See Figure 8.3.

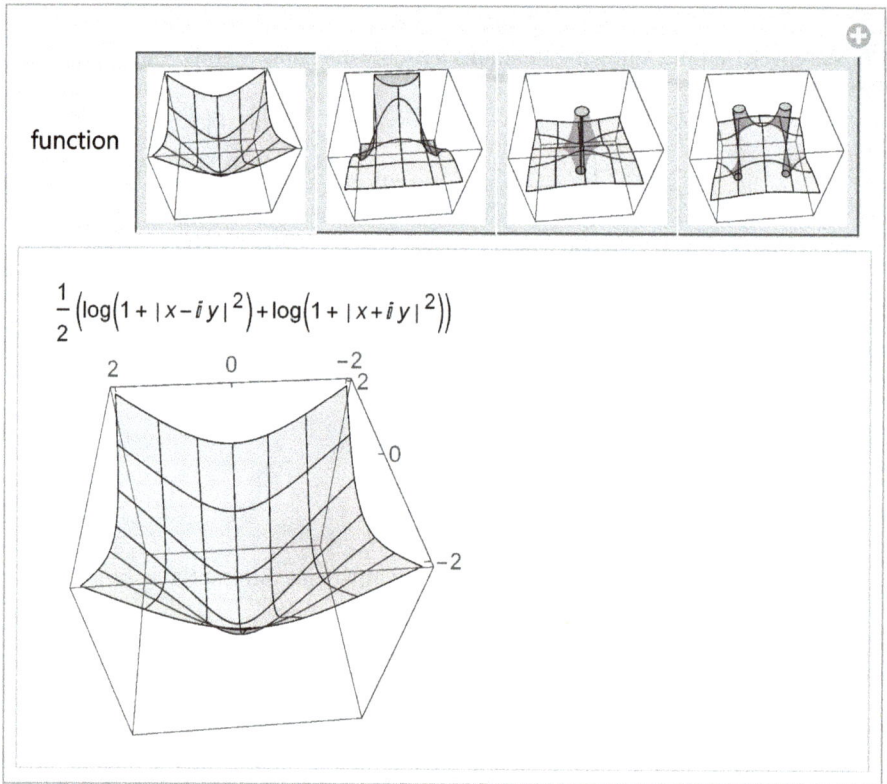

Figure 8.3: Solution for problem 8.3.

By a similar code, certain power and radical functions are introduced,

```
functions={u[x,y]/.f[t_]→Table[t^n,{n,{1,2,5,9}}],
u[x,y]/.f[t_]→Table[√t,{n,{2,3,5,9}}]};
```

and plotted.

```
Table[
icons=Table[function→Plot3D[function,{x,-2,2},
{y,-2,2},PlotStyle→{Gray,Opacity[.15]},PlotPoints→40,
ViewPoint→{-1.3,.1,1.5},ImageSize→68,Mesh→3,
MeshStyle→{Blue,Red},Ticks→None,BoxRatios→{1,1,1}],
{function,functions}];
Manipulate[
Plot3D[function,{x,-2,2},{y,-2,2},PlotPoints→40,
PlotStyle→{Gray,Opacity[.15]},ImageSize→175,Mesh→3,
ViewPoint→{-1.3,.1,1.5},BoxRatios→{1,1,1},
Ticks→{{-2,0,2},{-2,0,2},None},
PlotLabel→Style[function,Black,10]],
{{function,icons[[1,1]]},icons,Setter}],
{k,Length[functions]}]
```

See Figure 8.4.

8.2 Laplace equations on rectangles

8.4. We find a solution to the following boundary value problem to Laplace equation defined on a rectangle [64, p. 149]:

$$\begin{cases} \nabla^2_{x,y}u(x,y) = 0, & 0 < x < a, \quad 0 < y < b, \\ u(0,y) = c, & u(a,y) = u(x,0) = u(x,b) = 0. \end{cases}$$

8.4

```
Clear[a,b,c,x,y]
```

We define the problem.

```
dom=ImplicitRegion[0≤x≤a&&0≤y≤b,{x,y}];
harmonic=Laplacian[u[x,y],{x,y}]==0;
bc={u[0,y]==c,u[x,0]==u[a,y]==u[x,b]==0};
```

We look for an analytical solution.

```
sol=DSolve[{harmonic,bc},u[x,y],{x,y}∈dom]/.K[1]→n//
Flatten
```

$$\{u[x,y] \to \sum_{n=1}^{\infty} -\frac{2(-1+(-1)^n)c\,\text{Csch}[\frac{a\pi n}{b}]\text{Sin}[\frac{\pi y n}{b}]\text{Sinh}[\frac{\pi(a-x)n}{b}]}{\pi n}\}$$

We select the active part of the first 50 terms,

```
solf:=Activate[u[x,y]/.sol/.∞ →50]
```

{

,

$$\frac{1}{2}\left(\sqrt{x-iy}+\sqrt{x+iy}\right)$$

}

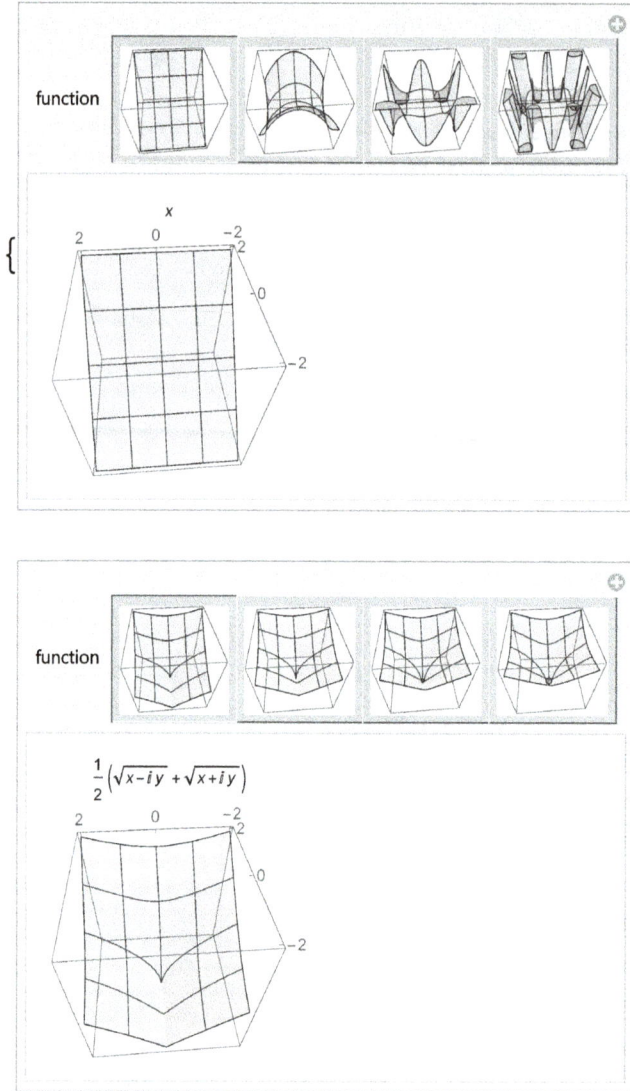

Figure 8.4: Certain power and radical functions.

set the parameters, and plot the solution.

```
Block[{a=1,c=4,b=4},
Plot3D[solf,{x,0,a},{y,0,b},Mesh→5,PlotRange→All,
PlotPoints→50,PlotStyle→{Gray,Opacity[.2]},
ImageSize→180,Ticks→{{0,a},{0,b},{0,c}}]]
```

See Figure 8.5.

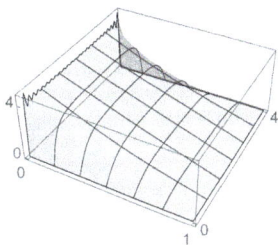

Figure 8.5: Analytical solution for problem 8.4.

8.5. We find a solution to the following boundary value problem to Laplace equation defined on a rectangle:

$$\begin{cases} \nabla^2_{x,y} u(x,y) = 0, & 0 < x < a, \quad 0 < y < b, \\ u(0,y) = c\,y(b-y), & u(a,y) = u(x,0) = u(x,b) = 0. \end{cases}$$

8.5

```
Clear[a,b,c,x,y,n]
```

The domain of integration is a rectangle.

```
dom=ImplicitRegion[0<x≤a&&0<y≤b,{x,y}];
harmonic=Laplacian[u[x,y],{x,y}]==0;
bc={u[0,y]==cy(b-y),u[x,0]==u[a,y]==u[x,b]==0};
```

The analytical solution of the problem follows.

```
sol=DSolve[{harmonic,bc},u[x,y],{x,y}]/.{K[1]→n}//
Flatten
```

$$\{u[x,y]\to \sum_{n=1}^{\infty} -\frac{2(-2+2(-1)^n)b^2 c\, \text{Csch}[\frac{a\pi n}{b}]\text{Sin}[\frac{n\pi y}{b}]\text{Sinh}[\frac{n\pi (a-x)}{b}]}{\pi^3 n^3}\}$$

We consider a partial sum of the series,

```
solf:=Activate[sol[[1,2]]]/.{∞ →50}]
```

introduce some values of the parameters, and plot the figure of the problem.

```
Block[{a=2,c=10,b=5},
Plot3D[solf,{x,0,a},{y,0,b},PlotPoints→25,Mesh→5,
PlotStyle→{Gray,Opacity[.2]},PlotRange→All,
ImageSize→175,Ticks→{{0,a},{0,b},{0,60}}]]
```

See Figure 8.6.

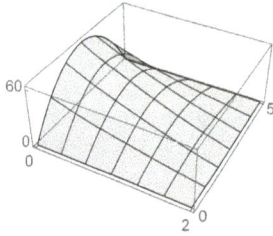

Figure 8.6: Analytical solution for problem 8.5.

8.6. We look for the solution to the following boundary value problem for the Laplace equation:

$$\begin{cases} \nabla^2_{x,y} u(x,y) = 0, & x \in [0,10], \quad y \in [0,30], \\ u(0,y) = 0, \quad u(10,y) = 0, \quad u(x,0) = -\sin(x\pi/30) = -u(x,30). \end{cases}$$

8.6

```
Clear[a,b,x,ξ,y,υ]
```

We define the problem:

```
eqncond={∇²{x,y}u[x,y]==0,u[0,y]==0,u[10,y]==0,
u[x,0]==-Sin[xπ/30],u[x,30]==Sin[xπ/30]};
```

and solve it. Because the analytical solution is given as a series, a partial sum is considered [56, p. 110].

```
sol=DSolve[eqncond,u,{x,y}]/.{∞ →m,K[1]→n}//First
```

$$\{u\rightarrow Function[\{x,y\},$$
$$\sum_{n=1}^{m}\left(-\frac{1}{\pi-9\pi n^2}9(-1)^n\sqrt{3}\,Csch[3\pi\,n]n\,Sin[\tfrac{\pi x n}{10}]\,Sinh[\tfrac{\pi(30-y)n}{10}]\right.$$
$$+\frac{1}{\pi-9\pi n^2}9(-1)^n\sqrt{3}\,Csch[3\pi\,n]n\,Sin[\tfrac{\pi x n}{10}]Sinh\,[\tfrac{\pi y n}{10}])]\}$$

The number of terms in the partial sum is restricted to 100.

```
asol[x_,y_]:=Activate[u[x,y]/.sol/.m→100]
```

The partial sum is plotted with two options Automatic and All.

```
Plot3D[asol[x,y],{x,y}∈Rectangle[{0,0},{10,30}],
Mesh→3,PlotStyle→{Gray,Opacity[.2]},
Ticks→{5{0,2},5{0,6},Automatic},ImageSize→175,
PlotLabel→Style[#,Bold,Black,12],
PlotRange→#]&/@{Automatic,All}
```

See Figure 8.7.

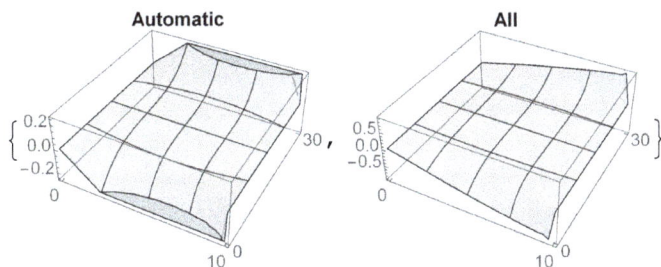

Figure 8.7: Plot of the partial sum of the solution for problem 8.6.

The partial sum is plotted in a dynamical way.

```
Manipulate[
Plot3D[asol[x,y],{x,0,a},{y,0,b},Mesh→3,
Ticks→{5{0,1,2},5{0,3,6},Automatic},ImageSize→175,
PlotStyle→{Gray,Opacity[.2]}],{a,5,10},{b,10,30},
SaveDefinitions→True]
```

See Figure 8.8.

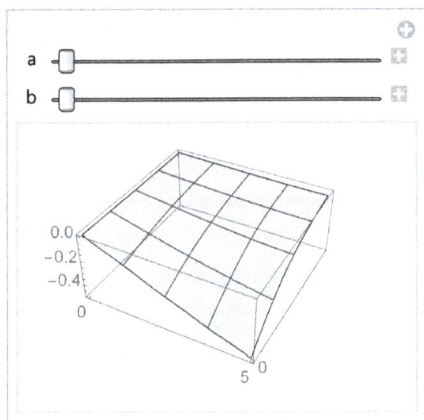

Figure 8.8: The partial sum in a dynamical way for problem 8.6.

8.7. We find a solution to the following Laplace equation on a rectangle with Dirichlet conditions:

$$\begin{cases} \nabla^2_{x,y}u(x,y) = 0, & x \in [0,10], \quad y \in [0,30], \\ u(0,y) = \sin(\frac{y\pi}{30}), \quad u(10,y) = -\sin(y(1-\frac{y}{30})), \\ u(x,0) = u(x,30) = 0. \end{cases}$$

8.7

```
Clear[m,n,x,ξ,y,υ]
```

We define the problem

```
eqncond={V²_{x,y}u[x,y]==0,u[0,y]==Sin[y π/30],
u[10,y]==-Sin[y(30-y)/30],u[x,0]==u[x,30]==0};
```

and solve it by the power of *Mathematica*.

```
sol=DSolve[eqncond,u,{x,y}]/.{∞ →m,K[1]→n}//First
```

$\{u \rightarrow \text{Function}[\{x,y\}, \sum_{n=1}^{m} -\sqrt{\frac{\pi}{15}}\text{Csch}[\frac{\pi n}{3}]\text{Sin}[\frac{\pi n}{2}]\text{Sin}[\frac{\pi y n}{30}]$

$(\text{Cos}[\frac{900+\pi^2 n^2}{120}](\text{FresnelS}[\frac{-30+\pi n}{2\sqrt{15}\pi}]-\text{FresnelS}[\frac{30+\pi n}{2\sqrt{15}\pi}])$

$+(-\text{FresnelC}[\frac{-30+\pi n}{2\sqrt{15}\pi}]+\text{FresnelC}[\frac{30+\pi n}{2\sqrt{15}\pi}])\text{Sin}[\frac{900+\pi^2 n^2}{120}])$

$\text{Sinh}[\frac{\pi x n}{30}]]\}$

We consider the partial sum of 100 terms.

```
asol[x_,y_]:=Activate[u[x,y]/.sol/.m→100]
```

and plot it.

```
{Plot3D[asol[x,y],{x,y}∈Rectangle[{0,0},{10,30}],
Mesh→3,PlotStyle→{Gray,Opacity[.2]},
Ticks→{5{0,2},5{0,6},Automatic},ImageSize→160,
PlotLabel→Style[#,Black,11],PlotRange→#]
&/@{Automatic,All}
```

See Figure 8.9.

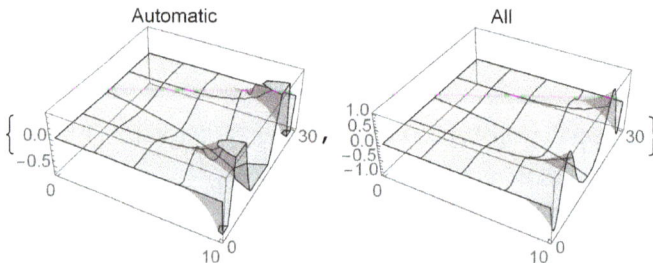

Figure 8.9: Plot of the partial sum of the solution for problem 8.7.

8.8. Let us find a solution to the following Laplace equation with three boundary conditions:

$$\begin{cases} \nabla^2_{x,y}u(x,y) = 0, & x \in [0,10], \quad y \in [0,30], \\ u(0,y) = -1, \quad u(10,y) = 1/2, \quad u(x,0) = -\cos(x\,\pi/30)) = -u(x,30). \end{cases}$$

8.8

```
Clear[a,b,x,y,m,n]
```

We define the problem

```
eqncond={V²{x,y}u[x,y]==0,u[0,y]==-1,u[10,y]=1/2,
u[x,0]==-Cos[x π/30],u[x,30]==Cos[x π/30]};
```

and look for an analytic solution.

```
sol=DSolve[eqncond,u,{x,y}]/.{∞ →m,K[1]→n}//First
```

$$\{u \rightarrow \text{Function}[\{x,y\},$$
$$\sum_{n=1}^{m} \left(\frac{2(-1+(-1)^n)\text{Csch}[\frac{\pi n}{3}]\text{Sin}[\frac{\pi y n}{30}]\text{Sinh}[\frac{\pi(10-x)n}{30}]}{\pi n} \right.$$
$$-(-1+(-1)^n)\frac{\text{Csch}[\frac{\pi n}{3}]\text{Sin}[\frac{\pi y n}{30}]\text{Sinh}[\frac{\pi x n}{30}]}{\pi n}$$
$$+\frac{\text{Csch}[3\pi n](90n-45(-1)^n n)\text{Sin}[\frac{\pi x n}{10}]\text{Sinh}[\frac{\pi(30-y)n}{10}]}{5(\pi-9\pi n^2)}$$
$$\left. -\frac{3\text{Csch}[3\pi n](-6n+3(-1)^n n)\text{Sin}[\frac{\pi x n}{10}]\text{Sinh}[\frac{\pi y n}{10}]}{\pi(-1+9n^2)} \right)]\}$$

We set the partial sum of 10 terms

```
asol[x_,y_]:=Activate[u[x,y]/.sol/.m→10]
```

and plot it.

```
{Plot3D[asol[x,y],{x,y}∈Rectangle[{0,0},{10,30}],##,
Ticks→{5{0,1,2},15{0,1,2},{-1,0,1}},ImageSize→160],
Manipulate[
Plot3D[asol[x,y],{x,0,a},{y,0,b},##,ImageSize→160,
Ticks→{5{0,1,2},15{0,1,2},Automatic}],{a,5,10},
{b,10,30},SaveDefinitions→True,DefaultDuration→30]}
&[PlotStyle→{Gray,Opacity[.2]},PlotPoints→50,
Mesh→3]
```

See Figure 8.10.

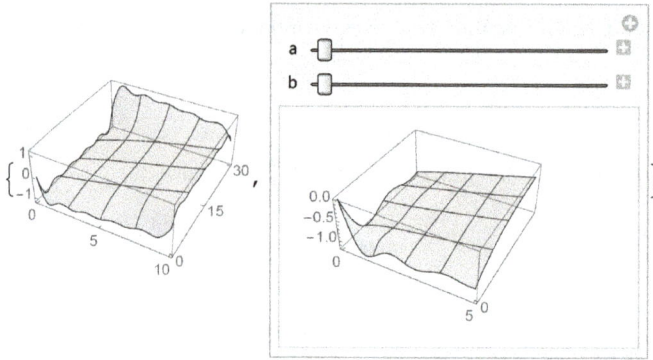

Figure 8.10: Plot of the partial sum of the solution for problem 8.8.

8.9. We find a solution to the Dirichlet problem to Laplace equation defined on a rectangle [64, p. 149],

$$\begin{cases} \nabla^2_{x,y} u(x,y) = 0, & 0 < x < a, \quad 0 < y < b, \\ u(0,y) = c\sin(\frac{\pi y}{b}), & u(a,y) = 0, \quad u(x,0) = d\sin(\frac{\pi x}{a}), \quad u(x,b) = 0. \end{cases}$$

8.9

```
Clear[a,b,c,d,x,y]
```

We introduce the problem:

```
dom=ImplicitRegion[0<x≤a&&0<y≤b,{x,y}];
cond={DirichletCondition[u[x,y]==c Sin[π y/b],x==0],
DirichletCondition[u[x,y]==0,x==a∨y==b],
DirichletCondition[u[x,y]==d Sin[π x/a],y==0]};
```

and try finding an analytical solution.

```
sol=DSolveValue[{Laplacian[u[x,y],{x,y}]==0,cond},
u[x,y],{x,y}]
```

$$\sum_{K[1]=1}^{(x)} 0$$

Two boundary conditions are not satisfied, so the solution returned by the system is not correct.

Therefore, we try finding a numerical solution of this problem.

```
Block[{a=2,b=5,c=10,d=20}, (* Set the parameters *)
soln=NDSolveValue[{Laplacian[u[x,y],{x,y}]==0,cond},u,
{x,y}∈dom]; (* Find the numerical solution *)
Plot3D[soln,{x,0,a},{y,0,b},PlotPoints→40,
PlotStyle→{Gray,Opacity[.15]},ImageSize→175,Mesh→5,
Ticks→Outer[Times,{1,2.5,10},{0,2}]]]
```

See Figure 8.11.

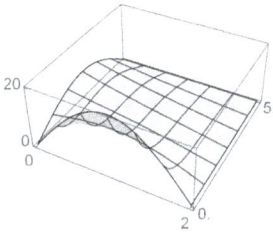

Figure 8.11: Numerical solution for problem 8.9.

We check the boundary conditions.

```
Block[{a=2,b=5,c=10,d=20},
{Plot[(soln-c Sin[π y/b])/.x→0,{y,0,5},##],
Plot[(soln-d Sin[π x/a])/.y→0,{x,0,2},##],
Plot[soln/.x→a,{y,0,5},##],
Plot[soln/.y→b,{x,0,2}]}&[PlotStyle→Black,
ImageSize→175]]
```

See Figure 8.12.

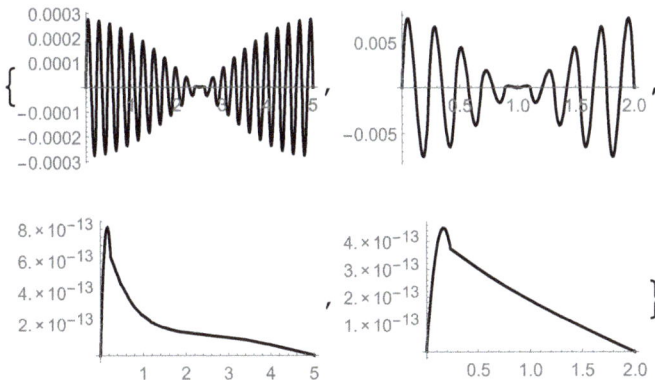

Figure 8.12: Check of the boundary conditions for problem 8.9.

8.10. Let us find a solution of the Neumann problem to the Laplace equation defined on a rectangle [64, p. 150],

$$\begin{cases} \nabla^2_{x,y} u(x,y) = 0, & 0 < x < a, \quad 0 < y < b, \\ \partial_n u(0,y) = \beta, \quad \partial_n u(a,y) = 0, \quad \partial_n u(x,0) = \alpha, \quad \partial_n u(x,b) = 0. \end{cases}$$

8.10

```
Clear[a,α,b,x,y]
```

We define the domain of the problem

```
dom=ImplicitRegion[0≤x≤a&&0≤y≤b,{x,y}];
```

and try finding an analytical solution.

```
β=a α/b
DSolve[{Laplacian[u[x,y],{x,y}]
+NeumannValue[α,y==0]+NeumannValue[0,y==b]
+NeumannValue[β,x==0]+NeumannValue[0,x==a]},u,
{x,y}∈dom]  (* No closed-form solution is returned *)
```

We add a Dirichlet condition to get a unique solution and plot it.

```
Block[{a=10,α=4,b=6},
sol=NDSolveValue[{Laplacian[u[x,y],{x,y}]
+NeumannValue[α,y==0]+NeumannValue[0,y==b]
+NeumannValue[β,x==0]+NeumannValue[0,x==a],
u[x,0]==Sin[2x]},u,{x,y}∈dom];
Plot3D[sol[x,y],{x,0,a},{y,0,b},PlotPoints→25,Mesh→5,
PlotRange→All,Ticks→Outer[Times,{5,3,-15},{0,2}],
PlotStyle→{Gray,Opacity[.2]},ImageSize→{180,160}]]
```

See Figure 8.13.

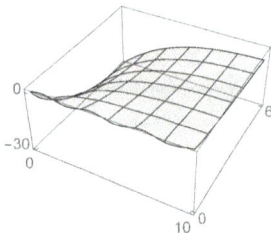

Figure 8.13: Numerical solution for problem 8.10.

8.3 Laplace equations on arbitrary domains

8.11. We find the solution to the Neumann problem of the Laplace equation defined on a disk [64, p. 150],

$$\nabla^2_{x,y} u(x,y) = 0, \quad x^2 + y^2 \leq 4, \quad \partial_n u(x,y) = x + y.$$

8.11

```
Clear[x,y]
```

Here is the domain of the problem.

```
dom=Assuming[{x,y}∈Reals,
ImplicitRegion[x²+y² ≤4,{x,y}]];
```

The uniqueness of the solution requires an extra condition, so we add a Dirichlet condition.

```
eqncond={Laplacian[u[x,y],{x,y}]==
NeumannValue[x+y,x²+y²==4],
DirichletCondition[u[x,y]==0,y==0]};
```

We look for an analytical solution.

```
DSolveValue[eqncond,u[x,y],{x,y}∈dom]; (* No
closed-form solution is returned *)
```

Then we look for a numerical solution

```
sol=NDSolveValue[eqncond,u,{x,y}∈dom];
```

and plot it.

```
Plot3D[sol[x,y],{x,-2,2},{y,-2,2},PlotPoints→25,
Mesh→3,ImageSize→175,PlotRange→All,
RegionFunction→Function[{x,y,z},First[dom]],
Ticks→Outer[Times,{2,2,5},{-1,0,1}],
PlotStyle→{Gray,Opacity[.2]}]
```

See Figure 8.14.

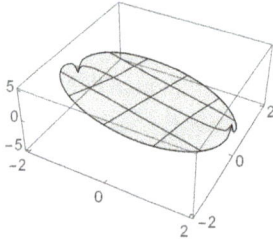

Figure 8.14: Numerical solution for problem 8.11.

8.12. We find a solution to the following Dirichlet problem for the elliptic partial differential equation in polar coordinates,

$$\begin{cases} \frac{1}{r}\frac{\partial}{\partial r}(r\frac{\partial u(r,t)}{\partial r}) + \frac{1}{r^2}\frac{\partial^2 u(r,t)}{\partial t^2} = 0, & -\pi \le t \le \pi, \quad 0 \le r \le 1, \\ u(1,t) = \sin(t). \end{cases}$$

8.12

```
Clear[t,r]
```

We set the problem

eqncond={$\frac{1}{r}\frac{\partial}{\partial r}$(r$\frac{\partial u[r,t]}{\partial r}$) + $\frac{1}{r^2}\frac{\partial^2 u[r,t]}{\partial t^2}$==0,u[1,t]==Sin[t]};

and look for an analytical solution.

```
DSolve[eqncond,u,{r,t}];  (* No closed-form solution
is returned *)
```

Now we look for a numerical solution

```
sol=NDSolve[eqncond,u,{r,.00001,1},{t,-π,π},
PrecisionGoal→4];
```

and plot it

```
Plot3D[u[r,t]/.First[sol],{r,.0001,1},{t,-π,π},
PlotStyle→{Gray,Opacity[.2]},Mesh→3,ImageSize→175,
Ticks→{{0,.5,1},π{-1,0,1},{-1,0,1}}]
```

See Figure 8.15.
    We pass to the polar coordinates and plot the figure.

```
Plot3D[u[r,t]/.sol/.{r→ √x² + y²,t→ArcTan[x,y]}
//Evaluate,{x,y}∈Disk[],PlotPoints→50,ImageSize→175,
PlotStyle→{Gray,Opacity[.2]},PlotRange→All,Mesh→2,
Ticks→Outer[Times,{1,1,1},{-1,0,1}]]
```

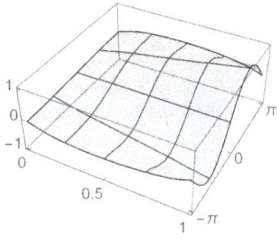

Figure 8.15: Numerical solution for problem 8.12.

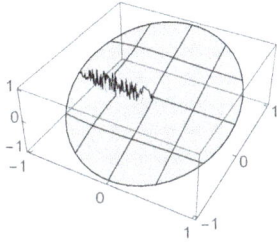

Figure 8.16: Numerical solution in polar coordinates for problem 8.12.

See Figure 8.16.

We also give a dynamical plot.

```
Manipulate[
Plot3D[u[r,t]/.sol/.r→Sqrt[x²+y²]//Evaluate,
{x,y}∈Disk[],PlotStyle→{Gray,Opacity[.2]},Mesh→3,
PlotRange→All,ImageSize→200,
Ticks→{{-1,1},{-1,0,1},Automatic}],{t,-π,π},
SaveDefinitions→True]
```

See Figure 8.17.

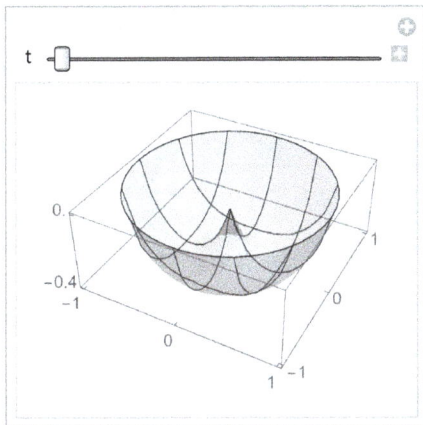

Figure 8.17: Dynamical numerical solution in polar coordinates for problem 8.12.

8.13. Let us find a solution to the following Dirichlet problem for the elliptic partial differential equation in polar coordinates:

$$\begin{cases} \frac{1}{r}\frac{\partial}{\partial r}\left(r\frac{\partial u(r,t)}{\partial r}\right) + \frac{1}{r^2}\frac{\partial^2 u(r,t)}{\partial t^2} = 0, & -\pi \le t \le \pi, \quad 0 \le r \le 1, \\ u(1,t) = t^2/2. \end{cases}$$

8.13

```
Clear[t,r,u]
```

We set the problem

```
eqncond={1/r ∂/∂r (r ∂u[r,t]/∂r) + 1/r² ∂²u[r,t]/∂t²==0,u[1,t]==t²/2};
```

and look for an analytical solution.

```
asol=DSolve[eqncond,u,{r,t}]//Flatten
{u→Function[{r,t}, π²/6 +∑∞_K[1]=1 (2(-1)^K[1] r^K[1] Cos[tK[1]])/K[1]² ]}
```

```
analyticalu[r_,t_]:=Activate[u[r,t]/.asol]/.{K[1]→n,
∞ →50}
analyticalu[r,t]
π²/6 +PolyLog[2,-e^{-it}r]+PolyLog[2,-e^{it}r]
```

Remark. We note that

$$\frac{\pi^2}{6}+\text{PolyLog}[2,-e^{-it}r]+\text{PolyLog}[2,-e^{it}r]=\frac{\pi^2}{6} + \sum_{k=1}^{\infty}\frac{(-1)^k 2\cos[k\,t]r^k}{k^2}$$

Thus the series in the right-hand side of analyticalu[r,t] converges uniformly and absolutely to a real number whenever $r \le 1$. □

We plot the solution.

```
Plot3D[analyticalu[r,t]/.{r→Sqrt[x²+y²],
t→ArcTan[x,y]}//Evaluate,{x,y}∈Disk[],Mesh→3,
PlotPoints→50,ImageSize→175,PlotRange →All,
Ticks→Outer[Times,{1,1,5},{-1,0,1}],
PlotStyle→{Gray,Opacity[.2]}]
```

See Figure 8.18.

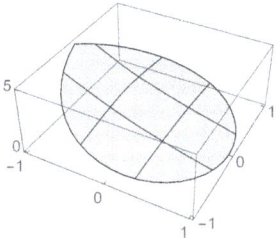

Figure 8.18: Analytical solution in polar coordinates for problem 8.13.

```
Manipulate[
Plot3D[analyticalu[r,t]/.{r→Sqrt[x²+y²],
t→ArcTan[x,y]}//Evaluate,{x,y}∈Disk[],Mesh→2,
PlotPoints→25,PlotStyle→{Gray,Opacity[.2]},
Ticks→Outer[Times,{1,1,1},{-1,0,1}],
PlotRange→All,ImageSize→175],{t,0,π},
SaveDefinitions→True]
```

See Figure 8.19.

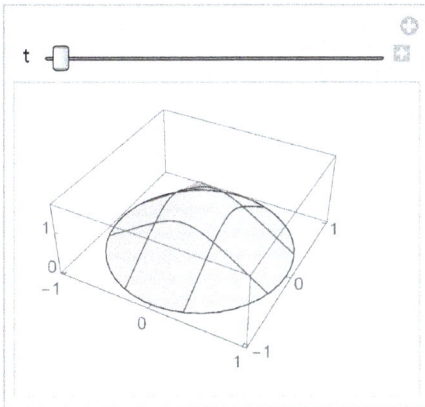

Figure 8.19: Dynamical solution in polar coordinates for problem 8.13.

We study the same problem by numerical methods

```
soln=NDSolveValue[eqncond,u,{r,.00001,1},{t,-π,π},
PrecisionGoal→4]
```

and plot the figures.

```
Plot3D[soln[r,t],{r,.0001,1},{t,-π,π},Mesh→3,
PlotStyle→{Gray,Opacity[.2]},ImageSize→175,
Ticks→{{0,1},π{-1,1},{0,5}}]
```

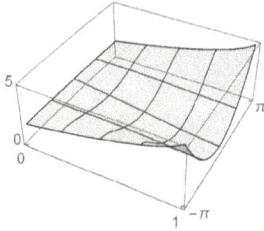

See Figure 8.20.

```
Plot3D[soln[r,t]/.{r→Sqrt[x²+y²],t→ArcTan[x,y]}
//Evaluate,{x,y}∈Disk[],Mesh→2,PlotPoints→50,
PlotStyle→{Gray,Opacity[.2]},PlotRange→All,
Ticks→Outer[Times,{1,1,5},{-1,0,1}],
ImageSize→175]
```

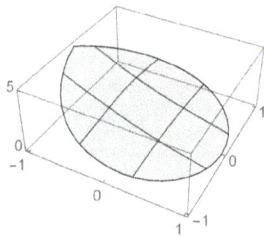

Figure 8.21: Numerical solution in polar coordinates for problem 8.13.

See Figure 8.21.

```
Manipulate[
Plot3D[soln[r,t]/.{r→Sqrt[x²+y²],t→ArcTan[x,y]}
//Evaluate,{x,y}∈Disk[],Mesh→3,ImageSize→175
PlotStyle→{Gray,Opacity[.2]},PlotRange→All,
Ticks→Outer[Times,{1,1,5},{-1,0,1}]],{t,-π,π},
SaveDefinitions→True]
```

See Figure 8.22.

8.14. We find a solution to the following elliptic partial differential equation in polar coordinates on a circular annulus:

$$\begin{cases} \frac{1}{r}\frac{\partial}{\partial r}(r\frac{\partial u(r,t)}{\partial r}) + \frac{1}{r^2}\frac{\partial^2 u(r,t)}{\partial t^2} = 0, & -\pi \le t \le \pi, \quad 1 \le r \le 2, \\ u(1,t) = 1 + \cos^2(t), & u(2,t) = 1 - \cos^2(t). \end{cases}$$

8.14

```
Clear[t,r]
```

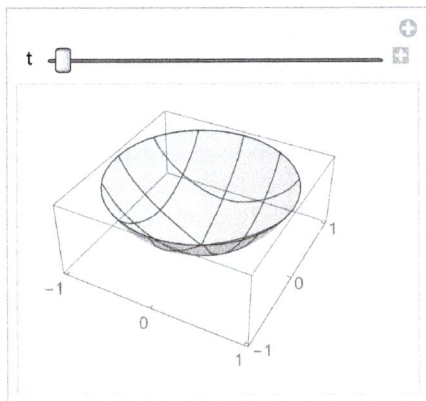

Dynamical numerical solution in polar coordinates for problem 8.13.

```
R=ParametricRegion[{r Cos[θ],r Sin[θ]},{θ,0,2π},
{r,1,2}];
RegionPlot[R,ImageSize→100]
(* Domain of integration of the problem *)
```

See Figure 8.23.

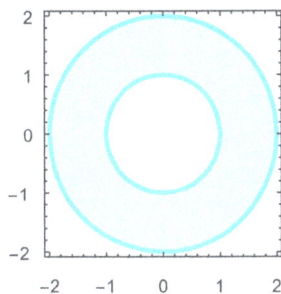

Figure 8.23: Annulus for problem 8.14.

We introduce the problem

$$eqncond=\{\frac{u^{(0,2)}[r,t]}{r^2} + \frac{u^{(1,0)}[r,t]+r\,u^{(2,0)}[r,t]}{r}==0,$$

```
u[1,t]==1+Cos[t]², u[2,t]==1-Cos[t]²};
sol=DSolveValue[eqncond,u[r,t],{r,t}]//Flatten
```

$$\left\{u[r,t] \rightarrow \left\{\begin{array}{ll} \frac{3}{2} - \frac{\text{Log}[r]}{\text{Log}[2]} & 1 \le r \le 2 \\ \text{Indeterminate} & \text{True} \end{array}\right.\right\}$$

and remark that the proposed solution is not correct because it does not satisfy any boundary condition.

Then we try finding a numerical solution

```
nsol=NDSolve[eqncond,u,{r,1,2},{t,-Pi,Pi}]//Flatten;
```

and plot it.

```
Plot3D[Evaluate[u[r,t]/.nsol/.{r→Sqrt[x²+y²],
t→ArcTan[x,y]}],{x,y}∈Annulus[{0,0},{1,2}],PlotRange→
All,PlotStyle→{Gray,Opacity[.2]},ImageSize→170,
Mesh→2,Ticks→Outer[Times,2{1,1,1},{-1,0,1}],
Exclusions→None]
```

See Figure 8.24.

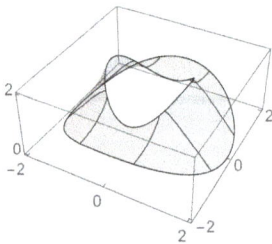

Figure 8.24: Numerical solution for problem 8.14.

8.15. Let us find a solution to the following Dirichlet problem to Laplace equation on a disk with holes:

$$\begin{cases} \nabla^2_{x,y}u(x,y) = 0, & 0 \le (x^2+y^2)^2 - 8(x^2-y^2) + c \,\&\&\, x^2+y^2 \le 25, \quad c \in \mathbb{R}, \\ u(x,y) = \begin{cases} 0, & x^2+y^2 = 25, \\ 100, & (x^2+y^2)^2 - 8(x^2-y^2) + c = 0. \end{cases} \end{cases}$$

8.15

```
Clear[x,y]
```

```
{dom=ImplicitRegion[0≤(x²+y²)²-8(x²-y²)+#&&x²+y² ≤25,
{x,y}];
boundary={DirichletCondition[u[x,y]==100,
(x²+y²)²-8(x²-y²)+#==0],
DirichletCondition[u[x,y]==0,x²+y²==25]};
DSolveValue[{∇²_{x,y}u[x,y]==0,boundary},u,{x,y}]; (* No
closed-form solution is returned *)
sol=NDSolveValue[{∇²_{x,y}u[x,y]==0,boundary},u,
{x,y}∈dom];
Plot3D[sol[x,y],{x,y}∈dom,PlotPoints→25,Mesh→2,
PlotStyle→{Gray,Opacity[.15]},ImageSize→140]}&/@
{3,0,-6}//Flatten
```

See Figure 8.25.

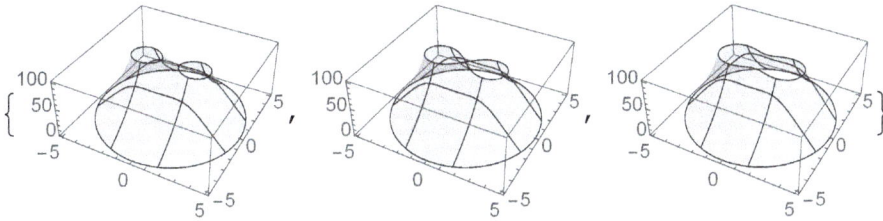

Figure 8.25: Numerical solution for problem 8.15.

Remark. Each innermost contour is a *lemniscate of Bernoulli*. □

8.16. We find a solution to the following Dirichlet and Neumann problem to the Laplace equation defined on a lemniscate:

$$\begin{cases} \nabla^2_{x,y} u(x,y) = 0, & \Omega = \{(x,y) \mid (x^2+y^2)^2 - (x^2-y^2) \le 0\} \subset \mathbb{R}^2, \\ u(x,y) = \begin{cases} \sin(5\pi xy), & x < 0, y < 0, \\ \cos(x+y), & x < 0, y > 0, \end{cases} & \partial_n u(x,y) = \begin{cases} 1, & x > 0, y > 0, \\ -1, & x > 0, y < 0. \end{cases} \end{cases}$$

8.16

```
Clear[x,y]
```

The domain of integration of the problem follows.

```
boundlemniscate=(x²+y²)²-(x²-y²);
Ω=ImplicitRegion[boundlemniscate≤0,{x,y}];
```

We look for a numerical solution

```
solve=NDSolveValue[{∇²_{x,y}u[x,y]==
NeumannValue[1,x>0&&y>0]+NeumannValue[-1,x>0&&y<0],
DirichletCondition[u[x,y]==Sin[5π x y],x<0&&y<0],
DirichletCondition[u[x,y]==Cos[x+y],x<0&&y>0]},
u,{x,y}∈ Ω];
```

and plot it.

```
Plot3D[solve[x,y],{x,y}∈ Ω,ImageSize→175,Mesh→3,
PlotStyle→Directive[Gray,Opacity[.15]],
Ticks→{{-1,1},.3{-1,1},{-.5,1}}]
```

See Figure 8.26.

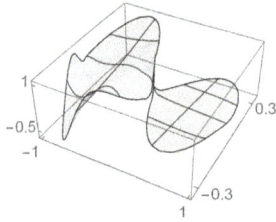

Figure 8.26: Numerical solution for problem 8.16.

8.17. We look for a solution to the Dirichlet and Neumann problem of the Laplace equation defined on the astroid in the first quadrant,

$$
\begin{cases}
\nabla^2_{x,y} u(x,y) = 0, & \Omega = \{(x,y) \mid x^{2/3} + y^{2/3} - 1 \le 0\} \subset \mathbb{R}^2, \\
u(x,y) = \begin{cases} 0, & x = 0, \\ 0, & y = 0, \end{cases} & \partial_n u(x,y) = -x^2 y^2 \text{ on the curvilinear arc.}
\end{cases}
$$

8.17

```
Clear[x,y]
```

```
boundastro={-x,-y,x^(2/3)+y^(2/3)-1};
```

We specify the region

```
Ω=ImplicitRegion[And@@(#≤0&/@boundastro),{x,y}];
```

and then the domain of integration

```
Show[RegionPlot[Ω],
ContourPlot[Evaluate[Thread[boundastro==0]],{x,0,1},
{y,0,1},ContourStyle→Black],ImageSize→130,
PlotRange→{{-.01,1},{-.01,1}}]
```

See Figure 8.27.

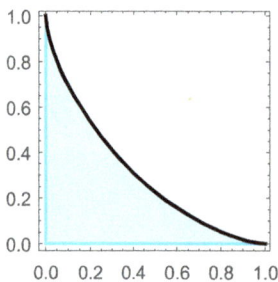

Figure 8.27: Domain of the solution for problem 8.17.

Then we solve the boundary value problem of Laplace equation on this domain.

```
nsolv=NDSolveValue[{V²_{x,y}u[x,y]==
NeumannValue[-x²y²,boundastro[[3]]==0],
DirichletCondition[u[x,y]==0,boundastro[[1]]==0],
DirichletCondition[u[x,y]==0,boundastro[[2]]==0]},
u,{x,y}∈ Ω,Method→{"FiniteElement","MeshOptions"→
{"BoundaryMeshGenerator"→"Continuation"}}];
```

The plot of the figure follows.

```
Plot3D[nsolv[x,y],{x,y}∈ Ω,Mesh→3,ImageSize→175,
PlotStyle→{Gray,Opacity[.2]},PlotRange→All,
Ticks→{{0,1},{0,1},{0,.003}}]
```

See Figure 8.28.

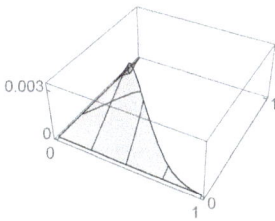

Figure 8.28: Numerical solution for problem 8.17.

8.18. One finds a solution of the Neumann and Dirichlet problem of the Laplace equation defined on an annulus [64, 10b, p. 150]. The Neumann conditions are considered successively,

$$\begin{cases} \nabla^2_{x,y}u(x,y) = 0, & 1 \le x^2 + y^2 \le 4, \\ u|_{x^2+y^2=1}(x,y) = 0, & \partial_{n\ x^2+y^2=4}\,u(x,y) = \begin{cases} x + y, \\ x^2 - y^2, \\ \sin(x)\cos(y). \end{cases} \end{cases}$$

8.18

```
Clear[x,y]
```

```
dom=Assuming[{x,y}∈Reals,  (* Domain of integration *)
ImplicitRegion[1≤x²+y² ≤4,{x,y}]];
RegionPlot[dom,ImageSize→100]
```

See Figure 8.29.

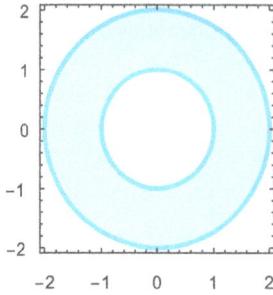

Figure 8.29: Domain of integration for problem 8.18.

We look for some solutions.

```
{eqncond={Laplacian[u[x,y],{x,y}]==
NeumannValue[#,x²+y²==4],
DirichletCondition[u[x,y]==0,x²+y²==1]};
DSolveValue[eqncond,u[x,y],{x,y}∈dom]; (* No
closed-form solution is returned *)
sol=NDSolveValue[eqncond,u,{x,y}∈dom]//Chop;
Plot3D[sol[x,y],{x,y}∈dom,PlotPoints→25,Mesh→5,
PlotRange→All,ImageSize→160,PlotStyle→{Gray,
Opacity[.2]},Ticks→Outer[Times,2{1,1,1},{-1,0,1}]]}
&/@{x+y,x²-y²,Sin[x]Cos[y]}//Flatten
```

See Figure 8.30.

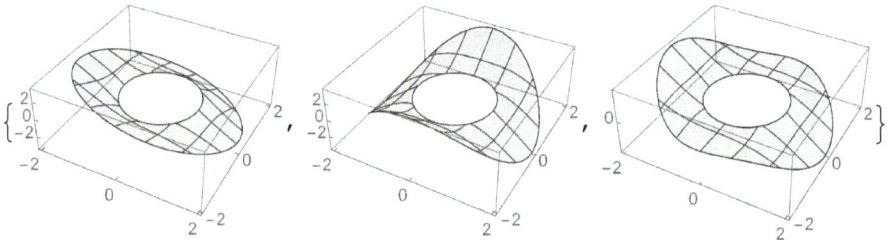

Figure 8.30: Solutions for problem 8.18.

8.4 Poisson equations on rectangles

8.19. We find a solution to the Neumann problem of a Poisson equation defined on a rectangle [64, 9, p. 150],

$$\begin{cases} \nabla^2_{x,y} u(x,y) = \alpha, & 0 < x < a, \quad 0 < y < b, \\ \partial_n u(x,y) = \beta & \text{on the boundary of } [0,a] \times [0,b]. \end{cases}$$

8.19

```
Clear[a,b,α,β,x,y]
```

We define the domain of integration.

```
dom=ImplicitRegion[0≤x≤a&&0≤y≤b,{x,y}];
```

```
DSolveValue[{Laplacian[u[x,y],{x,y}]==α
+NeumannValue[β,x==0||x==a||y==0||y==b]},u,{x,y}∈dom];
(* No closed-form solution is returned *)
```

```
a=1;α=-2;b=4;β=a b α/(2(a+b)); (* Set the
parameters *)
```

We add an extra condition to get a unique solution, and solve the problem numerically,

```
sol=NDSolveValue[{Laplacian[u[x,y],{x,y}]==α
+NeumannValue[β,x==0||x==a||y==0||y==b],
DirichletCondition[u[x,y]==0,x==0]},u,{x,y}∈dom];
```

and plot the solution.

```
Plot3D[sol[x,y],{x,0,a},{y,0,b},PlotPoints→50,
PlotStyle→{Gray,Opacity[.2]},Mesh→3,PlotRange→All,
Ticks→Outer[Times,{.5,2,1.1},{0,2}],ImageSize→175]
```

See Figure 8.31.

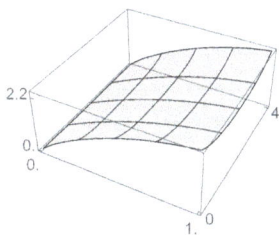

Figure 8.31: Numerical solution for problem 8.19.

8.5 Poisson equations on arbitrary domains

8.20. One finds a solution to the Dirichlet problem of the Poisson equation defined on the unitary disk,

$$\begin{cases} \nabla^2_{x,y}u(x,y) = \sin(x y), & x^2 + y^2 \le 1, \\ u(x,y)|_{x^2+y^2=1} = \sin(2\pi(x + y)). \end{cases}$$

8.20

```
Clear[x,y]
```

We look for the numerical solution of the problem,

```
sol=NDSolveValue[{Laplacian[u[x,y],{x,y}]==Sin[x y],
DirichletCondition[u[x,y]==Sin[2π(x+y)],x²+y²==1]},
u,{x,y}∈Disk[]];
```

and plot it.

```
Plot3D[sol[x,y],{x,y}∈Disk[],PlotPoints→50,Mesh→5,
PlotStyle→{Gray,Opacity[.2]},ImageSize→170,
Ticks→Outer[Times,{1,1,1},{-1,1}]]
```

See Figure 8.32.

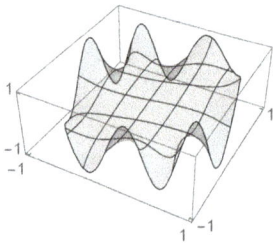

Figure 8.32: Numerical solution for problem 8.20.

8.21. We find the solutions to the Dirichlet problems of the Poisson equation successively defined on the unitary disk, on the ellipse centered at $(0,0)$ with semiaxes of lengths 1 and 2, and on an annular domain bounded by a circle and an ellipse, respectively,

$$
\begin{cases}
1. & \begin{cases} \nabla^2_{x,y}u(x,y) = \sqrt{|\sin(x+y)|}, & x^2+y^2 \le 1, \\ u(x,y)|_{x^2+y^2=1} = \exp(-xy) \end{cases} \\[2ex]
2. & \begin{cases} \nabla^2_{x,y}u(x,y) = \sqrt{|\sin(x+y)|}, & x^2+\frac{y^2}{4} \le 1, \\ u(x,y)|_{x^2+\frac{y^2}{4}=1} = \exp(-xy). \end{cases} \\[2ex]
3. & \begin{cases} \nabla^2_{x,y}u(x,y) = \sqrt{|\sin(x+y)|}, & x^2+y^2 \ge 1, \ \frac{x^2}{4}+\frac{y^2}{2} \le 1, \\ u(x,y)|_{x^2+y^2=1} = xy, \quad u(x,y)|_{\frac{x^2}{4}+\frac{y^2}{2}=1} = \exp(-xy). \end{cases}
\end{cases}
$$

8.21
1.

```
Clear[x,y]
```

We look for the numerical solution to the problem,

```
sol=NDSolveValue[{Laplacian[u[x,y],{x,y}]
==Sqrt[Abs[Sin[x+y]]],
DirichletCondition[u[x,y]==Exp[-x y],True]},u,
{x,y}∈Disk[]];
```

and plot it.

```
Plot3D[sol[x,y],{x,y}∈Disk[],PlotPoints→75,Mesh→3,
Ticks→{{-1,0,1},{-1,0,1},{1}},ImageSize→170,
PlotStyle→{Gray,Opacity[.2]}]
```

See Figure 8.33.

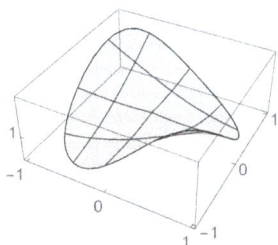

Figure 8.33: Solution for problem 8.21. 1.

2.

```
Clear[x,y]
```

We look for the numerical solution of the problem,

```
sol=NDSolveValue[{Laplacian[u[x,y],{x,y}]
==Sqrt[Abs[Sin[x+y]]],
DirichletCondition[u[x,y]==Exp[-x y],True]},
u,{x,y}∈Ellipsoid[{0,0},{1,2}]];
```

and plot it.

```
Plot3D[sol[x,y],{x,y}∈Ellipsoid[{0,0},{1,2}],
PlotPoints→75,Mesh→2,PlotStyle→{Gray,Opacity[.2]},
Ticks→{{-1,0,1},{-2,0,2},{1}},ImageSize→150,
PlotRange→All]
```

See Figure 8.34.

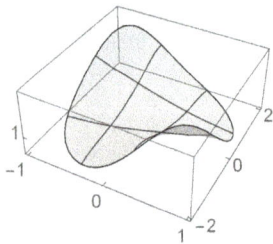

Figure 8.34: Solution for problem 8.21. 2.

3.

We define the domain of the problem

```
Clear[x,y]
dom=Assuming[{x,y}∈Reals,
ImplicitRegion[x²+y² ≥1&&x²/4+y²/2≤1,{x,y}]];
```

and plot it

```
RegionPlot[dom,AspectRatio→Automatic,ImageSize→140]
```

See Figure 8.35.

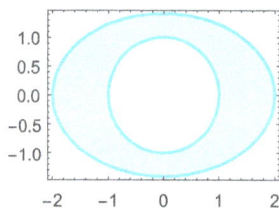

Figure 8.35: Domain for problem 8.21. 3.

We solve the problem on the annular domain.

```
sol=NDSolveValue[Laplacian[u[x,y],{x,y}
==Sqrt[Abs[Sin[x+y]]],
DirichletCondition[u[x,y]==x y,x²+y²==1],
DirichletCondition[u[x,y]==Exp[-x y],x²/4+y²/2==1],
u,{x,y}∈dom];
```

and plot the solution.

```
{Plot3D[sol[x,y],{x,y}∈dom,PlotPoints→75,
Mesh→2,PlotStyle→{Gray,Opacity[.2]}
Ticks→Outer[Times,{2,Sqrt[2],4},{-1,0,1}],
ImageSize→170,PlotRange→All]
```

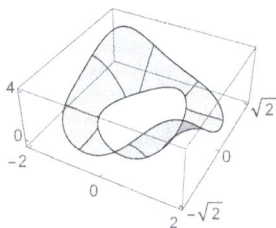

See Figure 8.36.

8.22. Let us find solutions to the Dirichlet problem of the Poisson equations defined on the unitary disk [64, 26, p. 154],

$$\begin{cases} \nabla^2_{x,y}u(x,y) = \begin{cases} (x^2 - y^2)\ln(1 + \sqrt{x^2 + y^2}), \\ (x^2 - y^2)\cos(1 + \sqrt{x^2 + y^2}), \\ (x^2 - y^2)\tan(1 + \sqrt{x^2 + y^2}), \\ (x^2 - y^2)\exp(1 + \sqrt{x^2 + y^2}), \end{cases} \quad x^2 + y^2 \leq 1, \\ u(x,y)|_{x^2+y^2=1} = 0. \end{cases}$$

8.22

```
Clear[p,x,y,u]
```

We numerically solve the problems and plot the solutions.

```
Manipulate[
sol=NDSolveValue[{Laplacian[u[x,y],{x,y}]
==(x²-y²) p[1+Sqrt[x²+y²]],
DirichletCondition[u[x,y]==0,True]},u,{x,y}∈Disk[]];
Plot3D[sol[x,y],{x,y}∈Disk[],PlotPoints∈75,Mesh→3,
PlotStyle∈{Gray,Opacity[.2]},ImageSize→175,
Ticks→Outer[Times,{1,1,.01},{-1,0,1}]],{p,{Log,Cos,
Tan,Exp},SaveDefinitions→True}]
```

See Figure 8.37.

8.23. We find a solution to the Dirichlet and Neumann problems to the Poisson equation [64, p. 154],

$$\begin{cases} \nabla^2_{x,y}u(x,y) = -200, & x^2 + y^2 \leq 1, \\ \partial_n u(x,y) = 100, & u(1,0) = 1. \end{cases}$$

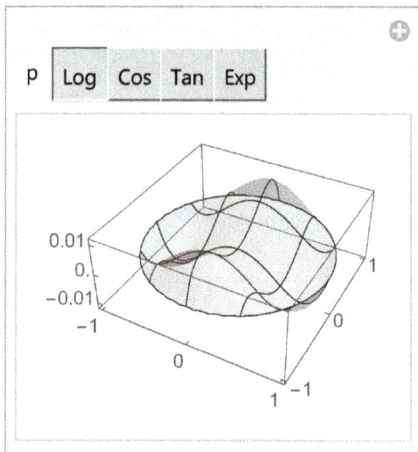

Figure 8.37: Solutions for problem 8.22.

8.23

```
Clear[x,y]
```

We solve the problem numerically,

```
sol=NDSolveValue[{Laplacian[u[x,y],{x,y}]==-200
+NeumannValue[100,True],
DirichletCondition[u[x,y]==1,x==1&&y==0]},u,
{x,y}∈Disk[]];
```

and plot its solution.

```
Plot3D[sol[x,y],{x,y}∈Disk[],PlotPoints→75,
ImageSize→160,Mesh→3,PlotRange→All,
Ticks→Outer[Times,{1,1,.01},{-1,0,1}]]
```

See Figure 8.38.

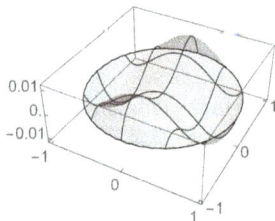

Figure 8.38: Solution for problem 8.23.

8.6 Laplace equation in higher dimension

8.24. We find a solution to the Dirichlet problem to the Laplace equation defined on a parallelepiped [64, p. 155],

$$\begin{cases} \nabla^2_{x,y,z} u(x,y,z) = 0, & 0 \le x \le a, \quad 0 \le y \le b, \quad 0 \le z \le c, \\ u(x,y,0) = x\,y, & u(x,y,c) = a\,x\,y \quad u(x,0,z) = x\,z, \\ u(x,b,z) = b\,x\,z, & u(0,y,z) = y\,z, \quad u(a,y,z) = c\,y\,z. \end{cases}$$

8.24

```
Clear[x,y,z]
```

```
a=1;b=2;c=3; (* Set the parameters *)
```

We introduce the domain of integration and plot it.

```
bound={x,a-x,y,b-y,z,c-z};
dom=ImplicitRegion[And@@(#≥0&/@bound),{x,y,z}];
```

We define the boundary conditions,

```
boundary={DirichletCondition[u[x,y,z]==x y,z==0],
DirichletCondition[u[x,y,z]==a x y,z==c],
DirichletCondition[u[x,y,z]==x z,y==0],
DirichletCondition[u[x,y,z]==b x z,y==b],
DirichletCondition[u[x,y,z]==y z,x==0],
DirichletCondition[u[x,y,z]==c y z,x==a]};
```

and try finding an analytical solution.

```
DSolveValue[{Laplacian[u[x,y,z],{x,y,z}]==0,
boundary},u,{x,y,z}∈dom]; (* No closed-form solution
is returned *)
```

Then we try finding a numerical solution, and plot its contour.

```
sol=NDSolveValue[{Laplacian[u[x,y,z],{x,y,z}]==0,
boundary},u,{x,y,z}∈dom]//Chop;
```

```
ContourPlot3D[sol[x,y,z],{x,0.,a},{y,0.,b},{z,0.,c},
PlotRange→All,Mesh→None,ImageSize→170,
Ticks→Outer[Times,.5{a,b,c},{0,1,2}],
ContourStyle→Directive[Gray,Opacity[.2]]]
```

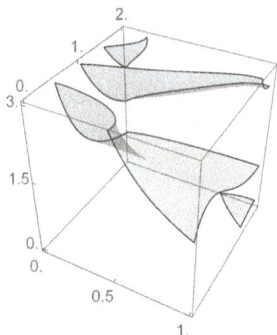

Figure 8.39: Contour of the solution for problem 8.24.

See Figure 8.39.

8.25. We find a solution to the Dirichlet problem of the Laplace equation defined on a parallelepiped [64, p. 156],

$$
\begin{cases}
\nabla^2_{x,y,z}u(x,y,z) = 0, & 0 \le x \le a, \quad 0 \le y \le b, \quad 0 \le z \le c, \\
u(x,y,z) = a\,x\,y(a-x)(b-y), & z = 0 \vee z = c, \\
u(x,y,z) = 0, & y = 0 \vee y = b \vee x = 0 \vee x = a.
\end{cases}
$$

8.25

```
Clear[x,y,z]
```

We introduce the domain of integration,

```
a=1.;α=5.;b=2.;c=3.;  (* Set the parameters *)
```

```
bound={x,a-x,y,b-y,z,c-z};
dom=ImplicitRegion[And@@(#≥0&/@bound),{x,y,z}];
```

and define the boundary conditions,

```
boundary={DirichletCondition[u[x,y,z]==α x y(a-x)(b-y),
z==0.||z==c],
DirichletCondition[u[x,y,z]==0.,y==0.||y==b],
DirichletCondition[u[x,y,z]==0.,x==0.||x==a]};
```

The system does not supply an analytical solution,

```
DSolveValue[{Laplacian[u[x,y,z],{x,y,z}]==0,boundary},
u,{x,y,z}];  (* No closed-form solution is returned *)
```

then we try finding a numerical solution,

```
sol=NDSolveValue[{Laplacian[u[x,y,z],{x,y,z}]==0,
boundary},u,{x,y,z}∈dom]//Chop;
```

and plot the contour of the solution.

```
ContourPlot3D[sol[x,y,z],{x,0.,a},{y,0.,b},{z,0.,c},
Ticks→{{.1,1},{.1,2},{.1,.8}},PlotRange→All,
Mesh→None,Contours→5,ImageSize→175,
ContourStyle→Directive[Gray,Opacity[.2]]]
```

See Figure 8.40.

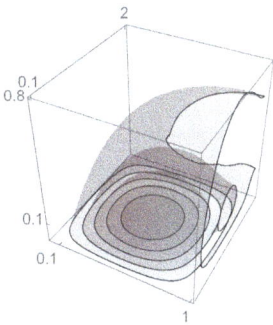

Figure 8.40: Contour of the solution for problem 8.25.

8.26. Let us find a solution to the Dirichlet problem to the Laplace equation defined on a parallelepiped [64, p. 156],

$$\begin{cases} \nabla^2_{x,y,z} u(x,y,z) = 0, & 0 \le x \le a, \quad 0 \le y \le b, \quad 0 \le z \le c, \\ u|_{x=0 \wedge x=a \wedge y=0 \wedge y=b \wedge z=c}(x,y,z) = 0, & u(x,y,0) = a. \end{cases}$$

8.26

```
Clear[x,y,z]
```

```
a=1;α=1;b=3;c=2; (* Set the parameters *)
```

We introduce the domain of integration.

```
bound={x,a-x,y,b-y,z,c-z};
dom=ImplicitRegion[And@@(#≥0&/@bound),{x,y,z}];
```

We define the boundary conditions,

```
eqncond={Laplacian[u[x,y,z],{x,y,z}]==
DirichletCondition[u[x,y,z]==0,x==0∨x==a∨y==0∨y==b
∨z==c],
DirichletCondition[u[x,y,z]==a,z==0]};
```

and try finding an analytical solution.

```
DSolveValue[eqncond,u,{x,y,z}]; (* No closed-form
solution is returned *)
```

If not, then one finds a numerical solution,

```
sol=NDSolveValue[eqncond,u,{x,y,z}∈dom]//Chop;
```

and plots the contour of the solution.

```
ContourPlot3D[sol[x,y,z],{x,0,a},{y,0,b},{z,0,c},
PlotRange→All,Mesh→None,Contours→2,PlotPoints→50,
ContourStyle→Directive[Gray,Opacity[.2]],
ImageSize→160]
```

See Figure 8.41.

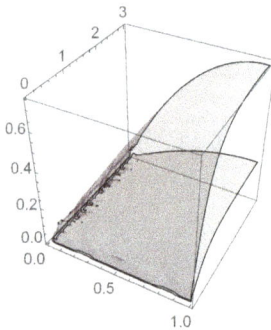

Figure 8.41: Contour of the solution for problem 8.26.

8.7 Poisson equations in higher dimensions

8.27. We find a solution to the Dirichlet problem of the Poisson equation defined on a parallelepiped [64, p. 157],

$$\begin{cases} \nabla^2_{x,y,z} u(x,y,z) = \alpha x + \beta y, & 0 \le x \le a, \quad 0 \le y \le b, \quad -c \le z \le c, \\ u|_{x=0 \wedge x=a \wedge y=0 \wedge y=b \wedge z=-c \wedge z=c}(x,y,z) = 0. \end{cases}$$

8.27

```
Clear[x,y,z]
```

```
a=1;α=5;b=2;β=4;c=5;  (* Set the parameters *)
```

We introduce the domain of integration.

```
dom=ImplicitRegion[Abs[x-a/2]≤a/2&&Abs[y-b/2]≤b/2
&&Abs[z]≤c,{x,y,z}];
```

We define the problem, i. e., equation and conditions,

```
eqncond={Laplacian[u[x,y,z],{x,y,z}]==α x+β y,
DirichletCondition[u[x,y,z]==0,x==0∧x==a∧y==0∧y==b
∧z==-c∧z==c]};
```

and try finding an analytical solution.

```
DSolveValue[eqncond,u,{x,y,z}];  (* No closed-form
solution is returned *)
```

If not, then one finds a numerical solution,

```
sol=NDSolveValue[eqncond,u,{x,y,z}∈dom]//Chop;
(* The system returned the message "No Dirichlet
Condition or Robin-type NeumannValue was specified
for [NoBreak]{u}[NoBreak]; the result is not unique up
to a constant." *)
```

and plot the contour of the solution.

```
ContourPlot3D[sol[x,y,z],{x,0.,a},{y,0.,b},{z,-c,c},
Contours→3,ContourStyle→Directive[Gray,Opacity[.2]],
PlotRange→All,Mesh→None,PlotPoints→50,
ImageSize→175]
```

See Figure 8.42.

8.28. Let us find a solution to the Dirichlet problem of the Poisson equation defined on the unit ball,

$$\begin{cases} \nabla^2_{x,y,z} u(x,y,z) = -(x^2 + y^2), & \{(x,y,z) \mid x^2 + y^2 + z^2 \leq 1\} \subset \mathbb{R}^3, \\ u|_{x+2y+z=1}(x,y,z) = \sin(2\pi(x+y+z)). \end{cases}$$

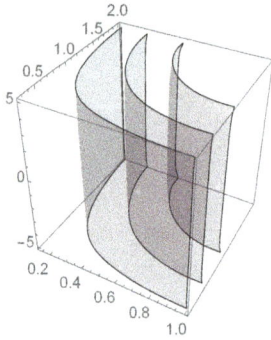

Figure 8.42: Contour of the solution for problem 8.27.

8.28

```
Clear[x,y,z]
```

We consider the equation, its boundary condition,

```
eqncond={Laplacian[u[x,y,z],{x,y,z}]==-(x²+y²),
DirichletCondition[u[x,y,z]==Sin[2π(x+y+z)],
x+2 y+z==1]};
```

and look for a numerical solution.

```
sol=NDSolveValue[eqncond,u,{x,y,z}∈Ball[]];
```

Here, we plot it.

```
ContourPlot3D[sol[x,y,z],{x,-1.,1},{y,-1.,1},{z,-1.,1},
PlotRange→All,Mesh→None,Contours→3,ImageSize→175,
Ticks→{{-.07,0,.08},{-.08,0,.07},{.92,1}},
ContourStyle→Directive[Opacity[.2],Gray]]
```

See Figure 8.43.

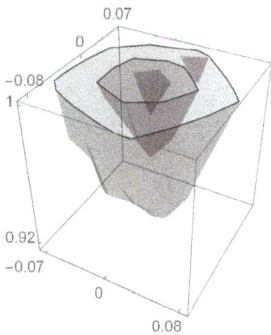

Figure 8.43: Contour of the solution for problem 8.28.

9 Parabolic partial differential equations

This chapter is dedicated to the next topics: 1D homogeneous parabolic equations, 1D inhomogeneous parabolic equations, 2D homogeneous parabolic equations, 2D inhomogeneous parabolic equations, the Burgers' equations, ansatz methods (tanh method, Hirota bilinear method, exp function method, and sine-cosine method), the Fisher equation, the Fitzhugh–Nagumo equation, the Calogero equation, the modified Klein–Gordon equation by ansatz methods, the double sine-Gordon equation by the tanh function method, and continuous dependence on a parameter.

A homogeneous parabolic partial differential equation with vanishing nonprincipal terms has one of the following forms:

$$\frac{\partial u(t,x)}{\partial t} = \frac{\partial^2 u(t,x)}{\partial x^2}, \quad (t,x) \in D \subset \mathbb{R}^2, \quad D \text{ nonempty, open, and connected,}$$

or, more generally, for $n \geq 1$,

$$\frac{\partial u(t,x_1,\ldots,x_n)}{\partial t} = \nabla^2_{x_1,\ldots,x_n} u(t,x_1,\ldots,x_n), \quad (t,x_1,\ldots,x_n) \in D \subset \mathbb{R}^{n+1},$$

and D is nonempty, open, and connected.

9.1 1D homogeneous parabolic equations

9.1. We introduce a 1D heat equation with a Gaussian initial condition,

$$\begin{cases} \partial_t u(t,x) = \partial_{x,x} u(t,x), & a,b \in \mathbb{R} \setminus \{0\}, \\ u(0,x) = b \exp(-a^2 x^2). \end{cases}$$

9.1

```
Clear[a,b,t,τ,x]
```

We solve the problem,

```
sol=Assuming[{a,b}∈Reals&&a≠0&&b≠0,
DSolve[{∂ₜu[t,x]=∂ₓ,ₓu[t,x],u[0,x]==b e^{-a²x²}},u,{t,x}]]
//Flatten} (* Solve the problem *)
```

$$\{u \to \text{Function}[\{t,x\}, \frac{b\, e^{-\frac{a^2 x^2}{1+4a^2 t}}}{\sqrt{1+4a^2 t}}]\} \quad (* \text{ Analytical solution } *)$$

and plot its solution.

https://doi.org/10.1515/9783111411392-011

```
Plot3D[u[t,x]/.sol/.{a→1,b→5},{t,0,5},{x,-5,5},
Mesh→{{.5,1,2.5,4},{-.75,.75}},PlotStyle→Directive[
Gray,Opacity[.2]],Ticks→{{0,5},5{-1,0,1},Automatic},
ImageSize→170,PlotLabel→#,PlotRange→#]
&/@{Automatic,All} (* Plot the solution *)
```

See Figure 9.1.

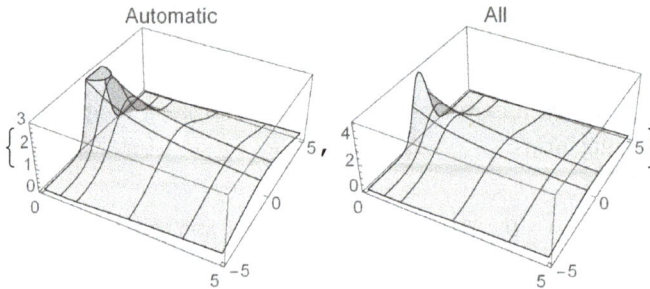

Figure 9.1: Analytical solution for problem 9.1.

9.2. We introduce an 1D heat equation with a periodic initial condition,

$$\begin{cases} \partial_t u(t,x) = a^2 \partial_{x,x} u(t,x), & a \in \mathbb{R}\setminus\{0\}, \\ u(0,x) = \sin(x). \end{cases}$$

9.2

```
Clear[a,t,τ,x,ξ]
```

We solve the problem

```
Assuming[a∈Reals&&a≠0,
DSolve[{∂_t u[t,x]==a²∂_{x,x}u[t,x],u[0,x]==Sin[x]},
u,{t,x}]]//Flatten//Simplify; (* Solve the problem *)
sol=u[t,x]/.%
sol=sol//FullSimplify
e^{-a²t}Sin[x]
```

and plot its solution for a particular value of the parameter.

```
{Plot3D[sol/.a→3,{t,-1,1},{x,0,2π},##,
ImageSize→175,Ticks→{{-1,0,1},π{0,1,2},Automatic}],
Manipulate[
Plot3D[sol/.a→3,{t,-1,τ},{x,0,ξ},##,ImageSize→160,
PerformanceGoal→"Quality",Boxed→False,Axes→None],
{τ,-.99,1},{ξ,.001,2π},SaveDefinitions→True]}
&[PlotStyle→Directive[Gray,Opacity[.2]],Mesh→3,
PlotRange→All]
```

See Figure 9.2.

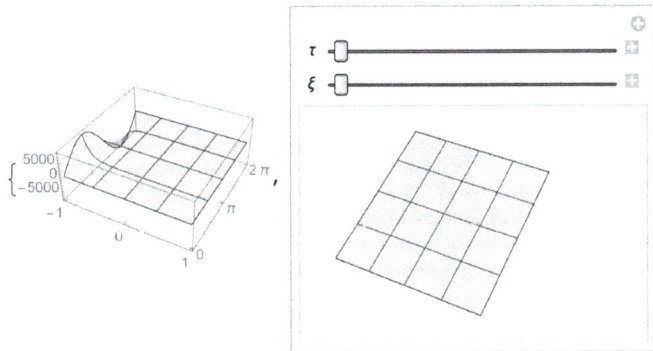

Figure 9.2: Dynamical representation of the solution for problem 9.2.

9.3. We introduce an 1D heat equation with discontinuous initial condition,

$$\begin{cases} \partial_t u(t,x) = \partial_{x,x} u(t,x), & c > 0, \\ u(0,x) = \begin{cases} c, & |x| < c, \\ 0, & |x| > c. \end{cases} \end{cases}$$

9.3

```
Clear[c,t,x,u]
```

```
eqncond={∂ₜu[t,x]==∂ₓ,ₓu[t,x], (* The problem *)
u[0,x]==Piecewise[{{c,Abs[x]<c},{0,Abs[x]>c}}]};
```

```
sol=Assuming[c>0,DSolve[eqncond,u,{t,x}]]//Flatten
(* Analytical solution *)
{u→Function[{t,x},½c(Erf[c-x/√2 t]+Erf[c+x/√2 t])]}
```

The solution satisfies the equation since

```
v[t_,x_]:=u[t,x]/.sol
∂ₜv[t,x]==∂ₓ,ₓv[t,x]
(* True *)
```

We plot the analytical solution for a particular value of the parameter.

```
Block[{c=1},
Plot3D[u[t,x]/.sol,{t,0,5},{x,-5,5},Mesh→5,
Ticks→{{0,5},5{-1,0,1},Automatic},ImageSize→170,
PlotStyle→Directive[Gray,Opacity[.2]],PlotRange→#,
PlotLabel→#]&/@{Automatic,All}]
```

See Figure 9.3.

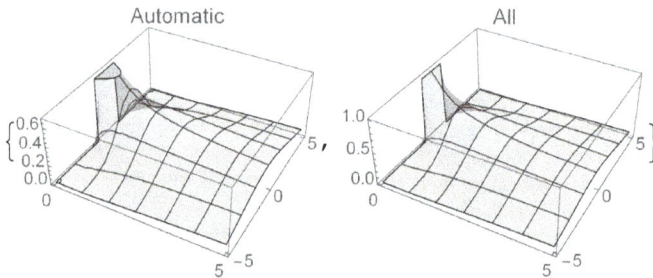

Figure 9.3: Representation of the analytical solution for problem 9.3.

9.4. Here, we study a 1D heat equation with initial and boundary conditions [64, p. 132],

$$\begin{cases} \partial_t u(t,x) = (1/r(x))\partial_x(p(x)\partial_x u(t,x)) + q(x)u(t,x), & t > 0, \quad 0 < x < l, \\ u(0,x) = x(l-x), \quad u(t,0) = 0, \quad u(t,l) = 0. \end{cases}$$

9.4

```
Clear[t,x]
l=10; (* Set the parameter *)
r[x_]:=1+Abs[x];p[x_]:=Sin[x];q[x_]:=Cos[x]
eqncond={∂ₜu[t,x]==(1/r[x])∂ₓ(p[x]∂ₓu[t,x])+q[x]u[t,x],
u[0,x]==x(l-x),u[t,0]==u[t,l]==0};
```

We try finding an analytical solution.

```
DSolve[eqncond,u,{t,x}]; (* No closed-form solution
is returned *)
```

In this case, we look for a numerical solution

```
sol=NDSolveValue[eqncond,u,{t,0,5},{x,0,1}];
```

and plot the figure.

```
Plot3D[sol[t,x],{t,0,5},{x,0,1},ImageSize→170,
Mesh→3,Ticks→{{0,5},5{0,1,2},Automatic},
PlotStyle→Directive[Gray,Opacity[.2]],PlotPoints→25,
PlotRange→#,PlotLabel→#]&/@{Automatic,All}
```

See Figure 9.4.

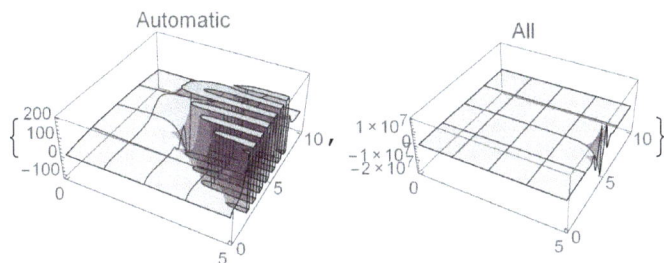

Figure 9.4: Numerical solution for problem 9.4.

9.5. Here, we study a 1D heat equation with initial and boundary conditions [64, p. 136],

$$\begin{cases} \partial_t u(t,x) = 2\partial_{x,x}u(t,x), & t \ge 0, \quad 0 < x < 10, \\ u(0,x) = x(10-x), & u(t,0) = 10\sin(t), \quad u(t,10) = 5\cos(t + \pi/2). \end{cases}$$

9.5

```
Clear[t,x]
```

We write the problem,

```
eqn={∂ₜu[t,x]==2∂ₓ,ₓu[t,x],u[0,x]==x(10-x),
u[t,0]==10Sin[t],u[t,10]==5Cos[t+Pi/2]};
```

look for an analytical solution,

```
sol=DSolve[%,u,{t,x}]/.{K[1]→n,∞ →50} (* Change the
index of summation and select only the first 50
terms *)
asol[t_,x_]:=Activate[u[t,x]/.sol]//Flatten;
```

and plot the solution.

```
{Plot3D[asol[t,x],{t,0,20},{x,0,10},PlotRange→All,
PlotStyle→Directive[Gray,Opacity[.15]],ImageSize→175,
Mesh→2,Ticks→Outer[Times,5{2,1,2},{0,1,2}]]}
```

See Figure 9.5.

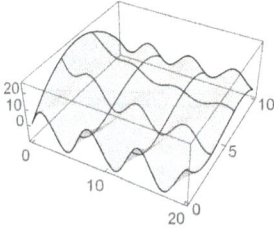

Figure 9.5: Analytical solution for problem 9.5.

9.6. We study the 1D heat equation with initial and periodic boundary conditions,

$$\begin{cases} \partial_t u(t,x) = (1/8)\partial_{x,x}u(t,x), & 0 \le t \le 10, \quad 0 \le x \le 1, \\ u(t,0) = \sin(2\pi\,t), & u(0,x) = 0, \quad \partial_x u(t,1) = 0. \end{cases}$$

9.6

```
Clear[t,x,ξ]
```

```
eqncond={∂ₜu[t,x]==(1/8)∂x,xu[t,x],u[t,0]==Sin[2π t],
u[0,x]==0,u^(0,1)[t,1]==0};
DSolve[%,u,{t,x}];(* No closed-form solution is
returned *)
```

We try finding a numerical solution and plot it.

```
sol=NDSolve[eqncond,u,{t,0,10},{x,0,1}];
```

```
{Plot3D[u[t,x]/.sol,{t,0,5},{x,0,1},Mesh→{2,5},
PlotStyle→{Gray,Opacity[.15]},PlotRange→All,
Ticks→{{0,5},{0,.5,1},{-.8,0,.9}},ImageSize→160],
Animate[(* A cross section *)
Plot[u[t,ξ]/.sol,{t,0,5},PlotStyle→Black,
Ticks→{2Range[5],.5Range[-2,2]},PlotRange→All,
ImageSize→140],{ξ,0,1},SaveDefinitions→True,
DefaultDuration→20,AnimationRunning→False,
ImageMargins→1,AnimationDirection→ForwardBackward]}
```

See Figure 9.6.

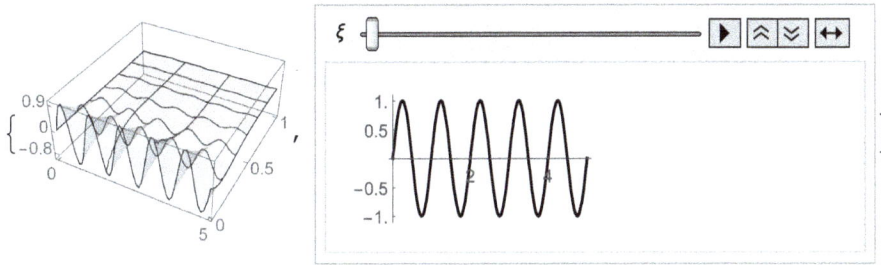

Figure 9.6: Numerical solution for problem 9.6.

9.7. We introduce a 1D heat equation with periodic boundary conditions,

$$\begin{cases} \partial_t u(t,x) = a^2 \partial_{x,x} u(t,x), & a > 0, \quad 0 \le t \le 1, \quad 0 \le x \le 2\pi, \\ u(0,x) = \sin(x), & u(1,x) = \cos(x). \end{cases}$$

9.7

```
Clear[a,t,x]
```

```
eqncond=Assuming[a>0,{∂ₜu[t,x]==a²∂ₓ,ₓu[t,x],
u[0,x]==Sin[x],u[1,x]==Cos[x]}]; (* The problem *)
```

```
DSolve[eqncond,u,{t,x}]; (* No closed-form solution
is returned *)
```

We look for a numerical solution,

```
a=1.;
sol=NDSolve[eqncond,u,{t,0,1},{x,0,2π}]//Flatten;
```

and plot it.

```
{Plot3D[u[t,x]/.sol,{t,0,1},{x,0,2π},ImageSize→175,
PlotStyle→{Gray,Opacity[.2]},Mesh→3,PlotPoints→50,
Ticks→{{0,1,2},π{0,1,2},Automatic}],
Manipulate[
Plot3D[Evaluate[u[t,x]/.sol],{t,0,τ},{x,0,ξ},
PlotRange→All,PlotStyle→{Gray,Opacity[.2]},
PerformanceGoal→"Quality",Mesh→3,ImageSize→150,
Boxed→False,Axes→None],{τ,.01,5.},{ξ,.01,2π},
SaveDefinitions→True]}
```

See Figure 9.7.

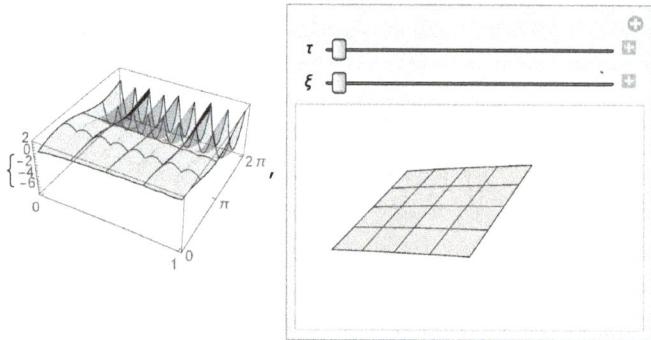

Figure 9.7: Plot of the numerical solution for problem 9.7.

9.8. We introduce a parabolic partial differential equation with constant coefficients,

$$4\partial_{t,t}u(t,x) - 12\partial_{t,x}u(t,x) + 9\partial_{x,x}u(t,x) = 0, \quad -7 \le t, x \le 7.$$

9.8

```
Clear[t,x]
```

We look for an analytical solution

```
DSolve[4D[u[t,x],t,t]-12D[u[t,x],t,x]+9D[u[t,x],x,x]
==0,u,{t,x}]; (* General solution *)
fn=u[t,x]/.%[[1]]/.{C[1][t_]→Sin[t²],
C[2][t_]→Log[Abs[t]]} (* Particular solution *)
```
$$t \, Log\left[Abs\left[\tfrac{3t}{2+x}\right]\right] + Sin\left[\left(\tfrac{3t}{2} + x\right)^2\right]$$

and plot it.

```
Plot3D[fn,{t,-7,7},{x,-7,7},ImageSize→175,Mesh→3,
Ticks→Outer[Times,{7,7,20},{-1,0,1}],PlotPoints→40,
PlotStyle→Directive[Gray,Opacity[.2]]]
```

See Figure 9.8.

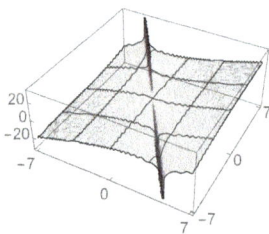

Figure 9.8: Analytical solution for problem 9.8.

9.9. We study a heat equation with Dirichlet conditions,

$$\begin{cases} 4\partial_{t,t}u(t,x) - 12\partial_{t,x}u(t,x) + 9\partial_{x,x}u(t,x) + \partial_t u(t,x) = 0, \\ u(t,0) = \tan(t), \quad u(0,x) = \tan(x), \quad 0 \le t, x \le 2\pi. \end{cases}$$

9.9

```
Clear[t,x]
```

```
DSolve[{4∂t,tu[t,x]-12∂t,xu[t,x]+9∂x,xu[t,x]
+∂tu[t,x]==0,u[t,0]==Tan[t],u[0,x]==Tan[x]},u,{t,x}];
(* No closed-form solution is returned *)
```

We look for a numerical solution

```
sol=NDSolve[{4∂t,tu[t,x]-12∂t,xu[t,x]+9∂x,xu[t,x]
+∂tu[t,x]==0,u[t,0]==Tan[t],u[0,x]==Tan[x]},u,{t,0,2π},
{x,0,2π}];
```

and plot it.

```
Plot3D[u[x,y]/.sol,{x,0,2π},{y,0,2π},Mesh→3,
Ticks→{π{0,1,2},πRange[0,2],Automatic},
PlotPoints→75,ImageSize→200,PlotRange→#,
PlotStyle→{Black,Opacity[.15]}]&/@{Automatic,All}
```

See Figure 9.9.

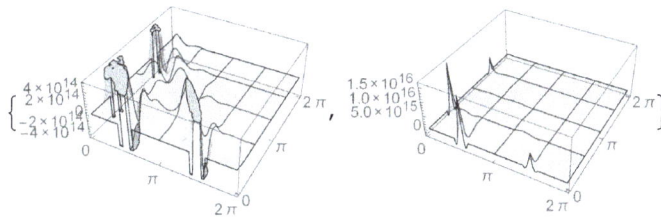

9.10. We study a heat equation with variable coefficients and periodic boundary conditions,

$$\begin{cases} t^2\partial_{t,t}u(t,x) - 2t\,x\,\partial_{t,x}u(t,x) + x^2\partial_{x,x}u(t,x) + \partial_t u(t,x) + \partial_x u(t,x) = 0, \\ u(t,-6) = u(t,6) = \sin(t\pi/6), \quad -5 \le t \le 5, \quad -6 \le x \le 6. \end{cases}$$

9.10

```
Clear[α,β,t,x]
```

We try finding an analytical solution.

```
DSolve[t²∂t,tu[t,x]-2t x∂t,xu[t,x]+x²∂x,xu[t,x]
+t∂tu[t,x]+x∂xu[t,x]==0,u,{t,x}]; (* No closed-form
solution is returned *)
```

Now we try finding a numerical solution,

```
sol=NDSolveValue[{t²∂t,tu[t,x]-2t x∂t,xu[t,x]
+x²∂x,xu[t,x]+∂tu[t,x]+∂xu[t,x]==0,
u[t,-6]==u[t,6]==Sin[tπ/6]},u,{t,-5,5},{x,-5,5}]
```

and plot it.

```
{Show[##,
ParametricPlot3D[{{5,x,sol[5,x]},{0,x,sol[0,x]},
{-5,x,sol[-5,x]}},{x,-6,6},PlotStyle→{{Black,Thick},
{Gray,Thick},{Black,Thick}}]],
Manipulate[
Show[##,
ParametricPlot3D[{{5,x,sol[5,x]},{α,x,sol[α,x]},
{-5,x,sol[-5,x]}},{x,-6,6},PlotStyle→{{Black,Thick},
{Gray,Thick},{Black,Thick}}],
ParametricPlot3D[{{t,6,sol[t,6]},{t,β,sol[t,β]},
{t,-6,sol[t,-6]}},{t,-5,5},PlotStyle→Black,Thick,
Gray,Thick,Black,Thick]],{α,-5,5},{β,-6,6},
SaveDefinitions→True]}&[
Plot3D[sol[t,x],{t,-5,5},{x,-6,6},Mesh→None,
Ticks→Outer[Times,{5,6,1},{-1,0,1}],
PlotStyle→{Gray,Opacity[.15]},ImageSize→175]]
```

See Figure 9.10.

9.2 1D inhomogeneous parabolic equations

9.11. Now we introduce a 1D inhomogeneous heat equation with an initial condition,

$$\begin{cases} \partial_t u(t,x) = 2\partial_{x,x}u(t,x) + b, & a,b,c,t,x \in \mathbb{R}, \quad b,c \neq 0, \\ u(0,x) = c\,\cos(a\,x). \end{cases}$$

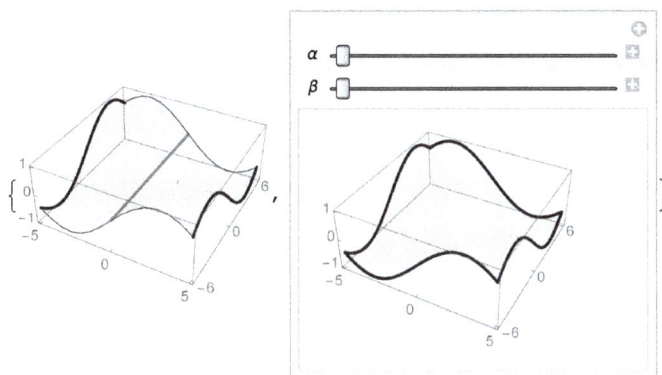

Figure 9.10: Numerical solution for problem 9.10.

9.11

```
Clear[a,b,β,c,γ,t,x]
```

We look for an analytical solution,

```
sol=DSolve[{∂ₜu[t,x]=2∂ₓ,ₓu(t,x)+b,u[0,x]==c Cos[a x]},
u,{t,x}]//Simplify//First (* Analytical solution *)
{u→Function[{t,x},b t+c e^{-2a²t}Cos[a x]]}
```

and plot it.

```
Manipulate[
Plot3D[u[t,x]/.sol/.{a→1,b→ β,c→ γ},{t,-5,5},
{x,-5,5},PlotRange→All,Mesh→2,ImageSize→180,
PlotStyle→Directive[Gray,Opacity[.2]],
Ticks→Outer[Times,{5,5,20000},{-1,0,1}]],{β,-1,1},
{γ,-1,1},SaveDefinitions→True]
```

See Figure 9.11.

9.12. Next, we study the boundary value problem of an inhomogeneous heat equation [64, p. 136],

$$\begin{cases} \partial_t u(t,x) - \partial_{x,x} u(t,x) = b\,x(l-x), & t \geq 0, \quad 0 \leq x \leq l, \\ u(0,x) = u(t,0) = u(t,l) = 0. \end{cases}$$

9.12

```
Clear[p,t,x]
```

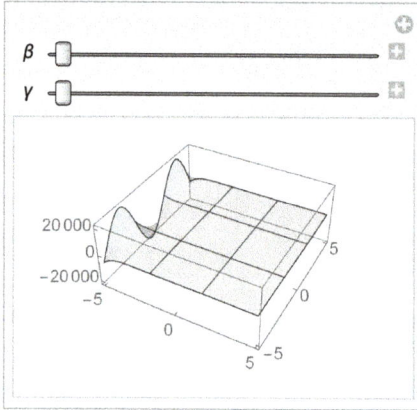

Figure 9.11: Analytical solution for problem 9.11.

```
Block[{b=2,l=4}, (* Set the parameters *)
eqncond={∂ₜu[t,x]-∂ₓ,ₓu[t,x]==b x(1-x),
u[0,x]==u[t,0]==u[t,l]==0};(* The problem *)]

DSolve[eqncond,u,{t,x}]; (* No closed-form solution is
returned *)
```

We try finding a numerical solution,

```
sol=NDSolveValue[eqncond,u,{t,0,5},{x,0,l}];
```

and plot it.

```
Plot3D[sol[t,x],{t,0,5},{x,0,l},ImageSize→175,Mesh→5,
Ticks→Outer[Times,{2.5,2,6},{0,1,2}],PlotRange→{{0,5},
{0,4},{0,13}},PlotStyle→{Gray,Opacity[.15]}]
```

See Figure 9.12.

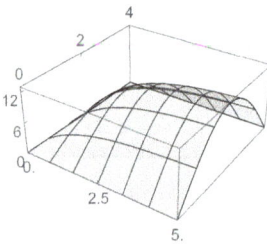

Figure 9.12: Numerical solution for problem 9.12.

Below we show a dynamical plot of the numerical solution.

```
Animate[
Plot3D[sol[t,x],{t,0,5p},{x,0,1},ImageSize→175,
Mesh→⌊5p⌋,PlotStyle→{Gray,Opacity[.15]},
Ticks→{{0,2.5p,5p},{0,2,4},Automatic},
PlotRange→{{0,5},{0,4},{0,13}}],{p,.03,1},
SaveDefinitions→True,DefaultDuration→25,
AnimationDirection→ForwardBackward,
AnimationRunning→False]
```

See Figure 9.13.

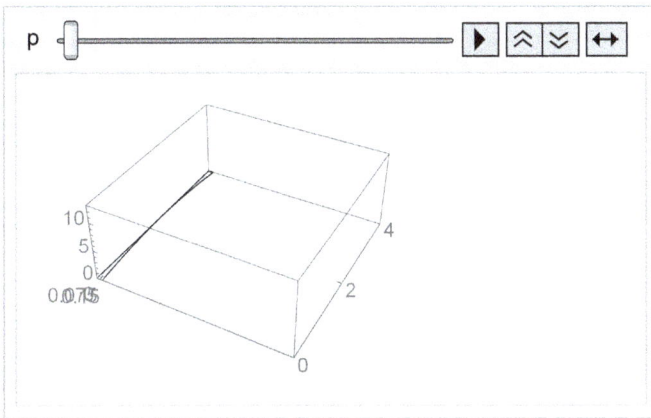

Figure 9.13: Dynamical plot of the numerical solution for problem 9.12.

9.3 2D homogeneous parabolic equations

9.13. We study a 2D boundary value problem to a heat equation,

$$\begin{cases} \partial_t u(t,x,y) = c^2 \nabla^2_{x,y} u(t,x,y), \quad a,b,c > 0, \quad t \ge 0, \\ (x,y) \in D = \{(x,y) \mid -a \le x \le a,\ -b \le y \le b\} \subset \mathbb{R}^2, \\ u(0,x,y) = 3xy + x^2 + y^2 \text{ and } \partial_n u(t,x,y) = 0 \text{ on } \partial D. \end{cases}$$

9.13

```
Clear[t,x,y]
```

```
a=b=c=1;  (* Set the parameters *)
```

We introduce the data of the problem

```
eqncond={∂_t u[t,x,y]-c²∇²_{x,y}u[t,x,y]==
NeumannValue[0,(-a≤x≤a&&(y==-b∨y==b))∨(-a≤x≤a&&y==b)
∨((x==-a∨x==a&&-b≤y≤b))∨(x==a&&-b≤y≤b)],
DirichletCondition[u[t,x,y]==3x y+x²+y²,t==0]};
```

and look for an analytical solution.

```
sol=DSolveValue[eqncond,u,{t,x,y}]; (* No closed-form
solution is returned *)
```

We numerically solve the problem,

```
nsol=NDSolveValue[eqncond,u,{t,0,5},{x,-a,a},{y,-b,b}];
```

and plot the solution.

```
ContourPlot3D[nsol[t,x,y],{t,0,5},{x,-a,a},{y,-b,b},
ContourStyle→Directive[Gray,Opacity[.2]],Mesh→None,
Contours→3,PlotRange→All,ImageSize→175]
```

See Figure 9.14.

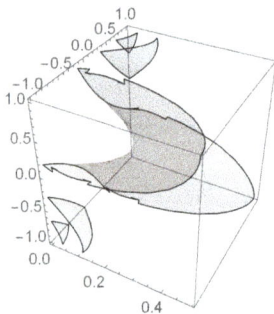

Figure 9.14: Contour of the numerical solution for problem 9.13.

9.14. Next, we study the 2D boundary value problem to a heat equation,

$$\begin{cases} \partial_t u(t,x,y) = a^2 \nabla^2_{x,y} u(t,x,y), \quad a > 0, \quad t \geq 0, \\ (x,y) \in D = \{(x,y) \mid x^2 + y^2 < 1\} \subset \mathbb{R}^2, \\ u(0,x,y) = 1 + x^2 + y^2, \quad \partial_n u(t,x,y) = 0 \text{ on } x^2 + y^2 = 1. \end{cases}$$

9.14

```
Clear[t,x,y]
```

We plot the domain of integration.

```
dom=ImplicitRegion[x²+y² ≤1,{x,y}];
```

```
a=2; (* Set the parameter *)
```

We write the data of the problem,

```
eqncond={∂ₜu[t,x,y]-a²∇²₍ₓ,y₎u[t,x,y]==
NeumannValue[0,x²+y²==1],
DirichletCondition[u[t,x,y]==1+x²+y²,t==0]};
```

and try finding an analytical solution.

```
sol=DSolve[eqncond,u,{t,x,y}]; (* No closed-form
solution is returned *)
```

We look for a numerical solution,

```
nsol=NDSolve[eqncond,u,{t,0,5},{x,y}∈dom,Method→
{"MethodOfLines","TemporalVariable"→t}]//First;
```

and plot it.

```
ContourPlot3D[u[t,x,y]/.nsol,{t,0,5},{x,-1,1},{y,-1,1},
PlotRange→All,PlotPoints→20,ImageSize→175,Mesh→
None,ContourStyle→Directive[Gray,Opacity[.2]]]
```

See Figure 9.15.

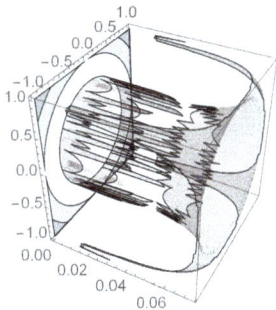

Figure 9.15: Contour plot of the numerical solution for problem 9.14.

9.15. We study the 2D heat equation on a disk with an initial value and Neumann condition [64, p. 137],

$$\begin{cases} \partial_t u(t,x,y) = a^2 \nabla^2_{xy} u(t,x,y), & a > 0, \quad t \geq 0, \\ (x,y) \in D = \{(x,y) \mid x^2 + y^2 \leq 1\} \subset \mathbb{R}^2, \\ u(0,x,y) = \ln(1 + \sqrt{x^2 + y^2}), & \partial_n u(t,x,y) = \sin(xy) \text{ on } x^2 + y^2 = 1. \end{cases}$$

9.15

```
Clear[t,x,y]
```

One shows the domain of integration.

```
dom=Disk[];
```

```
a=2; (* Set the parameter *)
```

We write the data of the problem,

```
eqncond={∂_t u[t,x,y]-a²∇²_{x,y}u[t,x,y]==
NeumannValue[Sin[x y],x²+y²==1],
DirichletCondition[u[t,x,y]==Log[1+√x² + y²],t==0]};
```

and look for an analytical solution.

```
sol=DSolve[eqncond,u,{t,x,y}]; (* No closed-form
solution is returned *)
```

Now we try finding a numerical solution,

```
nsol=NDSolve[eqncond,u,{t,0,5},{x,y}∈dom,Method→
{"MethodOfLines","TemporalVariable"→t}]//First;
```

and plot it.

```
Plot3D[u[t,x,y]/.nsol/.t→0,{x,-1.,1.},{y,-1.,1.},
RegionFunction→(#1²+#2²<1&),ImageSize→175,Mesh→3,
Ticks→Outer[Times,{-1,0,1},{1,1,.6}],
PlotStyle→{Gray,Opacity[.15]}]
```

See Figure 9.16.

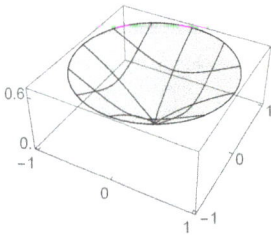

Figure 9.16: Plot of the numerical solution for problem 9.15 if $t = 0$.

We introduce a contour plot of the numerical solution.

```
ContourPlot3D[u[t,x,y]/.nsol,{t,0,5},{x,-1.,1.},
{y,-1.,1.},Ticks→Outer[Times,{5,1,1},{-1,0,1}],
ContourStyle→Directive[Gray,Opacity[.15]],Mesh→5,
PlotRange→All,ImageSize→175]
```

See Figure 9.17.

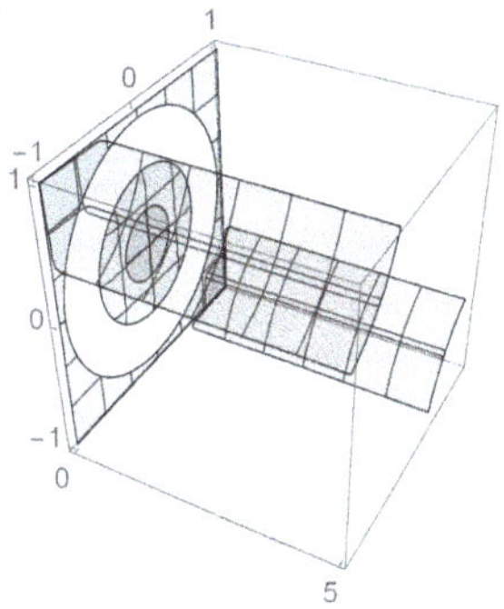

Figure 9.17: Contour plot of the numerical solution for problem 9.15.

9.16. We study the 2D boundary value problem to a heat equation [64, p. 137],

$$\begin{cases} \partial_t u(t,x,y) = a^2 \nabla^2_{x,y} u(t,x,y), & a > 0, \quad t \geq 0, \\ (x,y) \in D = \{(x,y) \mid x^2 + y^2 \leq 1\} \subset \mathbb{R}^2, \\ u(0,x,y) = \exp(-\sqrt{x^2 + y^2}), & \partial_n u(t,x,y) = \cos^2(x\,y) \text{ on } x^2 + y^2 = 1. \end{cases}$$

9.16

```
Clear[t,x,y]
```

The domain of integration is the unitary disk.
    We write the data of the problem,

```
Block[{a=2}, (* Set the parameter *)
eqncond={∂ₜu[t,x,y]-a²∇²_{x,y}u[t,x,y]==
NeumannValue[Cos[x y]²,x²+y²==1],
DirichletCondition[u[t,x,y]==Exp[-Sqrt[x²+y²]],t==0]}]
```

and look for an analytical solution.

```
sol=DSolve[eqncond,u,{t,x,y}]; (* No closed-form
solution is returned *)
```

We try finding a numerical solution,

```
nsol=NDSolve[eqncond,u,{t,0,5},{x,y}∈dom,Method→
{"MethodOfLines","TemporalVariable"→t}]//First;
```

and plot it for $t = 0$.

```
Plot3D[u[t,x,y]/.nsol/.t→0,{x,y}∈dom,Mesh→3,
Ticks→Outer[Times,{1,1,.8},{-1,0,1}],
PlotStyle→{Gray,Opacity[.15]},ImageSize→165]
```

See Figure 9.18.

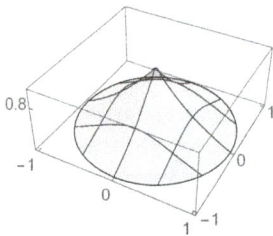

Figure 9.18: Plot of the numerical solution for problem 9.16 if t=0.

We introduce a contour plot of the numerical solution.

```
ContourPlot3D[u[t,x,y]/.nsol,{t,0,5},{x,-1.,1.},
{y,-1.,1.},Ticks→{{2,3},{-1,0,1},{-1,0,1}},Mesh→None,
PlotRange→All,ImageSize→165,ContourStyle→
Table[Directive[Gray,Opacity[.21-.03(k-1)]],{k,1,3}]]
```

See Figure 9.19.

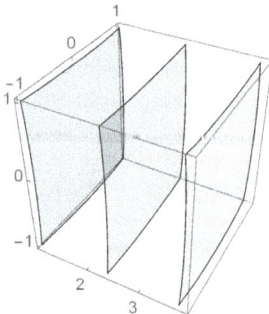

Figure 9.19: Contour plot of the numerical solution for problem 9.16.

9.4 2D inhomogeneous parabolic equations

9.17. We study a 2D inhomogeneous boundary value problem to a heat equation on a square [64, p. 141],

$$\begin{cases} \partial_t u(t,x,y) = \nabla^2_{x,y} u(t,x,y) + t\,x, & t \geq 0, \\ (x,y) \in D = \{(x,y) \mid 0 \leq x, y \leq 5\} \subset \mathbb{R}^2, \\ u(0,x,y) = x^2 - y^2. \end{cases}$$

9.17

```
Clear[t,x,y]
```

```
dom=ImplicitRegion[0≤x≤5&&0≤y≤5,{x,y}]; (* Domain of
integration *)
```

```
eqncond={∂_t u(t,x,y) == ∇²_{x,y} u(t,x,y) + t x,
u(0,x,y)==x² - y²} (* The problem *)
```

```
sol=DSolve[eqncond,u,{t,x,y}]; (* No closed-form
solution is returned *)
```

We look for a numerical solution and plot it for $t = 0$.

```
nsol=NDSolve[eqncond,u,{t,0,5},{x,y}∈dom,
Method→{"MethodOfLines","TemporalVariable"→t}]
//First;
```

```
Plot3D[u[t,x,y]/.nsol/.t→0,{x,y}∈dom,
Ticks→Outer[Times,5{1,1,4},{-1,0,1}],Mesh→3,
PlotStyle→{Gray,Opacity[.2]},ImageSize→165]
```

See Figure 9.20.

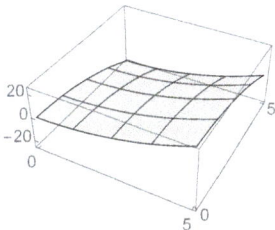

Figure 9.20: Numerical solution for problem 9.17 if $t = 0$.

We introduce a contour plot.

```
ContourPlot3D[u[t,x,y]/.nsol,{t,0,5},{x,-5,5},
{y,-5,5},Ticks→Outer[Times,5{1,1,1},{-1,0,1}],
Mesh→2,Contours→2,PlotRange→All,ImageSize→165,
ContourStyle→Table[Directive[Gray,
Opacity[.21-.03(k-1)]],{k,1,3}]]
```

See Figure 9.21.

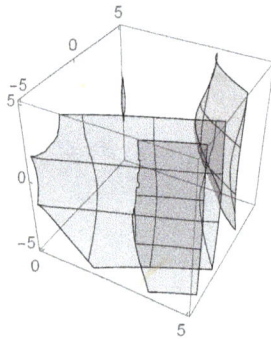

Figure 9.21: Contour plot for problem 9.17.

9.18. We study a 2D inhomogeneous initial value problem to a heat equation on a square [64, p. 141],

$$\begin{cases} \partial_t u(t,x,y) = \nabla^2_{x,y}u(t,x,y) + xy, & t \geq 0, \\ (x,y) \in D_2 = \{(x,y) \mid -5 \leq x,y \leq 5\} \subset \mathbb{R}^2, \\ u(0,x,y) = x^4 + y^4. \end{cases}$$

9.18

```
Clear[t,x,y]
```

```
dom=ImplicitRegion[-5≤x≤5&&-5≤y≤5,{x,y}]; (* Domain
of integration *)
```

```
eqncond={∂_t u[t,x,y] == ∇²_{x,y}u[t,x,y] + x y,
u(0,x,y)==x⁴ + y⁴} (* The problem *)
```

We look for an analytical solution.

```
sol=DSolve[eqncond,u,{t,x,y}]; (* No closed-form
solution is returned is returned *)
```

We look for a numerical solution and plot it for t=0.

```
nsol=NDSolve[eqncond,u,{t,0,5},{x,y}∈dom,
Method→{"MethodOfLines","TemporalVariable"→t}]
//First;
```

```
Plot3D[u[t,x,y]/.nsol/.t→0,{x,y}∈dom,Mesh→3,
Ticks→Outer[Times,5{1,1,200},{-1,0,1}],
PlotStyle→{Gray,Opacity[.2]},ImageSize→165]
```

See Figure 9.22.

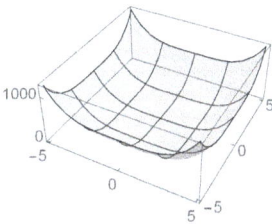

Figure 9.22: Numerical solution for problem 9.18 if t=0.

A contour plot is introduced.

```
ContourPlot3D[u[t,x,y]/.nsol,{t,0,5},{x,-5,5},
{y,-5,5},Mesh→2,Contours→2,PlotRange→All,
ContourStyle→Directive[Gray,Opacity[.2]],
ImageSize→165]
```

See Figure 9.23.

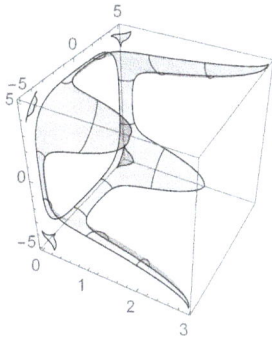

Figure 9.23: Contour plot for problem 9.18.

9.5 The Burgers' equations

9.5.1 The Burgers' equation in 1D

For a function $u(t, x)$ differentiable as many times as it is needed and diffusion coefficient d, the general form of the *Burgers' equation* in 1D space is

$$\partial_t u(t,x) + u(t,x)\partial_x u(t,x) = d\,\partial_{x,x} u(t,x), \quad d \in \mathbb{R}\setminus\{0\}.$$

9.19. We note that the solution to the Burgers' equation,

$$\partial_t u(t,x) + u(t,x)\partial_x u(t,x) = d\,\partial_{x,x} u(t,x), \quad d \in \mathbb{R}\setminus\{0\},$$

yielded by *Mathematica* is as follows.

9.19

```
Clear[t,τ,x,ξ]
```

We solve the Burgers' equation by *Mathematica* for a particular value of the diffusion coefficient

```
With[{d=-1},
sol=DSolve[∂_t u[t,x]+u[t,x]∂_x u[t,x]+d ∂_x,x u[t,x]==0,
u,{t,x}]]//Flatten
v[t_,x_]:= u[t,x]/.sol/.{C[1]→1,C[2]→1,C[3]→-1}
{u→Function[{t,x},-(C[1]+2C[2]²Tanh[t C[1]+x C[2]
+C[3]])/C[2]]}
```

and plot its particular solution.

```
{Plot3D[v[t,x],{t,-5,5},{x,-6,6},ImageSize→150,
Ticks→{5{-1,0,1},6{-1,0,1},{-3,-1,1}},Mesh→2,##],
Manipulate[
Plot3D[v[t,x],{t,-τ,τ},{x,-ξ,ξ},Mesh→2,ImageSize→175,
Ticks→{{-τ,0,τ},{-ξ,0,ξ},None},##],{τ,.01,5},{ξ,.01,6},
SaveDefinitions→True]}&[PlotStyle→{Gray,
Opacity[.15]}]
```

See Figure 9.24.

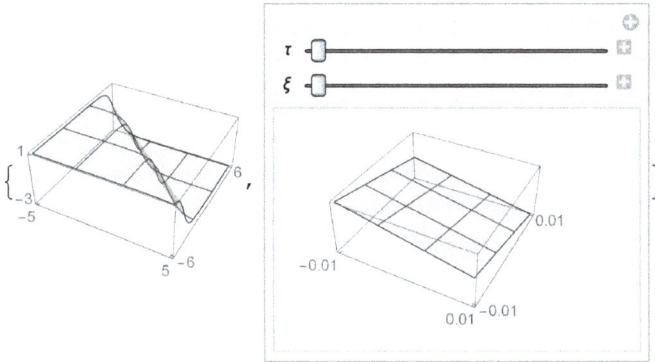

Figure 9.24: Plot by *Mathematica* of the analytical solution for problem 9.19.

9.20. Suppose that a list of functions is given

$$u(t,x) = \frac{\lambda + 2}{x + \lambda t + a},$$

$$u(t,x) = \frac{4x + 2a}{x^2 + a x + 2t + b},$$

$$u(t,x) = \frac{6(x^2 + 2t + a)}{x^3 + 6t x + 3a x + b},$$

$$u(t,x) = \frac{2\lambda}{1 + a \ \exp(-\lambda^2 t - \lambda x)},$$

$$u(t,x) = -\lambda + \frac{a(\exp(a(x - \lambda t)) - b)}{\exp(a(x - \lambda t)) + b},$$

$$u(t,x) = \lambda + \frac{a(\exp(a(x - \lambda t)) - b)}{\exp(a(x - \lambda t)) + b}$$

and we select from it the Burgers' functions, i. e., functions that satisfy the Burgers' equation. We put all the functions into a table mentioning which one is a Burgers' function and which one is not. In the end, we plot only the Burgers' functions, http://eqworld. ipmnet.ru/en/solutions/npde/npde1301.pdf.

9.20

```
Clear[a,b,λ,t,x]

listoffunctions=
{ λ+2/(x+λ t+a), 4x+2a/(x²+a x+2t+b), 6(x²+2t+a)/(x³+6t x+3a x+b), 2λ/(1+a Exp[-λ²t-λ x]),
 -λ + a(Exp[a(x-λt)]-b)/(Exp[a(x-λt)]+b), λ + a(Exp[a(x-λt)]-b)/(Exp[a(x-λt)]+b)};
(* List of functions *)
```

```
If[Length[listoffunctions]>0, (* If the list is
empty, then stop *)
header={"n","function u[t,x] " "," Burgers' function?
"};
triple={}; (* List of functions with answers *)
Do[u[t_,x_]=listofequations[[i]];
sol=-∂ₜu[t,x]+u[t,x]*∂ₓu[t,x]+∂ₓ,ₓu[t,x]//
FullSimplify;
If[sol===0,AppendTo[triple,{i,u[t,x],"yes"}],
AppendTo[triple,{i,u[t,x],"no"}]],
{i,Length[listofequations]}];
Grid[Join[{header},triple],Background→LightGreen,
Dividers→{{},{1→Thickness[3],2→Thickness[2],
-1→Thickness[3]}},Frame→ All,Alignment→Center],
Print["Empty list of equations"]]
```

These *Mathematica* commands generate Table 9.1.

Table 9.1: Certain functions tested as being Burgers' functions.

n	function	Burgers function?
1	$\lambda + \dfrac{2}{a+x+t\lambda}$	yes
2	$\dfrac{2a+4x}{b+2t+ax+x^2}$	yes
3	$\dfrac{6(a+2t+x^2)}{b+3ax+6tx+x^3}$	yes
4	$\dfrac{2\lambda}{a+ae^{-x\lambda-t\lambda^2}}$	yes
5	$\dfrac{a(-b+e^{a(x-t\lambda)})}{b+e^{a(x-t\lambda)}} - \lambda$	yes
6	$\dfrac{a(-b+e^{a(x-t\lambda)})}{b+e^{a(x-t\lambda)}} + \lambda$	no

```
listBurgers={}; (* List of Burgers' functions *)
Do[If[triple[[i,3]]==="yes",
AppendTo[listBurgers,triple[[i,2]]]],
{i,Length[triple]}];
listBurgers
```
$$\{\lambda+\frac{2}{a+x+t\lambda}, \frac{2a+4x}{b+2t+a\,x+x^2}, \frac{6(a+2t+x^2)}{b+3a\,x+6t\,x+x^3}, \frac{2\lambda}{1+a\,e^{-x\lambda-t\lambda^2}}, \frac{a(-b+e^{a(x-t\lambda)})}{b+e^{a(x-t\lambda)}} - \lambda\}$$

Here, we display all graphs in a condensed manner.

```
λ=1;a=1;b=2; (* Set the parameters *)
```

```
Table[listB→Plot3D[listB,{t,-5,5},{x,-6,6},
Axes→False,Boxed→False,Mesh→3,PlotStyle→{Gray,
Opacity[.15]},ImageSize→100],{listB,listBurgers}]
```

Below there are representations of the Burgers' functions.

```
Table[listB→Plot3D[listB,{t,-5,5},{x,-6,6},
Axes→False,Boxed→False,Mesh→3,ImageSize→100,
PlotStyle→{Gray,Opacity[.15]}],
{listB,listBurgers}]
```

See Figure 9.25.

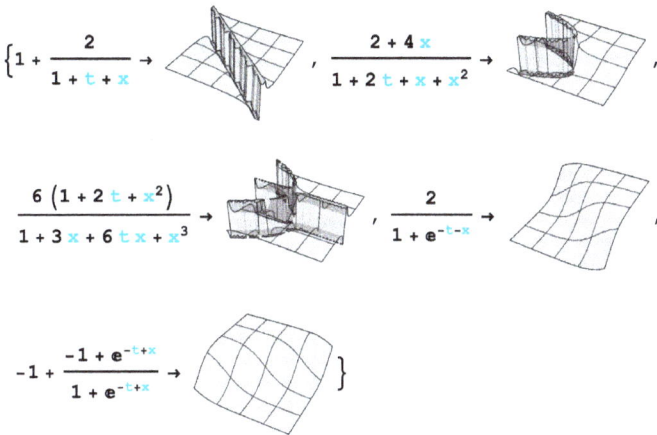

$$\left\{1 + \frac{2}{1 + t + x} \rightarrow \right.$$, $$\frac{2 + 4x}{1 + 2t + x + x^2} \rightarrow$$,

$$\frac{6\left(1 + 2t + x^2\right)}{1 + 3x + 6tx + x^3} \rightarrow$$, $$\frac{2}{1 + e^{-t-x}} \rightarrow$$,

$$-1 + \frac{-1 + e^{-t+x}}{1 + e^{-t+x}} \rightarrow \left. \right\}$$

Figure 9.25: Dynamical plot of the Burgers' functions for problem 9.20.

9.21. A solution for the Burgers' equation different from the above follows:

$$u(t,x) = \frac{e^{-\frac{x^2}{4t}}}{\sqrt{\pi t}\left(\frac{1}{2}\mathrm{erfc}(\frac{x}{2\sqrt{t}}) + \frac{1}{\sqrt{e}-1}\right)}, \quad t > 0, \quad x \in \mathbb{R}.$$

9.21

```
Clear[t,x]
```

The function

$$u[t\_,x\_] := \frac{e^{-\frac{x^2}{4t}}}{\sqrt{\pi t}(\frac{1}{2}\mathrm{Erfc}(\frac{x}{2\sqrt{t}}) + \frac{1}{\sqrt{e}-1})}$$

satisfies the Burgers' equation for d=1. Indeed,

```
∂ₜu[t,x]+u[t,x]∂ₓu[t,x]-∂ₓ,ₓu[t,x]//Simplify
0
```

We plot this Burgers' function.

```
Plot3D[u[t,x],{t,0,3},{x,-3,3},ImageSize→175,
Mesh→None,PlotRange→{{0,3},{-3,3},{0,2}},
PlotStyle→{Gray,Opacity[.15]},
Ticks→{{0,3},{-3,3},{0,2}}]
```

See Figure 9.26.

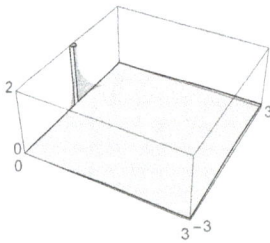

Figure 9.26: Plot of the function in problem 9.21.

This function is displayed for various instants with the commands shown below. Once generated, the sequence can be animated.

```
Manipulate[
Plot[u[.015k,x],{x,-3,3},PlotRange→{0,2.5},
Ticks→{Range[-3,3,3],Range[0,2,1]},PlotStyle→Black,
ImageSize→160,AxesLabel→{x,u}],{k,1,200}]
```

See Figure 9.27.

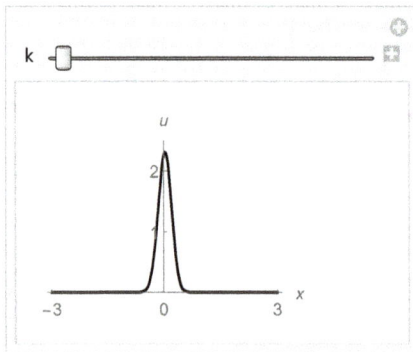

Figure 9.27: Dynamical plot of the function in problem 9.21.

9.5.2 The Burgers' equation in 2D

9.22. Here is an example of an initial and boundary values problem for a Burgers' e-quation in 2D [92, p. 234],

$$
\begin{cases}
\partial_t u(t,x,y) = d\,\nabla^2_{x,y} u(t,x,y) - u(t,x,y)(2\partial_x u(t,x,y) - \partial_y u(t,x,y)), \\
0 \le t \le l/2, \quad -l \le x,y \le l, \\
u(0,x,y) = \exp(-(x^2 + y^2)), \quad u(t,-l,y) = u(t,l,y), \quad u(t,x,-l) = u(t,x,l).
\end{cases}
$$

9.22

```
Clear[d,l,t,x,y]
```

We state the equation and the initial and boundary conditions,

```
eqn=∂ₜu[t,x,y]==d∇²_{x,y}u[t,x,y]
-u[t,x,y](2∂ₓu[t,x,y]-∂ᵧu[t,x,y]);
bound={u[0,x,y]==Exp[-(x²+y²)],u[t,-1,y]==u[t,1,y],
u[t,x,-1]==u[t,x,1]};
```

set the parameters, solve the problem,

```
l=3;d=.75;
sol=NDSolve[{eqn,bound},u,{t,0,l/2},{x,-1,1},{y,-1,1}];
```

and present the solution dynamically.

```
Animate[
Plot3D[u[t,x,y]/.sol,{x,-1,1},{y,-1,1},ImageSize→175,
Mesh→3,PlotRange→All,PlotStyle→{Gray,Opacity[.3]}],
{t,0,l/2},SaveDefinitions→True,DefaultDuration→25,
AnimationDirection→ForwardBackward,
AnimationRunning→False]
```

See Figure 9.28.

9.6 Ansatz methods

Large introductions to the ansatz methods may be found in many references. We only cite [82, Chapter 12] and [22, Appendix A].

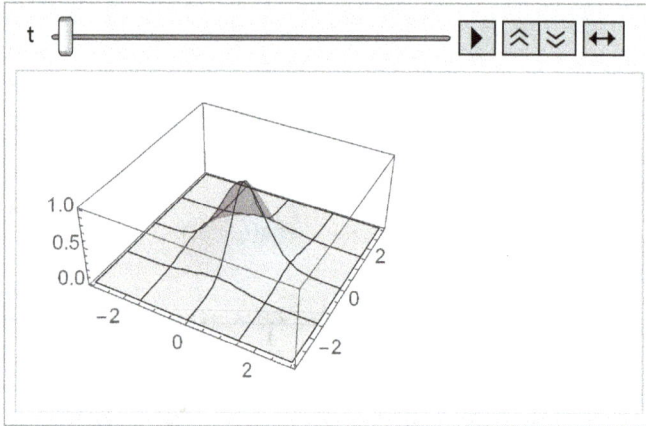

Figure 9.28: Dynamical plot of the solution for problem 9.22.

9.6.1 The tanh function method

This method assumes a traveling wave of the form $u(t, x) = w(\xi)$ with $\xi = k(x - vt)$, where v is the velocity of the wave and k is the wave number.

More assumptions are involved.

(a) We admit that the nonlinear partial differential equation is of the form

$$P(u, u_t, u_{t,t}, u_x, u_{t,x}, u_{x,x}, \dots) = 0. \tag{9.1}$$

(b) We admit that equation (9.1) has the wave variable $\xi = k(x - vt)$ so that

$$u(t, x) = w(\xi).$$

Then the partial derivatives there are as follows:

$$\frac{\partial u}{\partial t} = -kv\frac{dw}{d\xi}, \qquad \frac{\partial^2 u}{\partial t^2} = k^2v^2\frac{d^2 w}{d\xi^2}, \qquad \frac{\partial u}{\partial x} = k\frac{dw}{d\xi}, \tag{9.2}$$

$$\frac{\partial^2 u}{\partial t\partial x} = -k^2v\frac{d^2 w}{d\xi^2}, \qquad \frac{\partial^2 u}{\partial x^2} = k^2\frac{d^2 w}{d\xi^2}, \qquad \frac{\partial^3 u}{\partial x^3} = k^3\frac{d^3 w}{d\xi^3}, \tag{9.3}$$

$$\dots$$

By (9.2) and (9.3), equation (9.1) is transformed into an ordinary differential equation,

$$Q(w, w', w'', \dots) = 0. \tag{9.4}$$

(c) If equation (9.4) does not contain function w, then by integrating this equation and taking the constant of integration zero, the result is an ordinary differential equation whose order is diminished by 1.

(d) A new independent variable is introduced,

$$\zeta = \tanh(\xi),$$

as well as a new function $s(\zeta) = w(\xi)$. Then results that

$$\frac{dw}{d\xi} = (1 - \zeta^2)\frac{ds}{d\zeta}, \quad \frac{d^2 w}{d\xi^2} = (1 - \zeta^2)\left(-2\zeta\frac{ds}{d\zeta} + (1 - \zeta^2)\frac{d^2 s}{d\zeta^2}\right), \quad \text{etc.} \qquad (9.5)$$

(e) Now we introduce the ansatz that there exists a solution having the next representation,

$$s(\zeta) = a_0 + \sum_{n=1}^{m} a_n \zeta^n, \qquad (9.6)$$

or

$$s(\zeta) = a_0 + \sum_{n=1}^{m} a_n \zeta^n + \sum_{v=1}^{\mu} b_v \zeta^{-v}, \qquad (9.7)$$

where m and μ are positive integers that have to be found. By substituting (9.5), (9.6) or (9.7) into (9.4), there yields an equation in ζ. Usually, we balance the linear terms of the highest order with the highest degree of nonlinear terms according to the following formulas for the highest exponents of the function $s(\zeta)$ and its derivatives: $s(\zeta) \rightarrow m$, $s'(\zeta) \rightarrow m + 1$, $s''(\zeta) \rightarrow m + 2$, $s^{(n)}(\zeta) \rightarrow m + n$, and $s^n(\xi) \rightarrow nm$.

(f) This results in a polynomial equation in ζ that is identical null with coefficients k, v, and a_n, $n = \overline{0, m}$. Equating to zero, the coefficients of the like powers of ζ, and we look for consistent solutions of the resulted system of algebraic equations.

Remark. The tanh method is used by problems 9.28, 9.29, 10.19, 10.21, and 10.24. □

Remark. More details could be found in the pioneer works [47, 50], and [51], as well as in the later works [48, 49, 18], and [71, Section 39.3.4.]. □

9.6.2 The Hirota bilinear method

The *Hirota bilinear method* was introduced by Ryōgo Hirota in [28]. This method provides a direct method of finding N-soliton solutions to nonlinear evolutionary partial differential equations.

The Hirota bilinear operator, also called the *Hirota derivative*, is defined by [29],

$$D_x(f \cdot g) = \left(\frac{\partial}{\partial x} - \frac{\partial}{\partial x'} \right) f(x) \cdot g(x')|_{x'=x} = f'(x)g(x) - f(x)g'(x),$$

$$D_x^m = \left(\frac{\partial}{\partial x} - \frac{\partial}{\partial x'} \right)^m = \sum_{k=0}^{m} (-1)^k \binom{m}{k} \partial_x^{m-k} \partial_{x'}^{k} |_{x'=x}, \quad m \in \mathbb{N},$$

and

$$D_x^m D_t^n(f \cdot g) = \left(\frac{\partial}{\partial x} - \frac{\partial}{\partial x'} \right)^m \left(\frac{\partial}{\partial t} - \frac{\partial}{\partial t'} \right)^n f(t,x) \cdot g(t',x')|_{x'=x,\ t'=t} \tag{9.8}$$

Proposition 9.23. *The bilinear operator, just defined D, satisfies the next properties:*

$$D_x(f \cdot g) = f_x g - f g_x = -D_x(g \cdot f), \quad \text{(anticommutative)}$$

$$D_x^2(f \cdot g) = f_{xx} g - 2 f_x g_x + f g_{xx} = D_x^2(g \cdot f),$$

$$D_x D_t(f \cdot g) = f_{xt} g - f_t g_x - f_x g_t + f g_{xt} = D_x D_t(g \cdot f),$$

$$D_x D_t(f \cdot f) = 2(f f_{xt} - f_x f_t), \tag{9.9}$$

$$D_x^4(f \cdot g) = f_{xxxx} g - 4 f_{xxx} g_x + 6 f_{xx} g_{xx} - 4 f_x g_{xxx} + f g_{xxxx} = D_x^4(g \cdot f),$$

$$D_x^n(f \cdot f) = 0, \quad n \in \mathbb{N}, \quad odd.$$

This method assumes the existence of the *ancillary function f*, which is a logarithmic transformation of the form

$$u(t,x) = c \frac{\partial^2 \ln f(t,x)}{\partial x^2}, \tag{9.10}$$

where the constant c depends on the problem.

Function f in (9.10) is chosen of the form

$$f(t,x) = 1 + \sum_{n=1}^{\infty} e^n f_n(t,x), \tag{9.11}$$

where e is a formal expansion parameter and functions $f_1, f_2, \ldots$ yet to be found.

Proposition 9.24. *Let us suppose B is a bilinear application with $B(f \cdot f) = 0$, then the coefficients of like powers of e are obtained by the following set of equations:*

$$B(1 \cdot 1) = 0, \tag{9.12}$$

$$B(1 \cdot f_1 + f_1 \cdot 1) = 0, \tag{9.13}$$

$$B(1 \cdot f_2 + f_1 \cdot f_1 + f_2 \cdot 1) = 0, \tag{9.14}$$

$$B(1 \cdot f_3 + f_1 \cdot f_2 + f_2 \cdot f_1 + f_3 \cdot 1) = 0, \tag{9.15}$$

$$B(1 \cdot f_4 + f_1 \cdot f_3 + f_2 \cdot f_2 + f_3 \cdot f_1 + f_4 \cdot 1) = 0, \tag{9.16}$$

$$\cdots$$

Remark. It can be shown that if the nonlinear partial differential equation of evolution admits an N-soliton solution and f is a sum of exactly N simple exponential terms, then (9.11) is truncate at the $n = N$ term. Under such assumptions, system (9.12)–(9.16) is finite. □

Proposition 9.25. (a) *We assume that* $m = n = 1$ *in (9.8). Then*

$$D_x D_t (1 \cdot f_1) = D_x(-1 \cdot f_{1,t}) = f_{1,xt},$$
$$D_x D_t (f_1 \cdot 1) = D_x(f_{1,t} \cdot 1) = f_{1,xt},$$
$$D_x D_t (1 \cdot f_1 + f_1 \cdot 1) = 2f_{1,xt}.$$

(b) *We assume that* $m = 4$ *and* $n = 0$ *in (9.8). Then*

$$D_x^4 (1 \cdot f_1) = f_{1,xxxx},$$
$$D_x^4 (f_1 \cdot 1) = f_{1,xxxx},$$
$$D_x^4 (1 \cdot f_1 + f_1 \cdot 1) = 2f_{1,xxxx}.$$

(c) *By* (a) *and* (b), *we get that*

$$(D_x D_t + D_x^4)(1 \cdot f_1 + f_1 \cdot 1) = 2(f_{1,xt} + f_{1,xxxx}). \tag{9.17}$$

(d) Similar to (c), we have that

$$(D_x D_t + D_x^4)(1 \cdot f_2 + f_1 \cdot f_1 + f_2 \cdot 1) = 2(f_1 f_{1,xt} - f_{1,x} f_{1,t})$$
$$+ 2(f_{2,xt} + f_{2,xxxx} + f_1 f_{1,xxxx} - 4f_{1,x} f_{1,xxx} + 3f_{1,xx}^2). \tag{9.18}$$

Remark. The Hirota bilinear method is used by problems 9.30, 10.8, 10.9, 10.11, 10.10, and 10.29. □

Remark. The Hirota bilinear method is also discussed in, e. g., [26, 31, 30, 27], and [14]. □

9.6.3 The exp function method

This method assumes a traveling wave solution of the form $u(t, x) = w(\xi)$ with $\xi = k(x - vt)$. By this substitution, we suppose that the nonlinear partial differential equation is transformed into an ordinary differential equation. Afterwards, the method attempts to find solutions of the form

$$w(\xi) = \frac{\sum_{n=-d}^{c} a_n \exp(n\,\xi)}{\sum_{m=-q}^{p} b_m \exp(m\,\xi)}, \quad c, d, p, q \geq 0, \quad a_n, b_m \in \mathbb{R}.$$

A complete example of usage of this method is introduced in Section 10.1.2.

Remark. The exp function method is used by problems 10.13, 10.18, and 10.20. □

Remark. More details could be found in, e. g., [24]. □

9.6.4 The sine-cosine functions method

This method supposes that an equation (9.1) is given and by the wave variable $\xi = k(x - v\,t)$ is reduced to an ordinary differential of the form (9.4). We look for solutions of the forms $w(\xi) = \lambda \cos^\beta(\xi)$ or $w(\xi) = \lambda \sin^\beta(\xi)$ of equation (9.4). The parameter β is determined by balancing the exponents of each pair of sine or cosine functions. Then we gather all coefficients of the like powers in $\sin^k(\xi)$ or $\cos^k(\xi)$, where these coefficients have to vanish. The system of algebraic equations thus obtained helps us to find the parameters λ, k, and v.

Remark. The sine-cosine method is used by problems 10.22, 10.23, and 10.31. □

Remark. A presentation of this method may be found in, e. g., in [81, 2.1] and [97]. □

9.7 The Fisher equation

The *Fisher equation with convection* term is given below [25, (61)],

$$\partial_t u(t,x) + \kappa\, u(t,x)\partial_x u(t,x) - \partial_{x,x} u(t,x) - u(t,x)\big(1 - u(t,x)\big) = 0,$$
$$t, x, \kappa \in \mathbb{R}. \tag{9.19}$$

For $\kappa = 0$, one has the original *Fisher equation without convection* term.
    The *generalized Fisher equation* is of the form [79, (8)],

$$\partial_t u(t,x) = \partial_{x,x} u(t,x) + u(t,x)\big(1 - u^2(t,x)\big) = 0, \quad t, x \in \mathbb{R}. \tag{9.20}$$

The *nonlinear Fisher equation* is of the form [79, (28)],

$$\partial_t u(t,x) = \partial_{x,x} u(t,x) + 2u(t,x)\big(1 - u^2(t,x)\big) + \mu\big(1 - u^2(t,x)\big) = 0,$$
$$|\mu| < 1, \quad t, x \in \mathbb{R}. \tag{9.21}$$

9.7.1 The Fisher equation without convection term by *Mathematica*

The Fisher equation without a convection term by *Mathematica* is also discussed in [25, (61)].

9.26. We study equation (9.19) with $\kappa = 0$ by the power of *Mathematica*.

9.26

```
Clear[t,x]
eqn=∂_t u[t,x]-∂_{x,x} u[t,x]-u[t,x](1-u[t,x])==0;  (* The
equation *)
```

We solve the Fisher equation by *Mathematica*, thus resulting altogether eight real and complex solutions.

```
sol=u[t,x]/.DSolve[eqn,u,{t,x}]//Flatten//
ComplexExpand//FullSimplify;
Length[sol]
8
```

Four solutions are real and all others are complex. We select the real solutions

```
solreals=Select[sol,FreeQ[#,Complex]&]/.C[3]→1;
(* Select the real solutions for a particular value
of the constant of integration *)
Length[solreals]
4
```

and successively plot these solutions.

```
If[Length[solreals]>0,
Plot3D[solreals[[#]],{t,-5,5},{x,-5,5},Mesh→ None,
Ticks→{5{-1,0,1},5{-1,0,1},{0,1}},ImageSize→160,
PlotStyle→{Gray,Opacity[.2]}]
&/@Range[Length[sol]]]
```

See Figure 9.29.
Then we plot them all together

```
If[Length[solreals]>0,
Plot3D[solreals,{t,-5,5},{x,-5,5},Mesh→None,
Ticks→Outer[Times,{5,5,1},{-1,0,1}],
ColorFunction→Function[{x,y,z},Hue[Abs[y+5]/10+z]],
PlotStyle→{LightBlue,Opacity[.2]},ImageSize→160]]
```

See Figure 9.30.

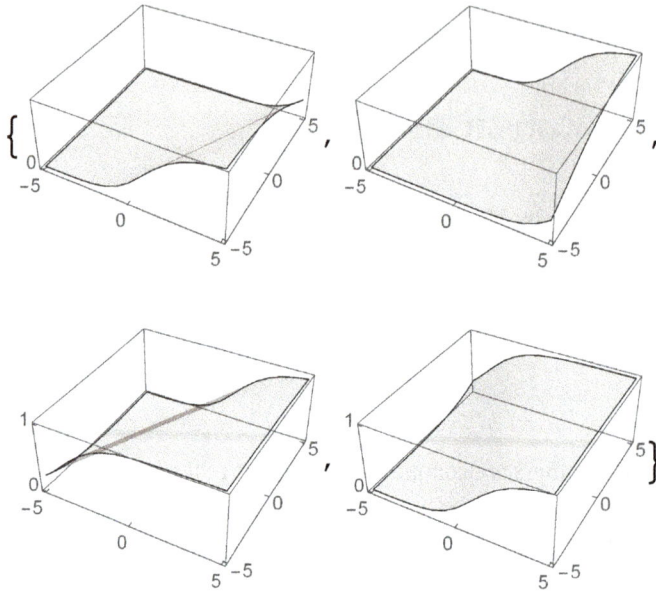

Figure 9.29: Successive plot of the real solutions for problem 9.26.

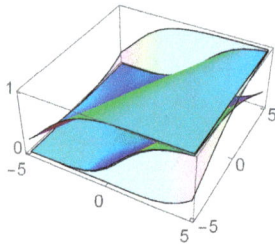

Figure 9.30: Plot of the real solutions for problem 9.26.

We insert a dynamical plot of the real solutions.

```
If[Length[solreals]>0,
Manipulate[
Plot3D[sol,{t,-τ,τ},{x,-ξ,ξ},Mesh→None,
Ticks→{{-τ,0,τ},{-ξ,0,ξ},{0,1}},ImageSize→185,
PlotStyle→Directive[Gray,Opacity[.1]],{τ,.01,5},
{ξ,.01,5},SaveDefinitions→True]]
```

See Figure 9.31.

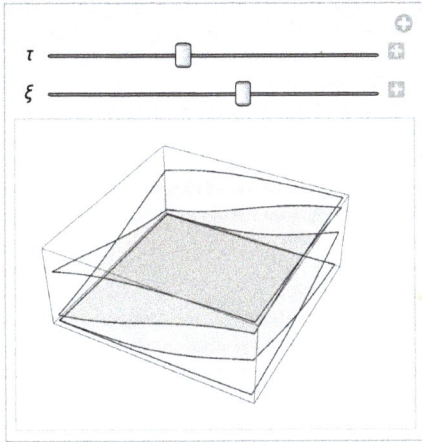

Figure 9.31: Dynamical plot of the real solutions for problem 9.26.

9.7.2 The Fisher equation with convection term by *Mathematica*

9.27. We study equation (9.19) by the power of *Mathematica*.

9.27 We solve the equation,

```
Clear[t,x,κ]
eqn=∂ₜu[t,x]+κ u[t,x]∂ₓu(t,x)-∂ₓ,ₓu[t,x]
-u[t,x](1-u[t,x])==0;  (* The equation *)

sol=u[t,x]/.DSolve[eqn,u,{t,x}]//Simplify//Flatten
Length[sol]
{1/2(1+Tanh[1/8(4t-2x κ+t κ²+8C[3])]),
1/2(1-Tanh[(x κ)/4-1/8t(4+κ²)+C[3]])}
2
```

We select Length[sol] particular solutions

```
sol=Table[sol[[k]]/.{C[3]→1,κ →2},{k,Length[sol]}]//
FullSimplify;
```

and plot them.

```
Plot3D[sol[[#]],{t,-7,5},{x,-5,5},Mesh→None,
Ticks→Outer[Times,{5,5,1},{-1,0,1}],
PlotStyle→{Gray,Opacity[.2]},ImageSize→160]&/@
Range[Length[sol]]
```

See Figure 9.32.

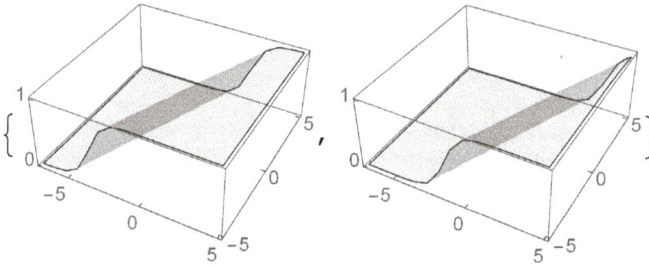

Figure 9.32: Plot of some particular solutions for problem 9.27.

9.7.3 The Fisher equation with a convection term by the tanh method

9.28. The Fisher equation with a convection term has the form

$$\partial_t u(t,x) + \kappa\, u(t,x)\partial_x u(t,x) - \partial_{x,x} u(t,x) - u(t,x)\big(1 - u(t,x)\big) = 0,$$
$$\kappa, t, x \in \mathbb{R} \quad \kappa \neq 0.$$

9.28 We introduce the equation,

```
Clear[t,x,k,κ]
eqn=∂_t u[t,x]+κ u[t,x]∂_x u[t,x]-∂_x,x u[t,x]
-u[t,x](1-u[t,x])==0;
```

We admit that the previous equation has a traveling wave variable $\xi = k(x - v\,t)$, and thus $u[t, x] = w[\xi]$. Then

```
∂_t u[t,x]=-k v w'[ξ]; ∂_x u[t,x]=k w'[ξ];
∂_x,x u[t,x]=k²w'[ξ]; (* and the equation *)
-k v w'[ξ]+κ k w[ξ]w'[ξ]-k²w''[ξ]-w[ξ]+w[ξ]²=0;
```

We introduce a new independent variable and the corresponding function,

```
ζ=Tanh[ξ]; w[ξ]=s[ζ]; w'[ξ]=(1-ζ²)s'[ζ]
w''[ξ]=-2ζ(1-ζ²)s'[ζ]+(1-ζ²)²s''[ζ] (* and the
equation *)
-k v (1-ξ²)s'[ζ]+κ k(1-ζ²)s[ζ]s'[ζ]
-k²(-2ζ(1-ζ²)s'[ζ]+(1-ζ²)²s''[ζ])-s[ζ]+s[ζ]²=0;
```

We apply the tanh method (see Section 9.6.1), consider polynomial $f[\zeta] = a_0 + \sum_{n=1}^{m} a_n \zeta^n$, m being a natural number, and substitute this finite sum into the equation

```
-k v (1-ξ²)f'[ζ]+κ k(1-ζ²)f[ζ]f'[ζ]
-k²(-2ζ(1-ζ²)f'[ζ]+(1-ζ²)f''[ζ])-f[ζ]+f[ζ]²=0;
```

By balancing the highest degree term $2m$ with the highest-order derivative $m + 2$, it follows that $m = 2$. Thus,

```
Clear[a,k,v,ζ]
f[ζ_]:=a₀+∑²ₙ₌₁aₙζⁿ
eqn=-k v(1-ζ²)f'[ζ]+κ k f[ζ](1-ζ²)f'[ζ]
-k²(-2ζ(1-ζ²)f'[ζ]+(1-ζ²)²f''[ζ])-f[ζ]+f[ζ]²//Simplify;

Series[eqn,{ζ,0,6}];
Clear[k]
Reduce[-a₀+a₀²-k v a₁+k κa₀a₁-2k²a₂==0&&
-a₁+2k²a₁+2a₀a₁+kκa₁²-2k v a₂+2kκa₀a₂==0&&
k v a₁-kκa₀a₁+a₁²-a₂+8k²a₂+2a₀a₂+3kκa₁a₂==0&&
-2k²a₁-kκa₁²+2k v a₂-2kκa₀a₂+2a₁a₂+2kκa₂²==0&&
-6k²a₂-3kκa₁a₂+a₂²==0&&kκa₂²==0,{a₀,a₁,a₂,k,v}];
```

The feasible solutions of this system of algebraic equations are as follows:

$$a_0 = \frac{1}{2}; a_1 = \pm\frac{1}{2}; a_2=0; k=-\frac{1}{2}\kappa\, a_1; v=\frac{4+\kappa^2}{2\kappa};$$

Thus, the solution is of the form

```
u[t,x]:=a₀+a₁ Tanh[k(x-v t)]
```

Because Tanh is an odd function, it is enough considering $a_1 = +\frac{1}{2}$. With this value for a_1, function u satisfies the equation:

```
∂ₜu[t,x]+κu[t,x]∂ₓu[t,x]-∂ₓ,ₓu[t,x]-u[t,x](1-u[t,x])
//FullSimplify
0
```

   We plot this solution for a particular value of parameter κ.

```
Block[{κ=1},
Plot3D[Evaluate[u[t,x]],{t,-10,10},{x,-10,10},
PlotStyle→{Gray,Opacity[.2]},ImageSize→175,
Ticks→Outer[Times,{10,10,1},{-1,0,1}],
Mesh→None]]
```

See Figure 9.33.

Remark. A new solution arises if we consider that $m = -1$, and thus

```
f[ζ_]:=a₀+a₋₁/ζ;
```

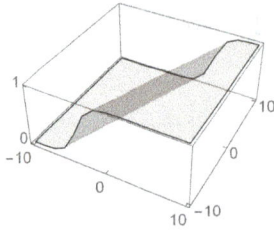

Then

```
eqn=-k v (1-ξ²)f'[ζ]+κ k(1-ζ²)f[ζ]f'[ζ]
-k²(-2ζ(1-ζ²)f'[ζ]+(1-ζ²)f''[ζ])-f[ζ]+f[ζ]²//Simplify
Series[eqn,{ζ,0,4}];
```

The following system of algebraic equations results in

```
Reduce[-2k²a₋₁-kκa²₋₁==0&&
k v a₋₁+a²₋₁-kκa₋₁a₀==0&&
-a₋₁+2k²a₋₁+kκa²₋₁a₀+2a₋₁a₀==&&
-k v a₋₁-a₀+kκa₋₁a₀+a²₀==0,{a₀,a₋₁,k,v}];
```

Its solutions follow

$$a_0 = \frac{1}{2} ; a_{-1} = \pm\frac{1}{2} ; k=-\frac{1}{2}\kappa a_{-1} ; v=\frac{4+\kappa^2}{2\kappa}$$

With these results, a single solution is obtained that reads

```
u[t_,x_]:=½-½Coth[¼κ(x-(4+κ²)/(2κ))]
∂ₜu[t,x]+κu[t,x]∂ₓu(t,x)-∂ₓ,ₓu[t,x]-u[t,x](1-u[t,x])
//Simplify (* The solution satisfies the equation *)
0
```

having the graph (for a particular value of κ),

```
Block[{κ=1},
Plot3D[Evaluate[u[t,x]],{t,-10,10},{x,-10,10},
Ticks→Outer[Times,{10,10,1},{-1,0,1}],Mesh→None,
PlotStyle→{Gray,Opacity[.2]},ImageSize→175,
PlotPoints→50]]
```

See Figure 9.34. □

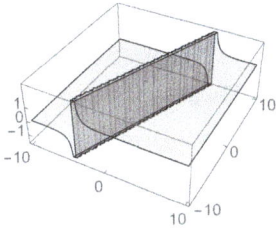

9.7.4 The generalized Fisher equation by the tanh method

This equation is also discussed in [79, (8)].

9.29. We study the generalized Fisher equation (9.20).

9.29 We introduce the equation under discussion,

```
Clear[t,x]
eq=∂_t u[t,x]-∂_x,x u[t,x]+u[t,x](1-u[t,x]²)==0;
```

We admit that this equation has a wave variable $\xi = k(x - v t)$, and thus $u(t, x) = w(\xi)$. We change the independent and the dependent variables. Then

```
ξ=k(x-vt);u[t,x]=w[ξ];∂_t u[t,x]=-k v w'[ξ];
∂_x,x u[t,x]=k²w''[ξ];
k v w'[ξ]+k²w''[ξ]+w[ξ]-w[ξ]³=0 (* The equation *)
```

Again, we change the independent and the dependent variable. Then

```
ζ=Tanh[ξ];w[ξ]=s[ζ];w'[ξ]=(1-ζ²)s'[ζ];
w''[ξ]=-2ζ(1-ζ²)s'[ζ]+(1-ζ²)²s''[ζ] (* the equation *)
k v (1-ζ²)s'[ζ]+k²(-2ζ(1-ζ²)s'[ζ]+(1-ζ²)²s''[ζ])+s[ζ]
-s[ζ]³=0
```

We apply the tanh method (see Section 9.6.1) and consider the polynomial $f[\zeta] = a_0 + \sum_{n=1}^{m} a_n \zeta^n$, where m is a natural number. We substitute this polynomial into the equation

```
k v (1-ζ²)f'[ζ]+k²(-2ζ(1-ζ²)f'[ζ]+(1-ζ²)²f''[ζ])+f[ζ]
-f[ζ]³=0
```

By balancing the highest-order derivative f'' with the highest degree f^3, it follows that $3m = m + 2$, and thus $m = 1$. Then

```
Clear[k,v,ζ,a]
f[ζ_]:=a₀+∑ₙ₌₁¹ aₙζⁿ;
eq=k v (1-ζ²)f'[ζ]+k²(-2ζ(1-ζ²)f'[ζ]+(1-ζ²)²f''[ζ])+f[ζ]
-f[ζ]³=0;
```

Further, we have that

```
Series[eq,{ζ,0,4}]//Simplify;
Reduce[-a₀+a₀³-k v a₁==0&&
-a₁+2k²a₁+3a₀²a₁==0&&
k v a₁+3a₀a₁²==0&&
-2k²a₁+a₁³==0,{a₀,a₁,k,v}];
```

We first consider the case,

```
(a₀==-1∨a₀==0∨a₀==1)&&a₁==0
```

Then one has the solutions:

```
1. Constant solution
u₁[t,x]=-1
2. Constant solution
u₂[t,x]=0
3. Constant solution
u₃[t,x]=1
```

All three solutions satisfy the equation. Indeed,

```
D[#,t]-D[#,x,x]-#(1-#²)]&/@{u₁[t,x],u₂[t,x],u₃[t,x]}
{0,0,0}
```

The plot of the constant solutions is as follows:

```
Plot3D[Evaluate[{u₁[t,x],u₂[t,x],u₃[t,x]}],{t,-3,3},
{x,-3,3},ImageSize→160,Mesh→None,PlotRange→All,
PlotStyle→Table[{Gray,Opacity[k]},{k,.2,.1,-.05}],
Ticks→Outer[Times,{3,3,1},{-1,0,1}],
PlotLegends→SwatchLegend["Expressions",
LegendLabel→"surfaces",
LegendFunction→(Framed[#,Background→LightBlue]&)]]
```

See Figure 9.35.

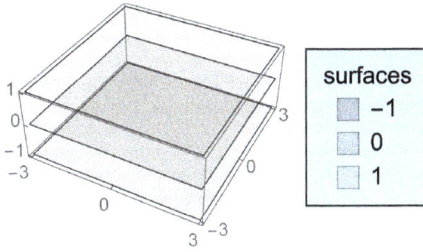

Figure 9.35: Plot of the constant solutions u for problem 9.29.

Now we consider the second case, namely

$$(a_0==-1/2 \vee a_0==0 \vee a_0==1/2) \&\&$$
$$(a_1==-\sqrt{1-3a_0^2} \vee a_1==\sqrt{1-3a_0^2}) \&\&$$
$$(k==-\frac{\sqrt{1-3a_0^2}}{\sqrt{2}} \vee k==-\frac{\sqrt{1-3a_0^2}}{\sqrt{2}}) \&\& v==-24k\, a_0 a_1$$

We put the coefficients into Table 9.2.

Table 9.2: Table of coefficients of solutions $u_4 - u_{15}$ for problem 9.29.

	a_0	a_1	k	v
1	−1/2	−1/2	−1/(2√2)	3/√2
2	−1/2	−1/2	1/(2√2)	−3/√2
3	−1/2	1/2	−1/(2√2)	−3/√2
4	−1/2	1/2	1/(2√2)	3/√2
5	0	−1	−1/√2	0
6	0	−1	1/√2	0
7	0	1	−1/√2	0
8	0	1	1/√2	0
9	1/2	−1/2	−1/(2√2)	−3/√2
10	1/2	−1/2	1/(2√2)	3/√2
11	1/2	1/2	−1/(2√2)	3/√2
12	1/2	1/2	1/(2√2)	−3/√2

Based on it, we define the functions:

$u_4[t\_,x\_]:=-\frac{1}{2}+\frac{1}{2}Tanh[\frac{1}{2\sqrt{2}}(x-\frac{3}{\sqrt{2}}t)]$

$u_5[t\_,x\_]:=-\frac{1}{2}-\frac{1}{2}Tanh[\frac{1}{2\sqrt{2}}(x+\frac{3}{\sqrt{2}}t)]$

$u_6[t\_,x\_]:=-\frac{1}{2}-\frac{1}{2}Tanh[\frac{1}{2\sqrt{2}}(x+\frac{3}{\sqrt{2}}t)]$ (* It coincides with u_5 *)

$u_7[t\_,x\_]:=-\frac{1}{2}+\frac{1}{2}Tanh[\frac{1}{2\sqrt{2}}(x-\frac{3}{\sqrt{2}}t)]$ (* It coincides with u_4 *)

$u_8[t\_,x\_]:=Tanh[\frac{1}{\sqrt{2}}x]$

$u_9[t\_,x\_]:=-Tanh[\frac{1}{\sqrt{2}}x]$

$u_{10}[t\_,x\_]:=-Tanh[\frac{1}{\sqrt{2}}x]$ (* It coincides with u_9 *)

$u_{11}[t\_,x\_]:=Tanh[\frac{1}{\sqrt{2}}x]$ (* It coincides with u_8 *)

$u_{12}[t\_,x\_]:=\frac{1}{2}+\frac{1}{2}Tanh[\frac{1}{2\sqrt{2}}(x+\frac{3}{\sqrt{2}}t)]$

$u_{13}[t\_,x\_]:=\frac{1}{2}-\frac{1}{2}Tanh[\frac{1}{2\sqrt{2}}(x-\frac{3}{\sqrt{2}}t)]$

$u_{14}[t\_,x\_]:=\frac{1}{2}-\frac{1}{2}Tanh[\frac{1}{2\sqrt{2}}(x-\frac{3}{\sqrt{2}}t)]$ (* It coincides with u_{13} *)

$u_{15}[t\_,x\_]:=\frac{1}{2}+\frac{1}{2}Tanh[\frac{1}{2\sqrt{2}}(x+\frac{3}{\sqrt{2}}t)]$ (* It coincides with u_{12} *)

Now we check that all these functions are indeed solutions,

```
D[#,t]-D[#,x,x]-#(1-#^2)&/@Table[u_j[t,x],{j,4,15}]
//Simplify
{0,0,0,0,0,0,0,0,0,0,0,0}
```

and plot them.

```
Plot3D[Evaluate[Table[u_j[t,x],{j,4,5}]],{t,-3,3},
{x,-3,3},ImageSize→160,Mesh→None,PlotRange→All,
PlotStyle→{{Gray,Opacity[.15]},{Gray,Opacity[.1]}},
Ticks→Outer[Times,{3,3,1},{-1,0,1}],
PlotLegends→SwatchLegend["Expressions",
LegendLabel→"surfaces u_4 - u_5",LabelStyle→11,
LegendFunction→(Framed[#,Background→{Gray,
Opacity[.4]}]&)]]
```

See Figure 9.36.

```
Plot3D[Evaluate[Table[u_j[t,x],{j,8,9}]],{t,-3,3},
{x,-3,3},ImageSize→160,Mesh→None,PlotRange→All,
PlotStyle→{{Gray,Opacity[.15]},{Gray,Opacity[.1]}},
Ticks→Outer[Times,{3,3,1},{-1,0,1}],
PlotLegends→SwatchLegend["Expressions",
LegendLabel→"surfaces u_8 - u_9",LabelStyle→11,
LegendFunction→(Framed[#,Background→{Gray,
Opacity[.4]}]&)]]
```

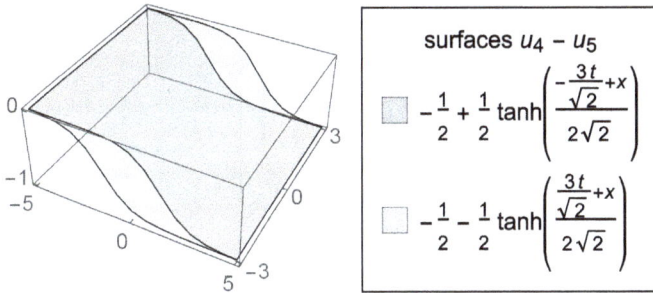

Figure 9.36: Plot of the solutions $u_4 - u_5$ for problem 9.29.

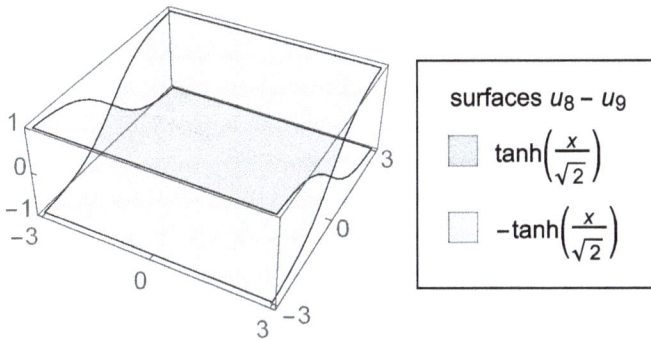

Figure 9.37: Plot of the solutions $u_8 - u_9$ for problem 9.29.

See Figure 9.37.

```
Plot3D[Evaluate[Table[u_j[t,x],{j,12,13}]],{t,-5,5},
{x,-3,3},ImageSize→160,Mesh→None,PlotRange→All,
PlotStyle→{{Gray,Opacity[.15]},{Gray,Opacity[.1]}},
Ticks→Outer[Times,{5,3,1},{-1,0,1}],
PlotLegends→SwatchLegend["Expressions",
LegendLabel→"surfaces u_12 - u_13",LabelStyle→11,
LegendFunction→(Framed[#,Background→{Gray,
Opacity[.4]}]&)]]
```

See Figure 9.38.

Now try finding a solution of the form,

```
Clear[a,ζ,k,v]
f[ζ_]:=a_0+∑_{n=-1}^{-1} a_n ζ^n
```

We repeat all the steps above, but this time we denote the solutions as q_j.

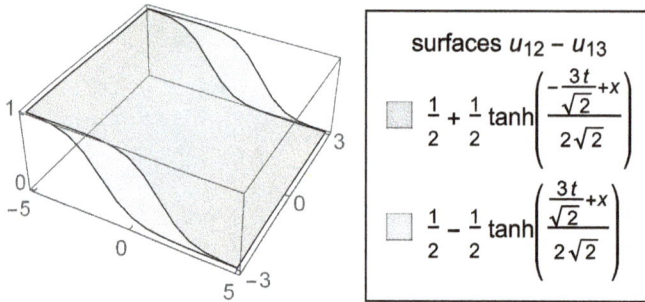

surfaces $u_{12} - u_{13}$

$$\frac{1}{2} + \frac{1}{2}\tanh\left(\frac{-\frac{3t}{\sqrt{2}}+x}{2\sqrt{2}}\right)$$

$$\frac{1}{2} - \frac{1}{2}\tanh\left(\frac{\frac{3t}{\sqrt{2}}+x}{2\sqrt{2}}\right)$$

Figure 9.38: Plot of the solutions $u_{12} - u_{13}$ for problem 9.29.

We first consider the case,

```
(a₀==-1∨a₀==0∨a₀==1)&&a₋₁==0
```

Then one has the solutions:

```
1. Constant solution
q₁[t,x]=-1
2. Constant solution
q₂[t,x]=0
3. Constant solution
q₃[t,x]=1
```

All three solutions satisfy the equation. Indeed,

```
D[#,t]-D[#,x,x]-#(1-#²)]&/@Table[qₙ[t,x],{n,3}]
{0,0,0}
```

Plot of the constant solutions:

```
Plot3D[Evaluate[Table[qₙ[t,x],{n,3}]],{t,-3,3},
{x,-3,3},ImageSize→160,Mesh→None,PlotRange→All,
PlotStyle→Table[{Gray,Opacity[k]},{k,.2,.1,-.05}],
Ticks→Outer[Times,{3,3,1},{-1,0,1}],PlotLegends→
SwatchLegend["Expressions",LegendLabel→"surfaces
q₁ - q₃",LabelStyle→11,LegendFunction→(Framed[
#,Background→{Gray,Opacity[.4]}]&)]]
```

See Figure 9.39.

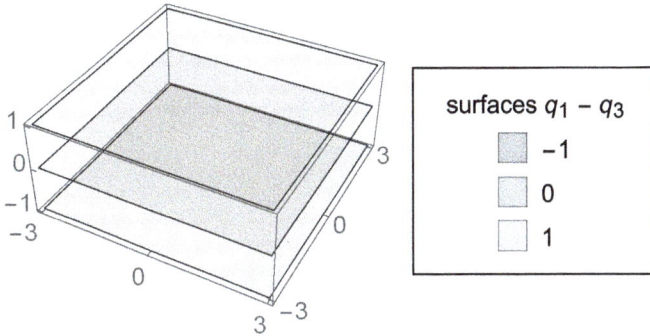

Figure 9.39: Plot of the constant solutions q for problem 9.29.

Further, one has

$q_4[t\_,x\_]:=-\frac{1}{2}+\frac{1}{2}\text{Coth}[\frac{1}{2\sqrt{2}}(x-\frac{3}{\sqrt{2}}t)]$

$q_5[t\_,x\_]:=-\frac{1}{2}-\frac{1}{2}\text{Coth}[\frac{1}{2\sqrt{2}}(x+\frac{3}{\sqrt{2}}t)]$

$q_6[t\_,x\_]:=-\frac{1}{2}-\frac{1}{2}\text{Coth}[\frac{1}{2\sqrt{2}}(x+\frac{3}{\sqrt{2}}t)]$ (* It coincides with q_5 *)

$q_7[t\_,x\_]:=-\frac{1}{2}+\frac{1}{2}\text{Coth}[\frac{1}{2\sqrt{2}}(x-\frac{3}{\sqrt{2}}t)]$ (* It coincides with q_4 *)

$q_8[t\_,x\_]:=\text{Coth}[\frac{1}{\sqrt{2}}x]$

$q_9[t\_,x\_]:=-\text{Coth}[\frac{1}{\sqrt{2}}x]$

$q_{10}[t\_,x\_]:=-\text{Coth}[\frac{1}{\sqrt{2}}x]$ (* It coincides with q_9 *)

$q_{11}[t\_,x\_]:=\text{Coth}[\frac{1}{\sqrt{2}}x]$ (* It coincides with q_8 *)

$q_{12}[t\_,x\_]:=\frac{1}{2}+\frac{1}{2}\text{Coth}[\frac{1}{2\sqrt{2}}(x+\frac{3}{\sqrt{2}}t)]$

$q_{13}[t\_,x\_]:=\frac{1}{2}-\frac{1}{2}\text{Coth}[\frac{1}{2\sqrt{2}}(x-\frac{3}{\sqrt{2}}t)]$

$q_{14}[t\_,x\_]:=\frac{1}{2}-\frac{1}{2}\text{Coth}[\frac{1}{2\sqrt{2}}(x-\frac{3}{\sqrt{2}}t)]$ (* It coincides with q_{13} *)

$q_{15}[t\_,x\_]:=\frac{1}{2}+\frac{1}{2}\text{Tanh}[\frac{1}{2\sqrt{2}}(x+\frac{3}{\sqrt{2}}t)]$ (* It coincides with q_{12} *)

We check that all these functions are indeed solutions:

```
D[#,t]-D[#,x,x]-#(1-#²)&/@Table[q_j[t,x],{j,4,15}]
//Simplify
{0,0,0,0,0,0,0,0,0,0,0,0}
```

```
Plot3D[Evaluate[#],{t,-1,1},{x,-1,1},ImageSize→160,
Mesh→None,PlotRange→All,PlotPoints→50,
PlotStyle→{Gray,Opacity[.2]},Ticks→{{-1,0,1},{-1,0,1},
{0}}]&/@Table[q_j[t,x],{j,4,5}]
```

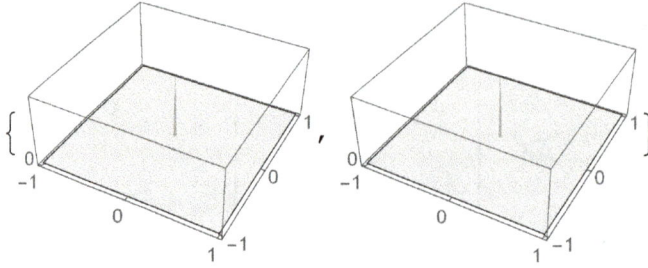

Figure 9.40: Plot of the solutions $q_4 - q_5$ for problem 9.29.

See Figure 9.40.

Remark. The peaks in Figure 9.40 appear due to the hyperbolic cotangent function. □

```
Plot3D[Evaluate[#],{t,-1,1},{x,-1,1},ImageSize→160,
PlotStyle→{Gray,Opacity[.2]},PlotPoints→50,
Mesh→None,PlotRange→All,
Ticks→{{-1,0,1},{-1,0,1},{0}}]&/@
Table[q_j[t,x],{j,8,9}]
```

See Figure 9.41.

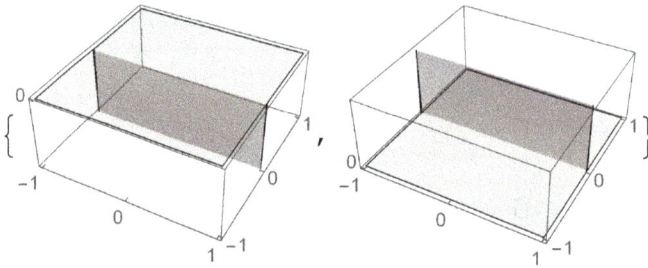

Figure 9.41: Plot of the solutions $q_8 - q_9$ for problem 9.29.

```
Plot3D[Evaluate[#],{t,-1,1},{x,-1,1},ImageSize→160,
Mesh→None,PlotRange→All,PlotStyle→{Gray,
Opacity[.2]},PlotPoints→50,Ticks→{{-1,0,1},{-1,0,1},
{0}}]&/@Table[q_j[t,x],{j,12,13}]
```

See Figure 9.42.

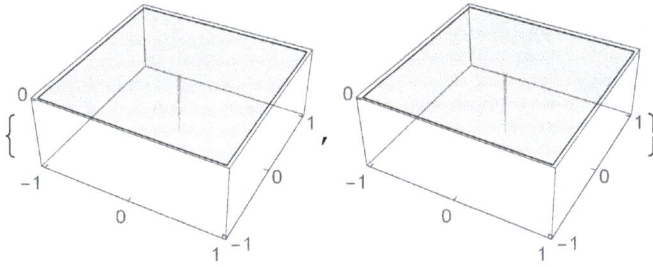

Figure 9.42: Plot of the solutions $q_{12} - q_{13}$ for problem 9.29.

Now we try finding a solution of the form:

```
Clear[a,ζ,k,v]
f[ζ_]:=a₋₁ζ⁻¹+a₀+a₁ζ, with a₋₁,a₁ ≠0
```

We repeat all the steps above, but this time we denote the solutions by y_j.
The results are shown in Table 9.3.

Table 9.3: Table of coefficients of solutions $y_1 - y_8$ for problem 9.29.

	a_{-1}	a_0	a_1	k	v
1	1/4	1/2	1/4	$1/(4\sqrt{2})$	$-3/\sqrt{2}$
2	1/4	1/2	1/4	$-1/(4\sqrt{2})$	$3/\sqrt{2}$
3	1/4	-1/2	1/4	$1/(4\sqrt{2})$	$3/\sqrt{2}$
4	1/4	-1/2	1/4	$-1/(4\sqrt{2})$	$-3/\sqrt{2}$
5	-1/4	1/2	-1/4	$1/(4\sqrt{2})$	$3/\sqrt{2}$
6	-1/4	1/2	-1/4	$-1/(4\sqrt{2})$	$-3/\sqrt{2}$
7	-1/4	-1/2	-1/4	$1/(4\sqrt{2})$	$-3/\sqrt{2}$
8	1/4	-1/2	-1/4	$-1/(4\sqrt{2})$	$3/\sqrt{2}$

Then one has the following solutions:

$$y_1[t\_,x\_]:=\frac{1}{4}\text{Coth}[\frac{1}{4\sqrt{2}}(x+\frac{3}{\sqrt{2}}t)]+\frac{1}{2}+\frac{1}{4}\text{Tanh}[\frac{1}{4\sqrt{2}}(x+\frac{3}{\sqrt{2}}t)]$$

$$y_2[t\_,x\_]:=\frac{1}{4}\text{Coth}[-\frac{1}{4\sqrt{2}}(x-\frac{3}{\sqrt{2}}t)]+\frac{1}{2}+\frac{1}{4}\text{Tanh}[-\frac{1}{4\sqrt{2}}(x-\frac{3}{\sqrt{2}}t)]$$

$$y_3[t\_,x\_]:=\frac{1}{4}\text{Coth}[\frac{1}{4\sqrt{2}}(x-\frac{3}{\sqrt{2}}t)]-\frac{1}{2}+\frac{1}{4}\text{Tanh}[\frac{1}{4\sqrt{2}}(x-\frac{3}{\sqrt{2}}t)]$$

$$y_4[t\_,x\_]:=\frac{1}{4}\text{Coth}[-\frac{1}{4\sqrt{2}}(x+\frac{3}{\sqrt{2}}t)]-\frac{1}{2}+\frac{1}{4}\text{Tanh}[-\frac{1}{4\sqrt{2}}(x+\frac{3}{\sqrt{2}}t)]$$

$$y_5[t\_,x\_]:=-\frac{1}{4}\text{Coth}[\frac{1}{4\sqrt{2}}(x-\frac{3}{\sqrt{2}}t)]+\frac{1}{2}-\frac{1}{4}\text{Tanh}[\frac{1}{4\sqrt{2}}(x-\frac{3}{\sqrt{2}}t)]$$

$$y_6[t\_,x\_]:=-\frac{1}{4}\text{Coth}[-\frac{1}{4\sqrt{2}}(x+\frac{3}{\sqrt{2}}t)]+\frac{1}{2}-\frac{1}{4}\text{Tanh}[-\frac{1}{4\sqrt{2}}(x+\frac{3}{\sqrt{2}}t)]$$

$$y_7[t\_,x\_]:=-\frac{1}{4}\text{Coth}[\frac{1}{4\sqrt{2}}(x+\frac{3}{\sqrt{2}}t)]-\frac{1}{2}-\frac{1}{4}\text{Tanh}[\frac{1}{4\sqrt{2}}(x+\frac{3}{\sqrt{2}}t)]$$

$$y_8[t\_,x\_]:=-\frac{1}{4}\text{Coth}[-\frac{1}{4\sqrt{2}}(x-\frac{3}{\sqrt{2}}t)]-\frac{1}{2}-\frac{1}{4}\text{Tanh}[-\frac{1}{4\sqrt{2}}(x-\frac{3}{\sqrt{2}}t)]$$

All the above defined functions are solutions. Indeed,

```
D[#,t]-D[#,x,x]-#(1-#²)&/@
Table[y_j][t,x],{j,1,8}]//Simplify
{0,0,0,0,0,0,0,0}
```

We plot them successively.

```
Plot3D[Evaluate[#],{t,-1,1},{x,-1,1},ImageSize→160,
PlotRange→All,PlotPoints→50,PlotStyle→{Gray,
Opacity[.2]},Ticks→{{-1,0,1},{-1,0,1},{0}}]&/@
Table[y_j[t,x],{j,1,4}]
```

See Figure 9.43.

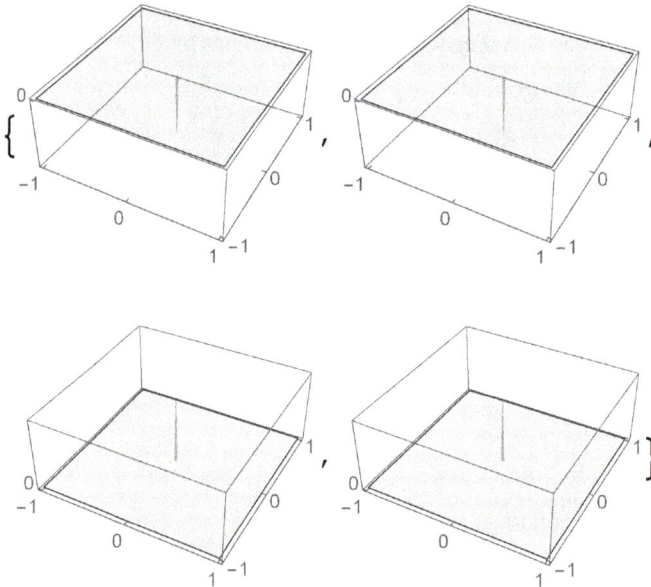

Figure 9.43: Plot of the solutions $y_1 - y_4$ for problem 9.29.

One can give the following representation of the solutions $y_5 - y_8$:

```
Plot3D[Evaluate[#],{t,-1,1},{x,-1,1},ImageSize→160,
PlotRange→All,PlotPoints→50,PlotStyle→{Gray,
Opacity[.2]},Ticks→{{-1,0,1},{-1,0,1},{0}}]
&/@Table[y_j[t,x],{j,5,8}]
```

See Figure 9.44.

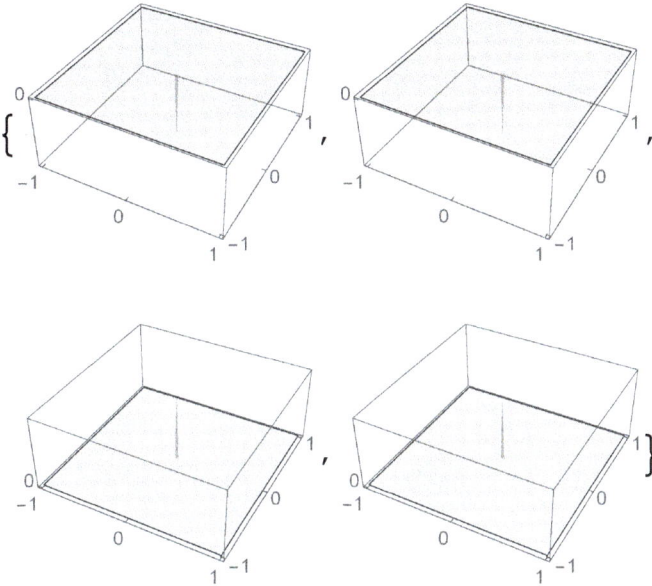

Figure 9.44: Plot of the solutions $y_5 - y_8$ for problem 9.29.

The last case in this investigation is generated by the next logical conditions.

$(a_{-1}==-1/2 \vee a_{-1}==1/2)\&\&a_0==0$
$\&\&a_1==1/27(25a_{-1}+40a_{-1}^3-128a_{-1}^5)$
$\&\&(k==-a_{-1}/\sqrt{2}\vee k==a_{-1}/\sqrt{2})\&\&v==0$

From here, this results in the following table of coefficients and list of solutions:

Table 9.4: Table of coefficients of solutions $y_9 - y_{12}$ for problem 9.29.

	a_{-1}	a_0	a_1	k	v
1	$\frac{1}{2}$	0	$\frac{1}{2}$	$\frac{1}{2\sqrt{2}}$	0
2	$\frac{1}{2}$	0	$\frac{1}{2}$	$-\frac{1}{2\sqrt{2}}$	0
3	$-\frac{1}{2}$	0	$-\frac{1}{2}$	$-\frac{1}{2\sqrt{2}}$	0
4	$-\frac{1}{2}$	0	$-\frac{1}{2}$	$\frac{1}{2\sqrt{2}}$	0

$y_9[t\_,x\_]:=\frac{1}{2}Coth[\frac{1}{2\sqrt{2}}x]+\frac{1}{2}Tanh[\frac{1}{2\sqrt{2}}x]$
$y_{10}[t\_,x\_]:=\frac{1}{2}Coth[-\frac{1}{2\sqrt{2}}x]+\frac{1}{2}Tanh[-\frac{1}{2\sqrt{2}}x]$
$y_{11}[t\_,x\_]:=-\frac{1}{2}Coth[\frac{1}{2\sqrt{2}}x]-\frac{1}{2}Tanh[\frac{1}{2\sqrt{2}}x]$
$y_{12}[t\_,x\_]:=-\frac{1}{2}Coth[-\frac{1}{2\sqrt{2}}x]-\frac{1}{2}Tanh[-\frac{1}{2\sqrt{2}}x]$

These functions are indeed solutions.

```
D[#,t]-D[#,x,x]-#(1-#²)&/@Table[y_j[t,x],{j,9,12}]
//Simplify
{0,0,0,0}
```

The graphs of the above functions follow below:

```
Plot3D[Evaluate[#],{t,-1,1},{x,-1,1},ImageSize→160,
PlotRange→All,PlotPoints→50,PlotStyle→{Gray,
Opacity[.2]},Ticks→{{-1,0,1},{-1,0,1},{0}}]
&/@Table[y_j[t,x],{j,9,12}]
```

See Figure 9.45.

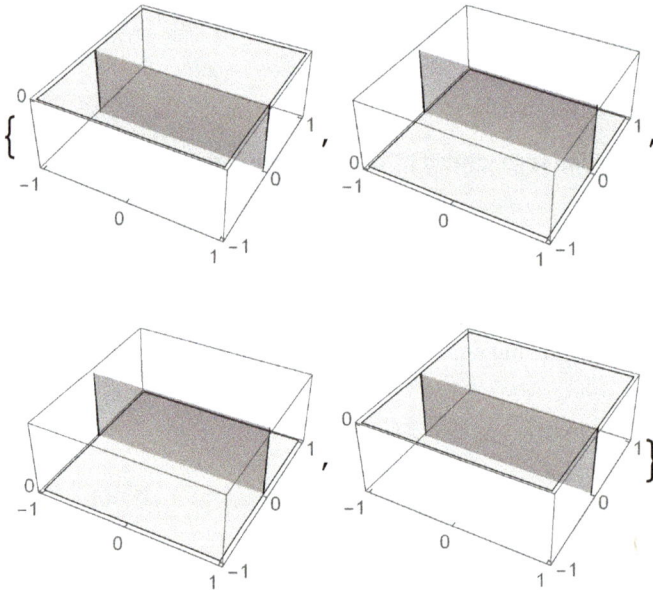

Figure 9.45: Plot of the solutions $y_9 - y_{12}$ for problem 9.29.

9.7.5 The Fisher equation with convection by the Hirota method

9.30. Below we study the existence of particular solutions of the Fisher equation with convection obtained by the Hirota method [25, (61)],

$$\partial_t u(t,x) + \kappa\, u(t,x)\partial_{x,x}u(t,x) - u(t,x)\bigl(1 - u(t,x)\bigr) = 0, \quad k,t,x \in \mathbb{R}.$$

9.30

```
Clear[a,b,c,d,κ,t,x]
```

One introduces the *ancillary function* and the associated solution:

```
f[t_,x_]:=1+Exp[a x+b t+c]
u[t_,x_]:=d D[Log[f[t,x]],x]
u[t,x]//ExpToTrig (* Function u under trigonometric
form *)
```

$$\frac{a\,d\,(\mathrm{Cosh}[c+b\,t+a\,x]+\mathrm{Sinh}[c+b\,t+a\,x])}{1+\mathrm{Cosh}[c+b\,t+a\,x]+\mathrm{Sinh}[c+b\,t+a\,x]}$$

We impose that this function satisfies the equation. Then

```
∂_tu[t,x]+κ u[t,x]∂_xu[t,x]-∂_x,xu[t,x]-u[t,x](1-u[t,x])
//FullSimplify
```

$$\frac{a\,d\,e^{a\,x+b\,t+c}}{(e^{a\,x+b\,t+c}+1)^3}$$

$$(-a^2+(a(a\,d\,\kappa+a+d)+b-2)e^{a\,x+b\,t+c} + (a\,d-1)e^{2(a\,x+b\,t+c)} + b-1)$$

In order for the coefficients to be a solution, they have to satisfy the system of algebraic equations:

```
Reduce[-1-a²+b==0&&-1+a d==0&&-2+b+a(a+d+a d κ)==0,
{a,b,c,d}]
```

i. e.,

```
a=-κ/2;b=1/4(4+κ²);d=-2/κ;
```

Based on these results, we plot a particular solution.

```
Block[{κ=-.1,c=0},
Plot3D[Evaluate[u[t,x]],{t,-10,10},{x,-5,5},Mesh→None,
PlotRange→All,Ticks→{10{-1,0,1},5{-1,0,1},{0,1}},
PlotStyle→{Gray,Opacity[.2]},ImageSize→175]]
```

See Figure 9.46.

9.31. We introduce another kink solution to the Fisher equation with convection term

$$\partial_t u(t,x) + \kappa\, u(t,x)\partial_x u(t,x) - \partial_{x,x} u(t,x) - u(t,x)(1 - u(t,x)) = 0,$$

$$\kappa, t, x \in \mathbb{R}, \quad \kappa \neq 0.$$

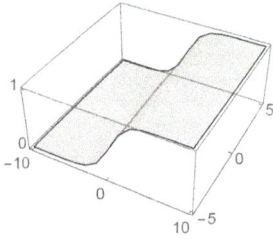

9.31

```
Clear[t,x,a,b,c,d,e,κ]
```

```
θ[t_,x_]:=a x+b t+c;  (* Ancillary function *)
u[t_,x_]:=(1+Tanh[θ[t,x]]/d+e Tanh[θ[t,x]])²  (* Form of the solution *)
```

We substitute this function in the equation to find the algebraic system of the parameters,

$$\partial_t u[t,x]+\kappa\, u[t,x]\partial_x u[t,x]-\partial_{x,x}u[t,x]-u[t,x](1-u[t,x])$$

Thus, we find the algebraic system of equations:

```
Reduce[(1+16a²-4b)(d-e)³==0&&
(-3-8a²+8b)(d-e)²(d+e)==0&&
(d-e)(4+(-3+8a²+4b)(d+e)²+16a κ)==0&&
(-2+d+e)(d+e)(2+d+e)==0,{a,b,c,d,e}]
```

For $\kappa\neq0$, we have that

```
a=-(1/(4κ));b=1/4(1+16a²);e=-d
```

Consequently,

```
θ[t_,x_]:=-(1/(4κ))x+1/4(1+16(1/(4κ))²)t+c
u[t_,x_]:=(1+Tanh[θ[t,x]]/d-d Tanh[θ[t,x]])²
```

This function satisfies the equation, thus it is a solution depending on certain parameters. We plot this solution.

```
Block[{κ=1,c=1/2,d=.1},
Plot3D[u[t,x],{t,25,500},{x,-10,0},PlotRange→All,
Ticks→{{25,500},{-10,0},{0,{2*10³⁴,"2*10³⁴"}}},
PlotStyle→{Gray,Opacity[.15]},PlotPoints→75,
Mesh→None,ImageSize→200]]
```

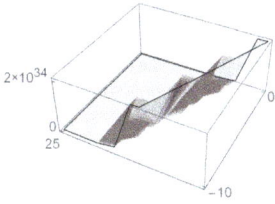

Figure 9.47: Particular solution for problem 9.31.

See Figure 9.47.

9.8 The Fitzhugh–Nagumo and the Calogero equations

9.8.1 The Fitzhugh–Nagumo equation

The *Fitzhugh–Nagumo* nonlinear partial differential *equation* is of the form [26, (39)], [25, (74)],

$$\partial_t u + k\,u\,\partial_x u - \partial_{x,x} u - u(1-u)(a-u) = 0, \quad a,k,t,x \in \mathbb{R}, \quad k \neq 0.$$

9.32. We study certain particular solutions of the Fitzhugh–Nagumo equation by the Hirota method:

$$\begin{cases} \partial_t u(t,x) + k\,u(t,x)\partial_x u(t,x) - \partial_{x,x} u(t,x) - u(t,x)(a - u(t,x)) = 0, \\ a,k,t,x \in \mathbb{R}, \quad a \neq 0, \quad k \neq 0. \end{cases}$$

9.32 We apply the Hirota method.

```
<<Notation`

Symbolize[a_]
Symbolize[b_]
Symbolize[c_]

Clear[a,b,c,d,k,p,t,x]

f[t_,x_]:=d+Exp[a₁t+b₁x+c₁]+Exp[a₂t+b₂x+c₂] (*
Ancillary function *)
u[t_,x_]:=p D[Log[f[t,x]],x] (* The form of the
solution *)
```

We introduce the equation,

```
eqn=∂ₜu[t,x]+k u[t,x]∂ₓu[t,x]-∂ₓ,ₓu[t,x]
-u[t,x](1-u[t,x])(a-u[t,x])//Simplify;
```

and transform it for further calculations,

$$\text{eqn1} = \frac{\text{eqn}}{p}\,(d+e^{a_1 t+b_1 x+c_1}+e^{a_2 t+b_2 x+c_2})^3 /.\{e^{a_1 t+b_1 x+c_1} \to y,$$
$$e^{a_2 t+b_2 x+c_2} \to z,\ e^{a_1 t+b_1 x+c_1+a_2 t+b_2 x+c_2} \to y\,z\}$$

We write eqn1 as a power series,

```
Series[eqn1,{y,0,4},{z,0,4}];
```

and select the system of algebraic equations finding the coefficients:

```
Assuming[k≠0&&p≠0&&a≠0,
Reduce[-a b₂d²+a₂b₂d²-b₂³d²==0&&
a b₂²d p-2a b₂d+a₂b₂d+b₂³d k p+b₂³d p+b₂³d==0&&
a b₂²p-a b₂-b₂³p²+b₂³p==0&&
2a b₂b₁d p-2 a b₁d+2a₁b₁d-a₂b₁d-2a b₂d-a₁b₂d+2a₂b₂d
+b₂b₁²d k p+b₂²b₁d k p+2b₂b₁d p-2b₁³d+3b₂b₁²d+3b₂²b₁d-2b₂³d==0&&
2a b₂b₁p+a b₂²p-a b₁+a₁b₁-a₂b₁-2a b₂-a₁b₂+a₂b₂+b₂b₁²k p
-2b₂²b₁k p+b₂³k p-3b₂b₁²p²+2b₂b₁p+b₂²p-b₁³+3b₂b₁²-3b₂²b₁+b₂³==0&&
a b₁²d p-2a b₁d+a₁b₁d+b₁³d k p+b₁²d p+b₁³d==0&&
a b₁²p+2a b₂b₁p-2a b₁+a₁b₁-a₂b₁-a b₂-a₁b₂+a₂b₂+b₁³k p-2b₂b₁²k p
+b₂²b₁k p-3b₂b₁²p²+b₁³p+2b₂b₁p+b₁³-3b₂b₁²+3b₂²b₁-b₂³==0&&
b₁(b₁p-1)(b₁p-a)==0,{a₁,b₁,c₁,a₂,b₂,c₂,d,p}]]
```

From here results a large list of feasible solutions.
    We only consider five solutions.
1.

```
Clear[a₁,b₁,c₁,a₂,b₂,c₂]
```

```
b₂=b₁;d=0;b₁≠0;p=1/b₁;
```

```
f[t_,x_]:=d+Exp[a₁t+b₁x+c₁]+Exp[a₂t+b₂x+c₂]
u[t_,x_]:=pD[Log[f[t,x]],x]
```

We check this solution.

```
∂ₜu[t,x]+k u[t,x]∂ₓu[t,x]-∂ₓ,ₓu[t,x]
-u[t,x](1-u[t,x])(a-u[t,x])//Simplify
0
```

Remark. We note that

```
u[t,x]//Simplify
1
```

so a solution to the Fitzhugh–Nagumo equation is the constant 1. ☐

2.

```
Clear[a₁,b₁,c₁,a₂,b₂,c₂]
```

```
a≠0;b₂=b₁;d=0;b₂≠0;p=1/b₂;
```

```
f[t_,x_]:=d+Exp[a₁t+b₁x+c₁]+Exp[a₂t+b₂x+c₂]
u[t_,x_]:=pD[Log[f[t,x]],x]
```

We check this solution.

```
∂ₜu[t,x]+ku[t,x]∂ₓu[t,x]-∂ₓ,ₓu[t,x]
-u[t,x](1-u[t,x])(a-u[t,x])//Simplify
0
```

```
u[t,x]//Simplify
a
```

Remark. This solution is also a constant. ☐

3.

```
Clear[a₁,b₁,c₁,a₂,b₂,c₂,a,k]
```

```
b₁=1/4(-k+Sqrt[-8+k²]);a₂=1/2(1-2a+2a₁+kb₁);b₂=0;d=0;
p=-k-2b₁;
```

```
f[t_,x_]:=d+Exp[a₁t+b₁x+c₁]+Exp[a₂t+b₂x+c₂]
u[t_,x_]:=pD[Log[f[t,x]],x]
```

We check this solution

```
∂ₜu[t,x]+ku[t,x]∂ₓu[t,x]-∂ₓ,ₓu[t,x]
-u[t,x](1-u[t,x])(a-u[t,x])//Simplify
0
```

and plot it for some values of the parameters.

```
Block[{a=4,k=-3,a₁=-1,c₁=-2,c₂=2},
Plot3D[Evaluate[u[t,x]],{t,-5,5},{x,-5,5},
Ticks→Outer[Times,{5,5,1},{-1,0,1}],Mesh→None,
PlotStyle→{Gray,Opacity[.2]},ImageSize→150]]
```

See Figure 9.48.

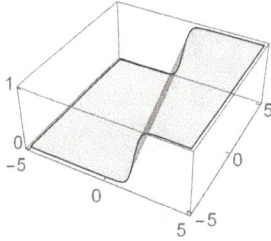

Figure 9.48: Kink solution for problem 9.32.

4.

```
Clear[a₁,b₁,c₁,a₂,b₂,c₂,a,k,d]
```

$$a_1=1/8(8a-4a^2+a^2k^2-a^2k\,\mathrm{Sqrt}[-8+k^2]);\quad b_1=\frac{2a-a^2-2a_1}{ak};$$
$$a_2=-\frac{a-2a^2+k\,b_1}{2a};\quad b_2=\frac{-1+2a-2a_2}{k};\quad p=-k-2\,b_2;$$

```
f[t_,x_]:=d+Exp[a₁t+b₁x+c₁]+Exp[a₂t+b₂x+c₂]
u[t_,x_]:=pD[Log[f[t,x]],x]
```

We check this solution

```
∂ₜu[t,x]+ku[t,x]∂ₓu[t,x]-∂ₓ,ₓu[t,x]
-u[t,x](1-u[t,x])(a-u[t,x])//Simplify
0
```

and plot it for some values of the parameters.

```
Block[{a=4,k=-3,c₁=2,c₂=0,d=0},
Plot3D[Evaluate[u[t,x]],{t,-5,5},{x,-5,5},
Ticks→{{-5,0,5},{-5,0,5},{1,4}},Mesh→None,
PlotStyle→{Gray,Opacity[.2]},ImageSize→150]]
```

See Figure 9.49.

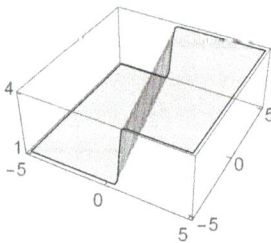

Figure 9.49: Another kink solution for problem 9.32.

5.

```
Clear[a₁,b₁,c₁,a₂,b₂,c₂,a,k]

b₁=¼(-a k-a Sqrt[-8+k²]);b₂=b₁/a;d=0;
a₂=(-a+a³+2a a₁-k b₁+a²k b₁)/(2a);p=1/b₂;

f[t_,x_]:=d+Exp[a₁t+b₁x+c₁]+Exp[a₂t+b₂x+c₂]
u[t_,x_]:=p D[Log[f[t,x]],x]
```

We check this solution

```
∂ₜu[t,x]+k u[t,x]∂ₓu[t,x]-∂ₓ,ₓu[t,x]
-u[t,x](1-u[t,x])(a-u[t,x])//Simplify
0
```

and plot it for some values of the parameters.

```
Block[{a=-4,k=-3,a₁=2;c₁=2,c₂=0},
Plot3D[Evaluate[u[t,x]],{t,-3,3},{x,-5,5},
Ticks→{{-3,0,3},{-5,0,5},{1,4}},Mesh→None,
PlotStyle→{Gray,Opacity[.2]},ImageSize→150]]
```

See Figure 9.50.

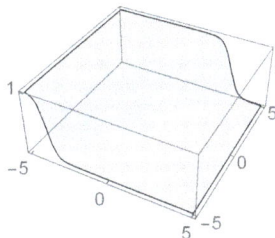

Figure 9.50: Antikink solution for problem 9.32.

9.8.2 The Calogero equation

The *Calogero* nonlinear partial differential *equation* is of the form [11, (1.1)], [25, (81)],

$$\partial_t u = \partial_{x,x,x} u + 3(u^2 \partial_{x,x} u + 3u(\partial_x u)^2 + u^4 \partial_x u), \quad t, x \in \mathbb{R}. \tag{9.22}$$

9.33. We introduce a particular solution to the Calogero equation (9.22).

9.33

```
Clear[t,x,k,δ]
```

A particular solution of the Calogero equation is [25, (85)],

$$u[t\_,x\_]:=\frac{\sqrt{k}}{2}\sqrt{1+Tanh[\frac{4k\,x+k^3\,t+\delta}{8}]}$$

where k and δ are constants.

We check whether the function just introduced u is a solution to equation (9.22). Indeed,

```
∂ₜu[t,x]-∂ₓ,ₓ,ₓu[t,x]-3(u[t,x]²∂ₓ,ₓu[t,x]+
3u[t,x]∂ₓu[t,x]²+u[t,x]⁴∂ₓu[t,x])//Simplify
0
```

We plot the figure some for particular values of the constants.

```
Block[{k=2,δ=0},
Plot3D[u[t,x],{t,-5,5},{x,-5,5},ImageSize→175,
Mesh→None,PlotStyle→{Gray,Opacity[.2]},
PlotPoints→25,PlotRange→All,
Ticks→Outer[Times,{5,5,1},{-1,0,1}]]]
```

See Figure 9.51.

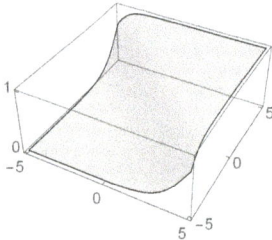

Figure 9.51: Particular solution for problem 9.33.

9.9 Modified Klein–Gordon equation by ansatz methods

We study the existence of a soliton solution to the modified Klein–Gordon equation, [71, Section 39.3.4.] or the sine-Gordon-like equation

$$\partial_{t,t}u(t,x) = \partial_{x,x}u(t,x) - u(t,x) + u^3(t,x). \tag{9.23}$$

In spite of the fact that the previous equation is a nonlinear hyperbolic equation, we inserted it here because we also use the ansatz method to solve it.

9.34. We try to solve equation (9.23) inspired by the ansatz methods.

9.34 1.

We applied the tanh method, Section 9.6.1 and [71, p. 1641].

```
Clear[k,t,x,v]
```

The modified Klein–Gordon equation is written by *Mathematica* under the form

```
eqn=∂t,tu[t,x]-∂x,xu[t,x]+u[t,x]-u[t,x]³;
```

and by the transformations

```
ξ=k(x-v t);u[t,x]=w[ξ];
```

it looks like

```
eqnw=k²v²w''[ξ]-k²w''[ξ]+w[ξ]-w[ξ]³;
```

and thus equation (9.23) is transformed in an ordinary differential equation.

According to the tanh method, we introduce another independent variable and dependent variable

```
ζ=Tanh[ξ];s[ζ]=w[ξ];
```

Then the equation is of the form

```
eqns=s[ζ]-s[ζ]³+k²(-1+v²)(-1+ζ²)(2ζz'[ζ]+(-1+ζ²)s''[ζ])
```

The tanh method supposes that there exists a solution of the form

$$s[\zeta]=\sum_{n=0}^{m}a_n\zeta^n$$

with *m* a natural number.

Substituting the function $s[\zeta]$ into eqns and balancing the term of the highest degree with the term of the highest order of derivative, we have that $3m = m + 2$, implying that $m = 1$. Thus, function s is an affine function in ζ, i. e.,

```
s[ζ_]:=a₀+a₁ζ;
```

We substitute it in the equation and find that

```
eqnζ=s[ζ]-s[ζ]³+k²(-1+v²)(-1+ζ²)(2ζs'[ζ]
+(-1+ζ²)s''[ζ]);
Series[eqnζ,{ζ,0,4}]
```

We try finding the parameters of this solution.

```
Reduce[a₀-a₀³==0&&a₁+2k²a₁-2k²v²a₁-3a₀²a₁==0&&a₀a₁²==0&&
2k²a₁-2k²v²a₁+a₁³==0, {a₀,a₁,k}]
(a₀==-1&&a₁==0)∨(a₀==0&&a₁==0)∨(a₀==1&&a₁==0)
∨(a₀==0&&(a₁==-1∨a₁==1)&&-1+v² ≠0&&
(k==-(1/Sqrt[-2+2v²])∨k==1/Sqrt[-2+2v²]))
```

If $a_1 = 0$, then there are three trivial solutions:

$$u[t,x]=\begin{cases} 1, \\ 0, \\ -1. \end{cases}$$

Suppose $a_1 \neq 0$. Then $a_0 = 0$, $a_1 = \pm 1$, and $k = \pm 1/\sqrt{-2 + 2v^2}$.

Function $u[t\_, x\_] := \tanh(1/\sqrt{-2 + 2v^2}(x - vt))$ and its opposite function are solutions of equation (9.23). We plot them:

```
Block[{v=1.5},
Plot3D[{u[t,x],-u[t,x]},{t,-7,7},{x,-5,5},Mesh→None,
PlotPoints→20,ImageSize→175,PlotStyle→{{Gray,
Opacity[.15]},{Gray,Opacity[.2]}},
Ticks→Outer[Times,{7,5,1},{-1,0,1}],PlotLegends→
SwatchLegend["Expressions",LegendFunction→
(Framed[#]&)]]]
```

See Figure 9.52.

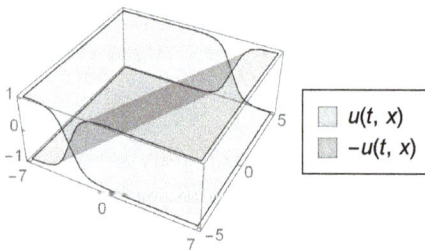

Figure 9.52: Plot of the solutions for problem 9.34.

We show that by a simple transformation the particular solution has another form:

$$(TrigToExp[u[t,x]]-\frac{e^{\frac{-tv+x}{Sqrt[2]Sqrt[-1+v^2]}}-e^{\frac{-(-tv+x)}{Sqrt[2]Sqrt[-1+v^2]}}}{e^{\frac{-tv+x}{Sqrt[2]Sqrt[-1+v^2]}}+e^{\frac{-(-tv+x)}{Sqrt[2]Sqrt[-1+v^2]}}})$$

```
//Simplify
0
```

We may conclude that there exists also an exp solution.

2.

We applied the exp function method, Section 9.6.3 and [71, p. 1643].

```
Clear[k,t,x,u,v]
```

We are reminded that the modified Klein–Gordon equation by *Mathematica* under the form

```
eqn=∂t,tu[t,x]-∂x,xu[t,x]+u[t,x]-u[t,x]³;
```

and by the transformations

```
ξ=k(x-v t);u[t,x]=w[ξ];
```

which looks like

```
eqnw=k²(v²-1)w''[ξ]+w[ξ]-w[ξ]³;
```

and thus equation (9.23) is transformed into an ordinary differential equation.

According to the exp function method, Section 9.6.3, we guess that the solution that we look for is of the form

$$w[\xi\_] := \frac{\sum_{n=-d}^{c} a_n \, e^{n\xi}}{\sum_{m=-q}^{p} b_m \, e^{m\xi}}$$

We substitute this function into equation eqnw and balance the highest-order linear term with the highest- order nonlinear term. Thus, it results that $c = p$. Similarly, we balance the lowest-order linear term with the lowest-order nonlinear term to get $d = q$. We choose $c = p = 1$ and $d = q = 1$. The solution that we look for looks like:

$$w[\xi\_] := \frac{a_1 \, e^{\xi} + a_0 + a_{-1} \, e^{-\xi}}{b_1 \, e^{\xi} + b_0 + b_{-1} \, e^{-\xi}}$$

Function $w[\xi]$ is substituted in the equation to find coefficients a_i, b_j, and parameters k and v.

We follow the next steps:

```
eqnw=k²(v²-1)w''[ξ]+w[ξ]-w[ξ]³//Simplify;
eqnw1=(b₁eξ+b₀+b₋₁e⁻ξ)³eqnw/.Table[enξ →qⁿ,{n,-3,5}]
//Expand//Simplify;
eqnw2=CoefficientList[Normal[Series[q³eqnw1,{q,0,4}]],
q]//Simplify
```

We transform this list into a system of algebraic equations and solve it. Because the resulted system of algebraic equations is rather complicated, we take certain initial values

$a_{-1}=-1;b_{-1}=1;a_1=1;b_1=1;a_0=0;b_0=0;$

then the system returns

$k=\pm\,\mathrm{Sqrt}[1/(-2+2v^2)],$

and thus the same solution as by the tanh function method is obtained.

3.

We apply the sine-cosine functions method; see Sect. 9.6.3 and [71, p. 1642].

`Clear[k,t,x,u,v]`

We are reminded that the modified Klein–Gordon equation by *Mathematica* under the form

`eqn=∂`$_{t,t}$`u[t,x]-∂`$_{x,x}$`u[t,x]+u[t,x]-u[t,x]`^{3}`;`

and by the transformations

$\xi=k(x-v\,t);\ u[t,x]=w[\xi];$

it looks like

`eqnw=k`^{2}`(v`^{2}`-1)w''[ξ]+w[ξ]-w[ξ]`^{3}`;`

and thus equation (9.23) is transformed into an ordinary differential equation.

According to the sine-cosine method, we may consider the traveling wave solutions of the form $w(\xi) = \lambda \sin^\beta(\xi)$ for $|\xi| \le \pi/|k|$.

`w[ξ_]:=λ Sin[ξ]`$^\beta$`;`
`eqnξ=k`^{2}`(v`^{2}`-1) D[w[ξ],ξ,ξ]+w[ξ]-w[ξ]`^{3}`//Simplify;`

Then

$2\,\mathrm{Sin}[\xi]^\beta+k^2(-1+v^2)\beta(-2+\beta+\beta(1-2\,\mathrm{Sin}[\xi]^2))\mathrm{Sin}[\xi]^{\beta-2}$
$-2\lambda^2\mathrm{Sin}[\xi]^{3\beta}=0$

We balance the exponents to get $3\beta = \beta - 2$, i. e., $\beta = -1$. Then we solve the system of algebraic equations to find k and λ.

`Reduce[-4k`^{2}`+4k`^{2}`v`^{2}`-2λ`^{2}`==0&&2+2k`^{2}`-2k`^{2}`v`^{2}`==0,{λ,k},Reals]`

The solutions are contained in Table 9.5.

Table of coefficients to the sine-cosine method for problem 9.34.

$|v| > 1$	$|\lambda| = \sqrt{2}$	$|k| = \sqrt{\frac{1}{-1+v^2}}$

We introduce a solution for some values of the parameters and plot it

```
u[t_,x_]:=λSin[Sqrt[1/(-1+v²)](x-v t)]⁻¹
Block[{λ=-Sqrt[2],v=2},
{Plot3D[u[t,x],##],Plot3D[-u[t,x],##]}
&[{t,-4,4},{x,-4,4},ImageSize→175,Mesh→None,
PlotPoints→45,PlotStyle→Directive[Gray,Opacity[.15]],
Ticks→Outer[Times,{4,4,10},{-1,0,1}]]]
```

See Figure 9.53.

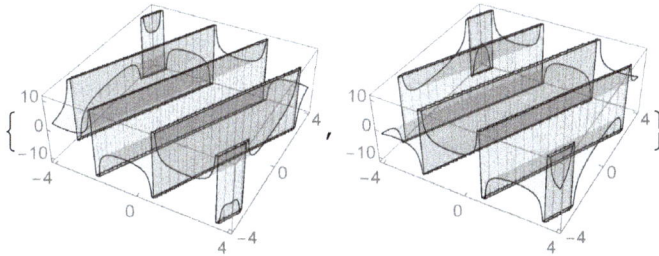

Figure 9.53: Plot of the solutions for problem 9.34.

9.10 The double sine-Gordon equation by the tanh function method

9.35. We study the simplest double sine-Gordon equation

$$\partial_{t,x} u(t, x) = \sin(u(t, x)) + \sin(2u(t, x)), \quad t, x \in \mathbb{R}$$

by the generalized tanh function method; see [18].

9.35

```
Clear[t,x,a,b]
```

Suppose u is a two variable function, i. e., $u(t, x)$. The simplest double sine-Gordon equation is of the form

```
∂t,xu[t,x]=Sin[u[t,x]]+Sin[2u[t,x]]
```

```
DSolve[∂t,xu[t,x]==Sin[u[t,x]]+Sin[2u[t,x]],u,{t,x}]
(* No closed-form solution is returned *)
```

We use the method in [18] and consider (f) in Theorem 1.7. Then

$$Sin[u]=(e^{iu}-e^{-iu})/(2i);Sin[2u]=(e^{2iu}-e^{-2iu})/(2i);$$
$$v=e^{iu}\Longrightarrow Sin[u]=(v-v^{-1})/(2i);Sin[2u]=(v^2-v^{-2})/(2i);$$
$$iu=Log[v];u=-iLog[v];v\in\mathbb{C};u\in\mathbb{R};$$

Then

$$\partial_t u=-i\frac{\partial_t v}{v};\partial_{t,x}u=-i(\frac{\partial_{t,x}v}{v}-\frac{\partial_t v\partial_x v}{v^2});$$

We substitute them in the equation

$$i(\frac{\partial_{t,x}v}{v}-\frac{\partial_t v\partial_x v}{v^2})-(v-v^{-1})/(2i)-(v^2-v^{-2})/(2i)=0;$$
$$2v\partial_{t,x}v-2\partial_t v\partial_x v-v^4-v^3+v+1=0;$$

We pass to the new variables

$$v[t,x]=z[\xi];\xi=k(x-\lambda t);$$

to get the equation

$$2\lambda z\,z''-2\lambda(z')^2+z^4+z^3-z-1=0;$$

We admit that the last equation has a solution as a polynomial in $f[\xi]$.

```
Clear[f]
z[ξ]:=a₀+∑ᵢ₌₁ᵐ aᵢf[ξ]ⁱ;
```

We substitute this solution into the last form of the equation and obtain that $m = 1$. Then we have

```
f[ξ]:=Sqrt[k] Tan[Sqrt[k](ξ+C[1])]
z[ξ]:=a|b f[ξ]
```

We substitute $z[\xi]$ into the equation and get a system of algebraic equations that try to solve it

```
Reduce[-1-a+a³+a⁴-2b²k²λ==0&&
-b Sqrt[k]+3a²b Sqrt[k]+4a³b Sqrt[k]+4a b k³ᐟ²λ==0&&
3a b²k+6a²b²k==0&&
b³k³ᐟ²+4a b³k³ᐟ²+4a b k³ᐟ²λ==0&&
b⁴k²+2b²k²λ==0,{a,b,λ,k}]
```

The following feasible solution results in the following:

```
b≠0;a=-1/2;λ=-b²/2;k=3/(4b²);
```

Then the solution of the double sine-Gordon equation is of the form:

```
v[t_,x_]:=-1/2+b Sqrt[3/(4b²)]Tan[Sqrt[3/(4b²)](3/(4b²)
(x+b²/2t)+c)]
Cos[u[t,x]]=-1/2+b Sqrt[3/(4b²)]Tan[Sqrt[3/(4b²)]
(3/(4b²)(x+b²/2t)+c)]
u[t_,x_]:=ArcCos[-1/2+b Sqrt[3/(4b²)]Tan[Sqrt[3/(4b²)]
(3/(4b²)(x+b²/2t)+c)]]
```

We plot the solution to the double sine-Gordon equation for two particular values of the parameters,

```
Block[{b=2,c=.219},
Plot3D[u[t,x],{t,-25,25},{x,-25,25},ImageSize→175,
Mesh→None,Ticks→Outer[Times,{25,25,3},{-1,0,1}],
PlotStyle→{Gray,Opacity[.2]}]]
```

See Figure 9.54.

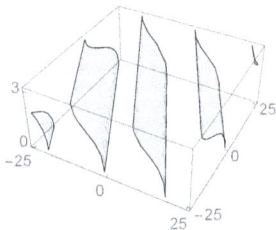

Figure 9.54: Particular solution for problem 9.35.

9.36. We study the double sine-Gordon equation

$$\partial_{t,x} u(t,x) = \lambda \sin(u(t,x)) + \mu \sin(2u(t,x)), \quad t, x, \lambda, \mu \in \mathbb{R}$$

by the method developed in [78].

9.36

```
Clear[t,x,λ,μ]
```

We introduce an independent variable ξ and a dependent variable w.

```
u[t,x]=w[ξ];ξ=k(x-v t)
```

Then

$$u_{t,x} = -k^2 v\, w''\,'[\xi]$$

We substitute in the equation and get

$$-w''\,'[\xi] = \frac{\lambda}{k^2 v}\, Sin[w[\xi]] + \frac{\mu}{k^2 v}\, Sin[2w[\xi]]$$

We multiply the previous equation by $w'[\xi]$ and get

$$-w'[\xi]w''\,'[\xi] = \frac{\lambda}{k^2 v}\, Sin[w[\xi]]w'[\xi] + \frac{\mu}{k^2 v}\, Sin[2w[\xi]]w'[\xi]]$$

By integration once, it results that

$$(w'[\xi])^2 = 2\frac{\lambda}{k^2 v}\, Cos[w[\xi]] + 2\frac{\mu}{k^2 v}\, Cos[w[\xi]]^2 + c - \frac{\mu}{k^2 v}$$

We write the right-hand side of the previous equation as

$$(a + b\, Cos[t])^2$$

Identifying the coefficients, we have that

$$b^2 = \frac{2\mu}{k^2 v}\;;\; 2\,a\,b = \frac{2\lambda}{k^2 v}\;;\; a^2 = c - \frac{\mu}{k^2 v}\;;$$

We try finding the coefficients a and b.

$$Reduce[b^2 == \frac{2\mu}{k^2 v}\,\&\&\,2\,a\,b == \frac{2\lambda}{k^2 v}\,\&\&\,a^2 == c - \frac{\mu}{k^2 v}, \{a,b\}]$$

A list of feasible solutions is of the form,

$$c = \frac{\lambda^2 + 2\mu^2}{2k^2 \mu v}\;;\; a = \frac{Sqrt[c\,k^2 v - \mu]}{k\,Sqrt[v]}\;;\; b = \frac{\lambda}{a\,k^2 v}$$

Remark. The constant of integration c depends on the coefficients λ and μ. □

Thus,

$$a = \frac{Sqrt[\lambda^2/\mu]}{Sqrt[2]\,k\,Sqrt[v]}\;;\; b = \frac{Sqrt[2]\lambda}{k\,Sqrt[v]\,Sqrt[\lambda^2]/\mu}\;;$$

and

$$w'[\xi] = \pm(a + b\, Cos[w[\xi]])$$

We solve the differential equation,

```
Clear[ξ,w]
DSolve[w'[ξ]==a+b Cos[w[ξ]],w,ξ]//Simplify;
```

and turn to the t, x variables

```
u[t_,x_]:=-2 ArcTan[(λ+2μ)/Sqrt[-λ²+4μ²]
Tanh[1/4(-((Sqrt[2]Sqrt[λ²/μ]
Sqrt[-λ²+4μ²](x-v t))/(k Sqrt[v]λ))-2 Sqrt[-λ²+4μ²] c)]]
```

where c is another integration constant and plot the figure for certain values of the parameters:

```
{Block[{c=-1,λ=.1,μ=5,k=1,v=5.5},
Plot3D[Evaluate[u[t,x]],{t,-3,1},{x,-2,2},
ImageSize→150,Ticks→{{-3,-1,1},{-2,0,2},{-1,0,1}},
Mesh→3,PlotStyle→{Gray,Opacity[.15]}]],
Block[{c=-1,λ=1,μ=1,k=.1,v=.5},
Plot3D[Evaluate[u[t,x]],{t,-3,1},{x,-2,2},
ImageSize→150,Ticks→{{-3,-1,1},{-2,0,2},{-2,0,2}},
Mesh→3,PlotStyle→{Gray,Opacity[.15]}]]}
```

See Figure 9.55.

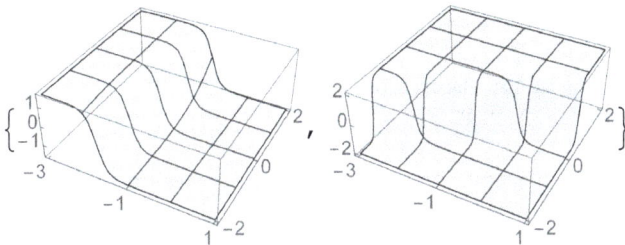

Figure 9.55: Particular solutions for problem 9.36.

Remark. Other methods of solving the double sine-Gordon equation are introduced in [78]. □

9.11 Continuous dependence on a parameter

9.37. Below we study the continuous dependence on a parameter of the solution to a simple heat equation,

$$\begin{cases} \partial_t u(t,x) = v\partial_{x,x}u(t,x), & 0 \le t \le 10, \quad -1 \le x \le 1, \quad v \in \mathbb{R}, \\ u(t,-1) = 0, \quad u(t,1) = t^2. \end{cases}$$

9.37

```
Clear[t,x,v]
eqncond={∂ₜu[t,x]==v∂ₓ,ₓu[t,x],u[t,-1]==0,u[t,1]==t²};
(* The problem *)
```

```
DSolve[eqncond,u,{t,x}] (* No closed-form solution
is returned *)
```

We try finding a numerical solution.

```
Block[{v=.002},
NDSolve[eqncond,u,{t,0,10},{x,-1,1}];
Plot3D[u[t,x]/.%,{t,0,10},{x,-1,1},ImageSize→175,
PlotStyle→{Gray,Opacity[.2]},Mesh→3,
Ticks→{{0,5,10},{-1,0,1},25000{0,1,2}}]]
```

See Figure 9.56.

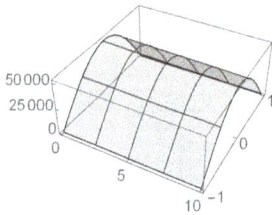

Figure 9.56: Plot of the solution for problem 9.37.

We turn to a dynamical approach seeing the continuous dependence on the parameter v. For it, we assume that function u also depends on v.

```
Clear[v,t,x,u]
sol=NDSolve[{∂ₜu[t,x,v]==v∂ₓ,ₓu[t,x,v],u[t,-1,v]==0,
u[t,1,v]==t²},u,{t,0,10},{x,-1,1},{v,0,.1}];
{Plot3D[u[t,x,v]/.sol/.v →.002,{t,0,10},{x,-1,1},
Mesh→3,ImageSize→175,PlotStyle→{Gray,Opacity[.2]},
Ticks→{{0,5,10},{-1,0,1},3500{0,1,2}}],
Manipulate[
Plot3D[Evaluate[u[t,x,v]/.sol],{t,0,10},{x,-1,1},
Mesh→3,PlotStyle→{Gray,Opacity[.15]},
Ticks→{5{0,1,2},{-1,0,1},Automatic},PlotRange→All,
ImageSize→200],{v,0,.1},SaveDefinitions→True]}
```

See Figure 9.57.

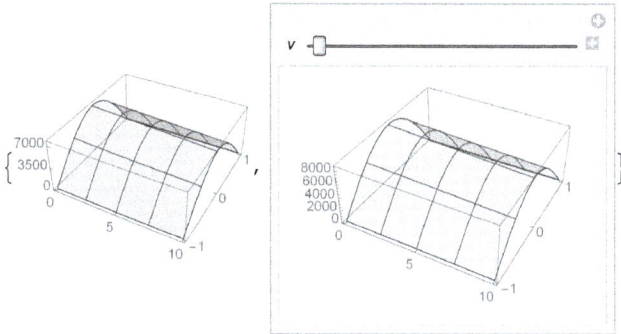

Figure 9.57: Dynamical plot of the solution for problem 9.37.

9.38. Below we study the continuous dependence of the solution on a parameter,

$$\begin{cases} \partial_t u(t,x) = v\,\partial_{x,x}u(t,x) + \sqrt{v}\,\partial_x u(t,x), \\ 0 \le t \le 10, \quad -1 \le x \le 1, \quad v > 0, \\ u(t,-1) = 0, \quad u(t,1) = t^2. \end{cases}$$

9.38

```
Clear[v,t,x,u]
eqncond={∂ₜu[t,x]==v∂ₓ,ₓu[t,x]+√v∂ₓu[t,x],u[t,-1]==0,
u[t,1]==t²};
DSolve[eqncond,u,{t,x}];
```

Because there is no closed-form solution, we set a particular value of v, look for a numerical solution, and plot it.

```
Block[{v=.02},
sol=NDSolve[eqncond,u,{t,0,10},{x,-1,1}];
Plot3D[u[t,x]/.sol,{t,0,10},{x,-1,1},Mesh→3,
ImageSize→175,PlotStyle→{Gray,Opacity[.2]},
Ticks→{5{0,1,2},{-1,0,1},60{0,1,2}}]]
```

See Figure 9.58.

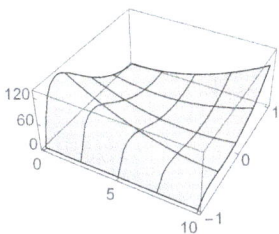

Figure 9.58: Plot of the solution for problem 9.38.

We turn to a dynamical approach seeing the continuous dependence on the parameter v. For it, we assume that function u also depends on v.

```
Clear[t,x,u,v]
eqncond={∂ₜu[t,x,v]==v∂ₓ,ₓu[t,x,v]+√v∂ₓu[t,x,v],
u[t,-1,v]==0,u[t,1,v]==t²};
DSolve[eqncond,u,{t,x,v}]; (* No closed-form
solution is returned *)
sol=NDSolve[eqncond,u,{t,0,10},{x,-1,1},{v,0,.3}]//
Flatten;
```

```
Animate[
Plot3D[Evaluate[u[t,x,v]/.sol],{t,0,10},{x,-1,1},
ImageSize→175,PlotStyle→{Gray,Opacity[.2]},
Ticks→{{0,5,10},{-1,0,1},Automatic},Mesh→3,
PlotRange→All],{v,0,.05},
SaveDefinitions→True,DefaultDuration→25,
AnimationRunning→False,
AnimationDirection→ForwardBackward]
```

See Figure 9.59.

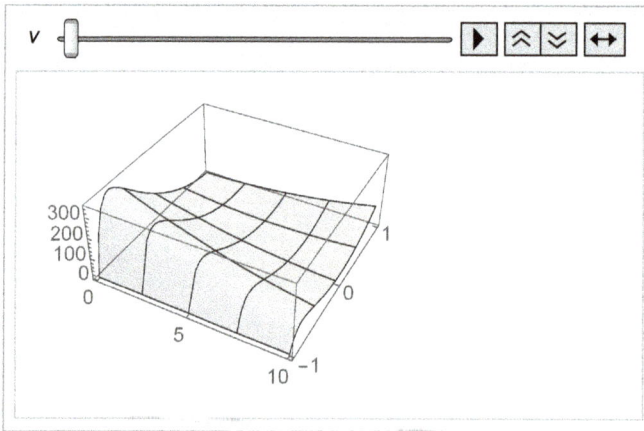

Figure 9.59: Dynamical plot of the solution for problem 9.38.

10 Third- and higher-order nonlinear partial differential equations

This chapter deals with the following evolution equations: the Korteweg–de Vries equations (algebraic solitons, soliton solutions to the Korteweg–de Vries equation, the modified Korteweg–de Vries equation, the potential Korteweg–de Vries equation, and the 2D Korteweg–de Vries equation), the Dodd–Bullough–Mikhailov equation, the Tzitzeica–Dodd–Bullough equation, the modified Kawahara equation, the Benjamin equation, the Kadomtsev–Petviashvili equation (single-soliton solution and the Hirota bilinear method), the Sawada–Kotera equation, and the Kaup–Kuperschmidt equation.

10.1 The Korteweg–de Vries equations

The *Korteweg–de Vries equation* introduced in [40] is a third-order nonlinear partial differential equation of the form

$$\partial_t u(t,x) + (a_1 + a_2 u(t,x))\partial_x u(t,x) + a_3 \partial_{x,x,x} u(t,x) = 0, \quad a_1, a_2, a_3, t, x \in \mathbb{R},$$

and models many scientific experiments, [82, Chapter 9 and Chapter 11]. By certain substitutions, the reduced forms of this equation are written as follows:

$$\partial_t u(t,x) + u(t,x)\partial_x u(t,x) + \partial_{x,x,x} u(t,x) = 0 \quad \text{and}$$
$$\partial_t u(t,x) - 6\,u(t,x)\partial_x u(t,x) + \partial_{x,x,x} u(t,x) = 0,$$

[67] and [68].

Of particular interest are solutions of this equation that are solitons.

The list of references on solitons is very large. We mention only a few titles: [16, 1, 22, 82], and [35].

10.1. We find a single-soliton solution to the reduced Korteweg–de Vries equation by *Mathematica*,

$$\partial_t u(t,x) + u(t,x)\partial_x u(t,x) + \partial_{x,x,x} u(t,x) = 0, \quad t, x \in \mathbb{R}.$$

10.1

```
Clear[t,x,ξ]

eqn=∂_t u[t,x]+u[t,x]∂_x u[t,x]+∂_{x,3}u[t,x];  (* The
equation *)
```

https://doi.org/10.1515/9783111411392-012

We solve the equation by the power of *Mathematica*,

```
sol=DSolve[eqn==0,u[t,x],{t,x}]//Flatten (* Analytical
solution *)
```
$$\{u[t,x]\to \frac{C[1]-8C[2]^3+12C[2]^3\text{Tanh}[t\,C[1]+x\,C[2]\ +C[3]]^2}{6C[2]}\}$$

select a particular solution,

```
z[t_,x_]:=u[t,x]/.sol/.{C[1]→1,C[2]→1,C[3]→1},
```

and plot it,

```
{Plot3D[z[t,x],{t,-10,10},{x,-8,8},ImageSize→175,
PlotStyle→Directive[Black,Opacity[.15]],Mesh→2,
PlotRange→All,Ticks→Outer[Times,{10,8,5},{-1,0,1}],
PlotPoints→25],
Animate[
Plot3D[z[t,x],{t,-10,10},{x,-ξ,ξ},ImageSize→175,
Axes→False,PlotStyle→Directive[Black,Opacity[.15]],
Mesh→2,PlotRange→All,Boxed→False,PlotPoints→25],
{ξ,.01,8},SaveDefinitions→True,DefaultDuration→25,
AnimationDirection→ForwardBackward,
AnimationRunning→False]}
```

See Figure 10.1.

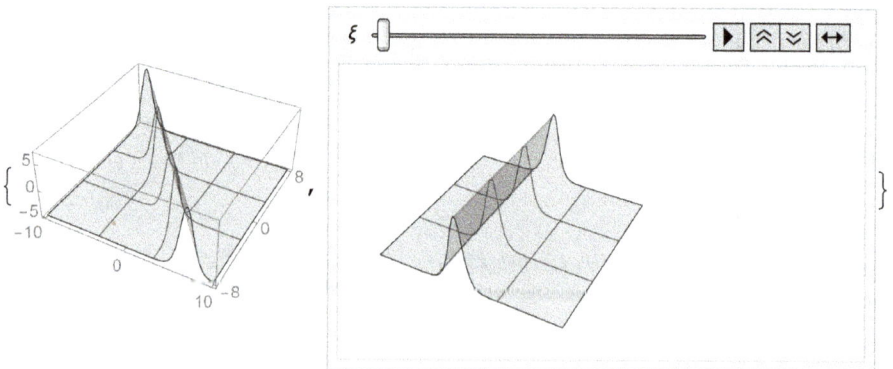

Figure 10.1: Plot of the particular solution for problem 10.1.

10.2. One checks that one-soliton + one-pole function, [12, p. 284], [70, Chapter 13, Section 4, p. 859], and [69],

$$\begin{cases} u(t,x) = -2p^2 \left(\cosh^{-2}(p\,z(t,x)) - \frac{\tanh^2(p\,z(t,x))}{(1+px)^2} \right)\left(1 - \frac{\tanh(p\,z(t,x))}{1+px}\right)^{-2}, \\ t,x,c,p \in \mathbb{R}, \\ z(t,x) = x - 4p^2 t - c \end{cases}$$

is a solution of the reduced Korteweg–de Vries equation of the form,

$$\partial_t u(t,x) - 6\,u(t,x)\partial_x u(t,x) + \partial_{x,x,x} u(t,x) = 0.$$

10.2

```
Clear[t,x,c,p]
```

```
z[t_,x_]:=x-4p²t-c;  (* Introduce functions z and u *)
u[t_,x_]:=-2p² (Cosh[p z[t,x]]⁻²-(1+p x)⁻²Tanh[p z[t,x]]²)
×(1-(1+p x)⁻¹Tanh[p z[t, x]])⁻²;
```

```
∂ₜu[t,x]-6u[t,x]∂ₓu[t,x]+∂₍ₓ,ₓ,ₓ₎u[t,x]==0//Simplify
(* Function u satisfies the equation *)
True
```

```
Block[{p=2,c=1}, (* Choose some particular values
and plot function u *)
Plot3D[u[t,x],{t,-.3,.3},{x,-6,6},ImageSize→160,
PlotStyle→Directive[Black,Opacity[.15]],PlotPoints→
50,PlotRange→#,PlotLabel→Style[#,Black,12],
Mesh→3]&/@{Automatic,All}]
```

See Figure 10.2.

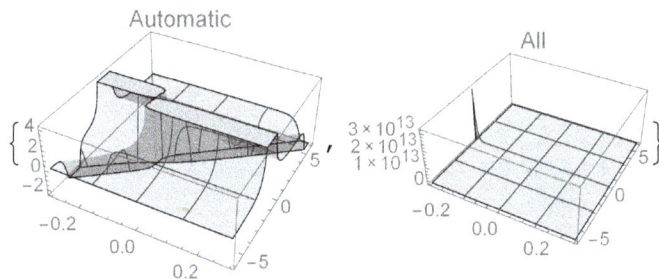

Figure 10.2: Plot of the particular solution for problem 10.2.

A dynamical plot follows:

```
{Manipulate[
Plot[u[t,x],{x,-6,6},Ticks→{{-6,0,6},Automatic},
ImageSize→160,PlotStyle→Black],{t,-.3,.3},
SaveDefinitions→True],
Manipulate[
Plot3D[Evaluate[u[t,x]],{t,-τ,τ},{x,-ξ,ξ},Mesh→2,
PlotStyle→Directive[Black,Opacity[.15]],Ticks→None,
ImageSize→160],{ξ,.0001,6},{τ,.0001,.3}]}
```

See Figure 10.3.

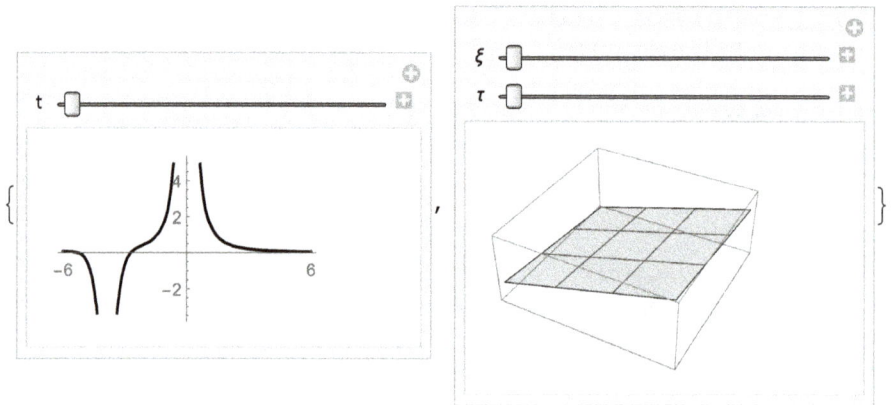

Figure 10.3: Dynamical plot of the solution for problem 10.2.

Remark. The case of n-soliton + one-pole function may be found in [12, p. 284 and p. 297]. □

10.3. Another problem for the Korteweg–de Vries equation is considered below:

$$\begin{cases} \partial_t u(t,x) = -\varepsilon u(t,x)\partial_x u(t,x) - \mu\partial_{x,x,x} u(t,x), & t,x \in \mathbb{R}, \\ \varepsilon = .002,\ \mu = .01,\ c = 1,\ d = 10,\ a = \sqrt{\varepsilon\, c/\mu}/2, \\ u(t,0) = 0,\ u(t,2) = 0,\ u(0,x) = 3c\,\text{sech}^2(a x + d). \end{cases}$$

10.3

```
Clear[t,x,ε,μ]
```

```
eqn=∂_t u[t,x]==-ε u[t,x]∂_x u[t,x]-μ∂_{x,3}u[t,x];
(* The equation *)
```

```
{ε,μ,c,d,a}={.002,.01,1,10,Sqrt[εc/μ]/2}; (* Set the
parameters *)

sol=NDSolve[{eqn,u[t,0]==0,u[t,2]==0,
u[0,x]==3c Sech[a x+d]²},u,{t,0,3},{x,0,2}];
(* Solve numerically the problem *)

{Plot3D[u[t,x]/.sol,{t,0,3},{x,0,2},Mesh→2,
PlotRange→Automatic,Ticks→{{0,3},{0,1,2},Automatic},
PlotStyle→Directive[Black,Opacity[.15]],
ImageSize→175,PlotRange→All,PlotPoints→50],
Plot[3c Sech[a x+d]²,{x,0,2},ImageSize→175,
PlotStyle→Black,PlotLabel→"Initial condition"]}
```

See Figure 10.4.

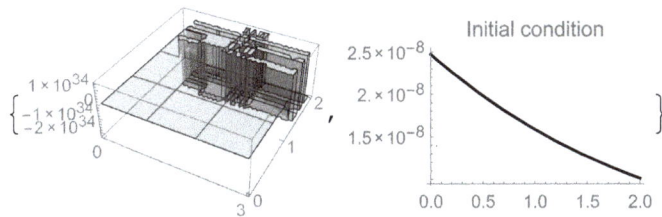

Figure 10.4: Plot of the solution for problem 10.3.

10.4. A problem with an initial condition for the Korteweg–de Vries equation is considered below:

$$\begin{cases} \partial_t u(t,x) + u(t,x)\partial_x u(t,x) + \delta^2 \partial_{x,x,x} u(t,x) = 0, & \delta = .022, \\ u(0,x) = \cos(\pi x). \end{cases}$$

10.4

```
Clear[a,t,x,δ]
```

We try finding a closed-form solution.

```
eqn=∂_t u[t,x]+u[t,x]∂_x u[t,x]+δ²∂_{x,3}u[t,x]; (* The
equation *)
DSolve[{eqn==0,u[0,x]==Cos[π x]},u[t,x],{t,x}] (* No
closed-form solution is returned *)
```

We solve the initial value problem numerically

```
Block[{δ=.022},
sol=NDSolve[{eqn==0,u[0,x]==Cos[π x]},u,{t,0,.55},
{x,0,2},MaxStepSize→.002];]
```

and plot its solution.

```
{Plot3D[u[t,x]/.First[sol],{t,0,.55},{x,0,2},Mesh→3,
PlotStyle→Directive[Black,Opacity[.15]],
Ticks→{{0,.55},{0,1,2},{-1,1,3}},PlotPoints→25,
ImageSize→175],
Plot[u[.44,x]/.First[sol],{x,0,2},ImageSize→160,
PlotStyle→Black,Ticks→{{0,1,2},{-1,0,1,2}},
Epilog→Text[Style["Cross section t=.44",Italic,11],
{1.28,1.2}]]}
Manipulate[
Plot[u[τ,x]/.First[sol],{x,0,2},ImageSize→175,
Ticks→{Range[0,2,.5],{-1,1}},PlotStyle→Black],
{τ,.001,.55},SaveDefinitions→True]
```

See Figure 10.5.

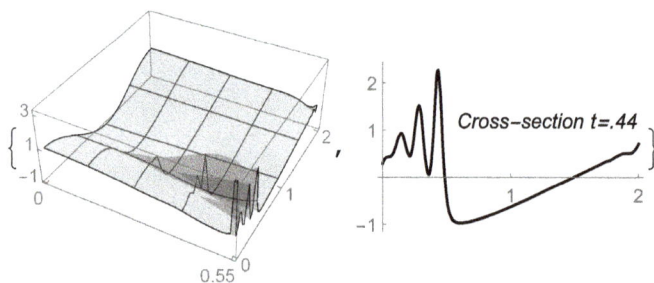

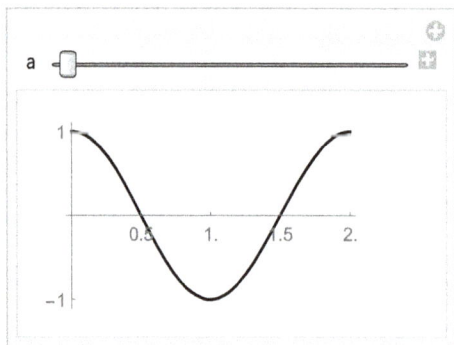

Figure 10.5: Plot of the solution for problem 10.4.

10.1.1 Algebraic solitons

Next, we introduce three algebraic (rational functions) soliton solutions of the reduced Korteweg–de Vries equation under the form

$$\partial_t u(t,x) - 6u(t,x)\partial_x u(t,x) + \partial_{x,x,x} u(t,x) = 0.$$

10.5. Our first algebraic soliton is considered here, [70, Chapter 13, Section 5] and [69],

$$u(t,x) = \frac{6x(x^3 - 24\,t)}{(x^3 + 12\,t)^2}$$

of the Korteweg–de Vries equation

$$u_t(t,x) - 6\,u(t,x)\,u_x(t,x) + u_{x,x,x}(t,x) = 0.$$

10.5

```
Clear[t,x,ξ]
```

```
u[t_,x_]:=6x(x³-24 t)/(x³+12 t)²  (* Define the function *)
```

```
∂ₜu[t,x]-6 u[t,x]∂ₓu[t,x]+∂₍ₓ,₃₎u[t,x]//Simplify
(* The function satisfies the equation *)
0
```

We plot the soliton.

```
Plot3D[Evaluate[u[t,x]],{t,-10,10},{x,-10,10},
Ticks→{10{-1,0,1},10{-1,0,1},Automatic},ImageSize→175,
PlotStyle→Directive[Black,Opacity[.15],Mesh→5,
PlotRange→#,PlotLabel→Style[#,Black]]
&/@{Automatic,All}
```

See Figure 10.6.

A dynamical plot of the soliton is presented below:

```
Manipulate[
Plot3D[Evaluate[u[t,x]],{t,-10,10},{x,-ξ,ξ},Mesh→3,
Axes→False,ImageSize→175,PlotPoints→50,
PlotStyle→Directive[Black,Opacity[.15]]],{ξ,.1,10},
SaveDefinitions→True]
```

See Figure 10.7.

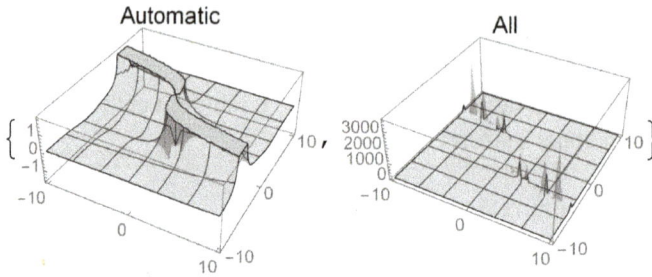

Figure 10.6: Plot of the first algebraic soliton in problem 10.5.

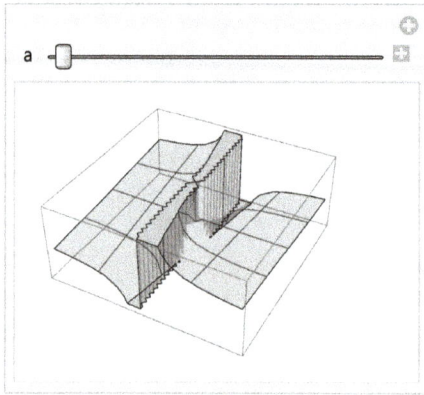

Figure 10.7: Dynamical plot of the soliton solution for problem 10.5.

10.6. Our second algebraic soliton solution is considered below, [70, Chapter 13, Section 5] and [69],

$$u(t, x) = -2\partial_{x,x} \ln(x^6 + 60x^3 t - 720t^2).$$

10.6

```
Clear[t,x]
```

```
u[t_,x_]:=-2 D[Log[x⁶+60 x³t-720 t²],x,x]  (* Define
the function *)
```

```
{u[t,x],
∂ₜu[t,x]-6u[t,x]∂ₓu[t,x]+∂₍ₓ,₃₎u[t,x]}//Simplify
(* The function is algebraic and satisfies the
equation *)
```

$$\left\{ \frac{12x(43200t^3 + 5400t^2 x^3 + x^9)}{(-720t^2 + 60t\,x^3 + x^6)^2}, 0 \right\}$$

A plot of the solution is presented next:

```
Plot3D[Evaluate[u[t,x]],{t,-10,10},{x,-10,10},
ImageSize→175,Mesh→3,PlotPoints→50,
Ticks→{10{-1,0,1},10{-1,0,1},None},
PlotStyle→Directive[Black,Opacity[.15]],
PlotLabel→Style[#,Black],PlotRange→#],
&/@{Automatic,All}
```

See Figure 10.8.

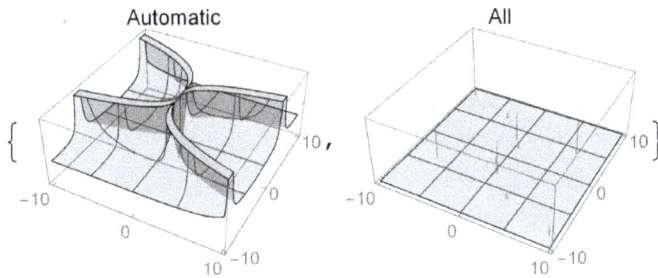

Figure 10.8: Plot of the solution for problem 10.6.

A dynamical plot of the solution is presented next:

```
Manipulate[
Plot3D[Evaluate[u[t,x]],{t,-10,10},{x,-ξ,ξ},
PlotStyle→Directive[Black,Opacity[.15]],
Boxed→False,Axes→False,PlotPoints→50,Mesh→3,
ImageSize→160],{ξ,.1,20},SaveDefinitions→True]
```

See Figure 10.9.

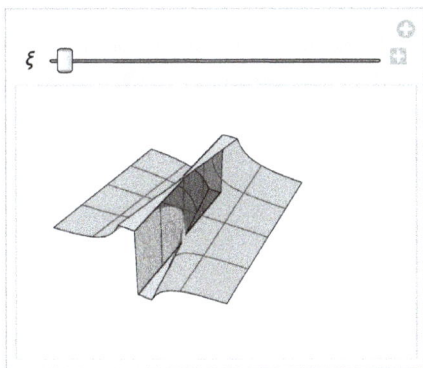

Figure 10.9: Dynamical plot of the solution for problem 10.6.

10.7. Our third algebraic soliton solution is considered now [35, p. 154],

$$u(t, x) = -\frac{6(6t\,x + x^4)}{(x^3 - 3t)^2}\,.$$

10.7

```
Clear[t,x,ξ]
u[t_,x_]:=-6(6t x+x⁴)/(x³-3t)²  (* Define the function *)
```

We check that this function is a solution of the reduced Korteweg–de Vries equation

```
∂ₜu[t,x]-3/2 u[t,x]∂ₓu[t,x]-1/4∂{x,3}u[t,x]//Simplify
(* The function satisfies the equation *)
0
```

and plot it.

```
{Plot3D[Evaluate[u[t,x]],{t,-10,10},{x,-10,10},##,
Ticks→{{-10,0,10},{-10,0,10},Automatic}],
Manipulate[
Plot3D[Evaluate[u[t,x]],{t,-10,10},{x,-ξ,ξ},##,
Boxed→False,Axes→False],{ξ,.1,20},
SaveDefinitions→True]}&[ImageSize→175,PlotPoints→50,
Mesh→3,PlotStyle→Directive[Black,Opacity[.15]]]
```

See Figure 10.10.

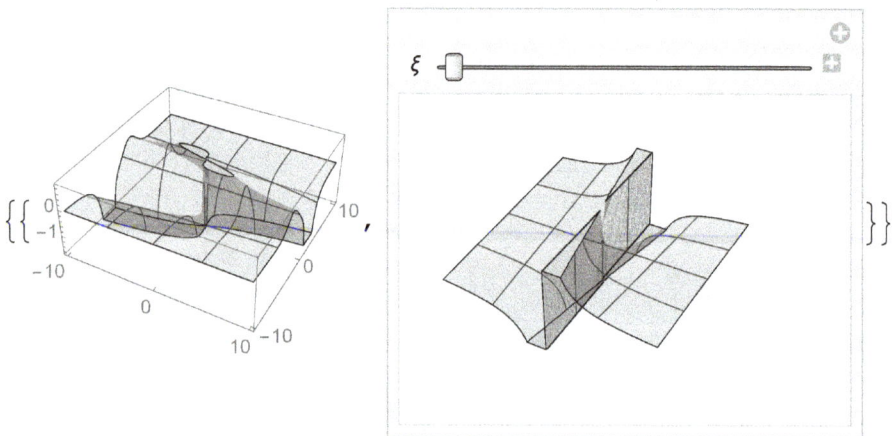

Figure 10.10: Plot of the solution for problem 10.7.

10.1.2 Soliton solutions to the Korteweg–de Vries equation

10.1.2.1 Single soliton solutions

10.8. Paper [40] contains an example of a single-soliton solution of the reduced Korteweg–de Vries equation,

$$u(t, x) = 3v \, \text{sech}^2\!\left(\frac{\sqrt{v}}{2}(x - v t) \right), \quad v > 0, \tag{10.1}$$

where v in (10.1) is the speed of the wave and $3v$ is the amplitude of the wave. Function (10.1) also appears in [69] and [71, p. 858].

10.8 We prove that function (10.1) is indeed a solution of the reduced Korteweg–de Vries equation.

```
Clear[t,x,k,v]
```

We introduce the function

```
u[t_,x_]:=3v Sech[Sqrt[v]/2(x-vt)]²
```

and check if it satisfies the reduced Korteweg–de Vries equation.

```
∂ₜu[t,x]+u[t,x]∂ₓu[t,x]+∂ₓ,ₓ,ₓu[t,x]
0
```

```
{Plot3D[u[t,x]/.v→1,{t,-5,5},{x,-10,15},Mesh→2,
PlotStyle→Directive[Black,Opacity[.15]],
ImageSize→175],
Manipulate[
Plot[u[t,x]/.v→1,{x,-10,15},ImageSize→155,
PlotStyle→Black,Ticks→{5Range[-1,3],{3}}],
{t,-5,5},SaveDefinitions→True]}
```

See Figure 10.11.

10.1.2.2 Two-soliton solutions

10.9. We show that the function, [90, (17.21), p. 583] and [39, p. 35],

$$u(t, x) = 12 \frac{k_1^2 u_1(t, x) + k_2^2 u_2(t, x) + 2(k_1 - k_2)^2 u_1(t, x) u_2(t, x)}{(1 + u_1(t, x) + u_2(t, x) + a \, u_1(t, x) u_2(t, x))^2}$$

$$+ 12 \frac{a \, u_1(t, x) u_2(t, x)(k_1^2 u_2(t, x) + k_2^2 u_1(t, x))}{(1 + u_1(t, x) + u_2(t, x) + a \, u_1(t, x) u_2(t, x))^2}$$

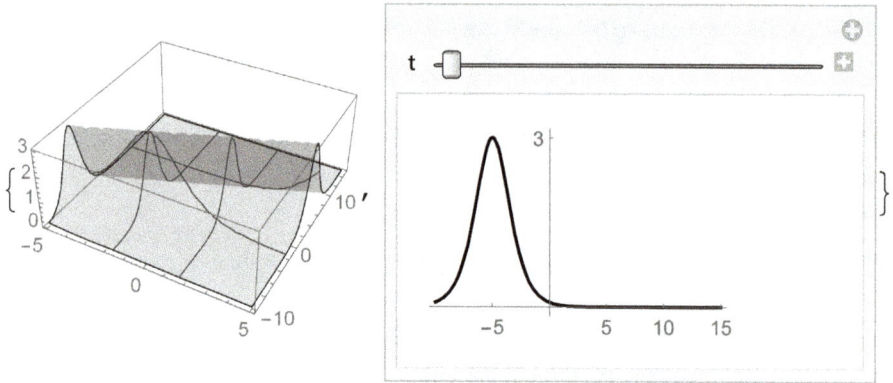

Figure 10.11: Plot of a particular solution to equation (10.1) and a cross-section through it.

where

$$u_1(t,x) = \exp(k_1^3 t - k_1 x), \quad u_2(t,x) = \exp(k_2^3 t - k_2 x), \quad a = \left(\frac{k_1 - k_2}{k_1 + k_2}\right)^2,$$

is a two-soliton solution of the reduced Korteweg–de Vries equation of the form

$$\partial_t u(t,x) + u(t,x)\partial_x u(t,x) + \partial_{x,x,x} u(t,x) = 0, \qquad (10.2)$$

10.9

```
Clear[k,t,x,ξ]
```

We introduce the solution

```
u[t_,x_]:=12(k₁²u₁[t,x]+k₂²u₂[t,x]+2(k₁-k₂)²u₁[t,x]u₂[t,x]
+a u₁[t,x]u₂[t,x](k₁²u₂[t,x]+k₂²u₁[t,x]))/
(1+u₁[t,x]+u₂[t,x]+a u₁[t,x]u₂[t,x])²
```

where

```
u₁[t_,x_]:=Exp[k₁³ t-k₁x], u₂[t_,x_]:=Exp[k₂³ t-k₂x],
a=(k₁-k₂/k₁-k₂)²
```

and check whether function u just defined satisfies the Korteweg–de Vries equation (10.2)

```
∂ₜu[t,x]+u[t,x]∂ₓu[t,x]+∂ₓ,ₓ,ₓu[t,x]==0//Simplify
True
```

We consider a particular solution

```
w[t_,x_]:=u[t,x]/.{k₁ →1,k₂ →1}
```

and plot it.

```
{Plot3D[w[t,x],{t,-12,8},{x,-10,10},ImageSize→175,
Mesh→2,Ticks→{{-12,8},10{-1,1},{0,2}},
PlotRange→{{-12,8},10{-1,1},{0,2}},PlotPoints→25,
PlotStyle→Directive[Black,Opacity[.15]]],
Manipulate[
Plot3D[w[t,x],{t,-12,18},{x,-ξ,ξ},ImageSize→165,
Boxed→False,Axes→False,Mesh→2,PlotPoints→25,
PlotStyle→Directive[Black,Opacity[.15]]],
{ξ,.1,10},SaveDefinitions→True]}
```

See Figure 10.12.

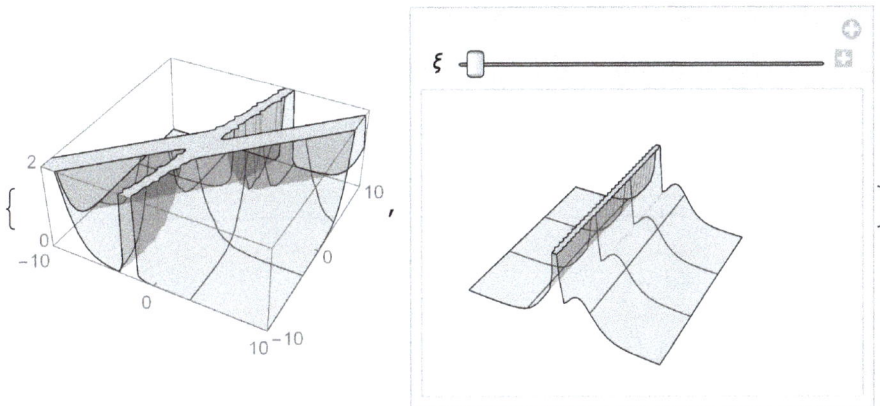

Figure 10.12: Plot of a particular solution for problem 10.9.

10.10. Another function is introduced [90, pp. 580–583],

$$u(t,x) = -2\partial_{x,x}(\ln(1 + b_1 \exp(u_1(t,x)) + b_2 \exp(u_2(t,x))$$
$$+ a\, b_1 b_2 \exp(u_1(t,x) + u_2(t,x))))$$

where

$$u_1(t,x) = k_1 x - k_1^3 t, \quad u_2(t,x) = k_2 x - k_2^3 t, \quad a = \left(\frac{k_1 - k_2}{k_1 + k_2}\right)^2$$

which is a two-soliton solution of the Korteweg–de Vries equation of the form

$$\partial_t u(t,x) - 6\,u(t,x)\partial_x u(t,x) + \partial_{x,x,x}u(t,x) = 0,$$

10.10

```
Clear[b,k,t,x,ξ]
```

Let us introduce

$$u_1[t\_,x\_]:=k_1x-k_1^3t;u_2[t\_,x\_]:=k_2x-k_2^3t;a=(\frac{k_1-k_2}{k_1+k_2})^2;$$

and

```
u[t_,x_]:=-2D[Log[1+b₁Exp[u₁[t,x]]+b₂Exp[u₂[t,x]]
+ab₁b₂Exp[u₁[t,x]+u₂[t,x]]],x,x]//Simplify
(* Define the solution *)
```

We check the solution

$$\partial_t u[t,x]-6u[t,x]\partial_x u[t,x]+\partial_{x,x,x}u[t,x]==0//Simplify$$
```
(* The equation is satisfied by the solution *)
True
```

Set a particular solution

$$w[t\_,x\_]:=u[t,x]/.\{k_1 \to 1,k_2 \to 2,b_1 \to 1,b_2 \to 1\}$$

and plot it.

```
{{Plot3D[Evaluate[w[t,x]],##,ImageSize→175,
PlotPoints→25],
Plot3D[Evaluate[w[t,x]],##,ImageSize→160,
ViewPoint→{0,0,8},PlotPoints→50]}&[{t,-20,20},
{x,-10,10},Mesh→2,Ticks→{{-20,20},{-10,10},{-.2,0}},
PlotStyle→Directive[Black,Opacity[.15]]],
{Manipulate[
Plot3D[Evaluate[w[t,x]],##,PlotPoints→25],
{ξ,.1,10},SaveDefinitions→True],
Manipulate[
Plot3D[Evaluate[w[t,x]],##,ViewPoint→{0,0,8},
PlotPoints→50],{ξ,.1,10},
SaveDefinitions→True]}&[{t,-20,20},{x,-ξ,ξ},
ImageSize→150,Boxed→False,Axes→False,Mesh→2,
PlotStyle→Directive[Black,Opacity[.15]]]}//Flatten
```

See Figure 10.13.

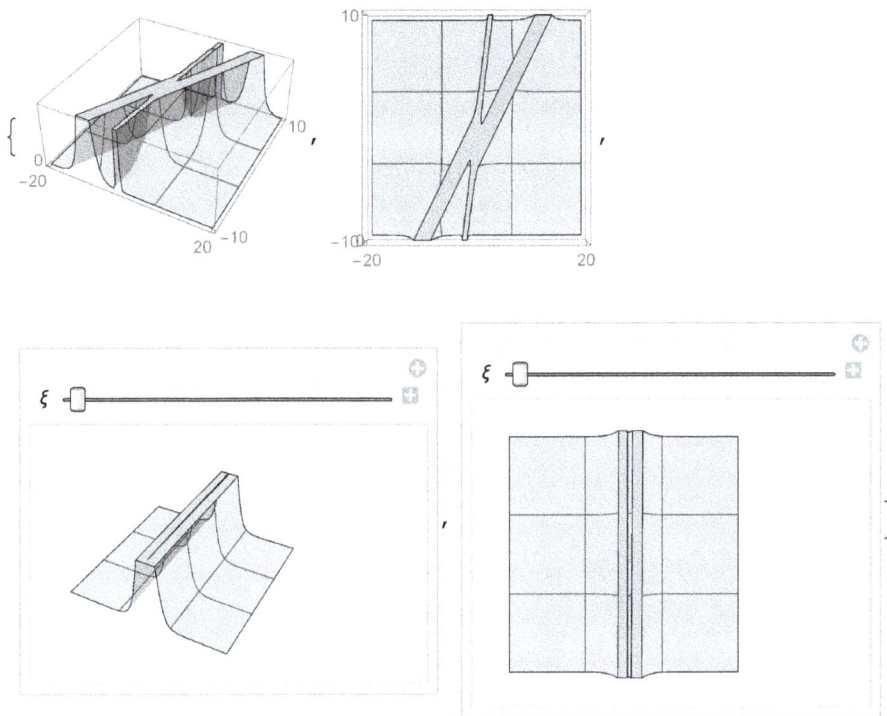

Figure 10.13: Plot of a particular solution for problem 10.10.

10.11. Next, we show by the Hirota bilinear method that the reduced Korteweg–de Vries equation.

$$\partial_t u(t,x) - 6\,u(t,x)\partial_x u(t,x) + \partial_{x,x,x} u(t,x) = 0,$$

has a two-soliton solution, [90, pp. 580–583] and [69].

10.11

```
<<Notation`
```

```
Symbolize[k_]
```

```
Clear[f,g,i,k,m,t,x,v]
```

We introduce the Korteweg–de Vries equation

```
kdv[t_,x_]:=D[u[t,x],t]+6u[t,x]D[u[t,x],x]+D[u[t,x],
{x,3}]
```

We start with the Hirota derivatives

```
hxm[f_,g_,m_]:=∑ᵢ₌₀ᵐ(-1)ᵐBinomial[m,i]D[f,{x,m-i}]
*D[g,{x,i}]
htm[f_,g_,m_]:=∑ᵢ₌₀ᵐ(-1)ᵐBinomial[m,i]D[f,{t,m-i}]
*D[g,{t,i}]
hxt[f_,g_]:=f D[g,x,t]-D[f,x]D[g,t]-D[f,t]D[g,x]
+D[f,x,t]g
```

We consider the ansatz for f_1, following (9.13),

```
f₁₁[t_,x_]:=Exp[k₁(x-v₁t)]; f₁₂[t_,x_]:=Exp[k₂(x-v₂t)]
f₁[t_,x_]:=f₁₁[t,x]+f₁₂[t,x]
b₁[f1_]:=hxt[1,f1]+hxt[f1,1]+hxm[1,f1,4]+hxm[1,f1,4]
(* Bilinear form *)
```

```
Reduce[b₁[f₁₁[t,x]]==0&&b₁[f₁₂[t,x]]==0,{v₁,v₂}]
(k₁==0&&(k₂==0‖v₂==k₂²))‖(k₂==0&&v₁==k₁²)‖
(v₁==k₁²&&v₂==k₂²)
```

We neglect the trivial cases and keep the cases $v_1 = k_1^2$ and $v_2 = k_2^2$.

In what follows, we consider the ansatz for f_2, (9.14),

```
f₂[t_,x_]:=a₁₂Exp[k₁(x-v₁t)+k₂(x-v₂t)]
f[t_,x_]:=1+f₁[t,x]+f₂[t,x]
b₂[f1_,f2_]:=hxt[1,f2]+hxt[f2,1]+hxm[1,f2,4]+
hxm[f2,1,4]+hxt[f1,f1]+hxm[f1,f1,4]
v₁=k₁²;v₂=k₂²;
b₂[f₁[t,x],f₂[t,x]]//Simplify
```

Then it follows that

```
a₁₂=(k₁-k₂)²/(k₁+k₂)²;
u[t_,x_]:=2D[Log[f[t,x]],x,x] (* This is the function
in the statement of the problem *)
```

Here, we insert a figure with a two-soliton solution of this problem.

```
Block[{k₁=1,k₂=Sqrt[2]}
Plot3D[Evaluate[u[t,x]],{t,0,1},{x,-7,12},Mesh→None,
PlotStyle→Directive[Black,Opacity[.15]],ImageSize→
175,Ticks→{{0,1},{-7,12},{0,.7}},PlotPoints→25]]
```

See Figure 10.14.

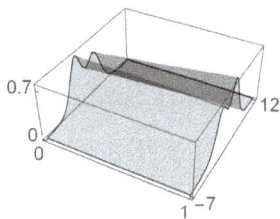

Figure 10.14: Plot of the two-soliton solution for problem 10.11.

Remark. This problem is stated in a simplified form in [39, p. 35]. □

A dynamical plot of the solution is introduced for a particular case.

```
k₁=1;k₂=Sqrt[2];
```

```
Manipulate[
Plot3D[Evaluate[u[t,x]],{t,0,1},{x,-7,ξ},Mesh→None,
PlotStyle→Directive[Black,Opacity[.15]],ImageSize→
175,Mesh→None,Ticks→{{0,1},{-7,N[ξ,2]},{0,.7}},
PlotPoints→25],{ξ,-6.8,12},SaveDefinitions→False,
SynchronousUpdating→False]
```

See Figure 10.15.

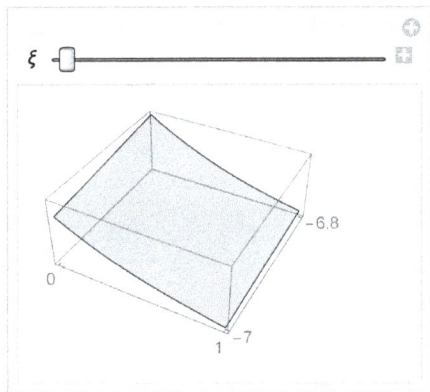

Figure 10.15: Dynamical plot of the two-soliton solution for problem 10.11.

10.1.2.3 Three-soliton solutions

10.12. Here, we search a three-soliton solution to the Korteweg–de Vries equation of the form

$$\partial_t u(t,x) + 6u(t,x)\partial_x u(t,x) + \partial_{x,x,x} u(t,x) = 0,$$

by the Hirota bilinear method, Section 9.6.2 and [26].

10.12

```
<<Notation`
```

```
Symbolize[k_]
```

```
Clear[a,b,i,m,f,g,t,x,k,v]
```

We introduce the Korteweg–de Vries equation

```
kdv[t_,x_]:=D[u[t,x],t]+6u[t,x]D[u[t,x],x]+D[u[t,x],
{x,3}]
```

We start with the Hirota derivatives

```
hxm[f_,g_,m_]:=∑ᵐᵢ₌₀(-1)ᵐBinomial[m,i]D[f,{x,m-i}]
*D[g,{x,i}]
htm[f_,g_,m_]:=∑ᵐᵢ₌₀(-1)ᵐBinomial[m,i]D[f,{t,m-i}]
*D[g,{t,i}]
hxt[f_,g_]:=f D[g,x,t]-D[f,x]D[g,t]-D[f,t]D[g,x]
+D[f,x,t] g
```

and discuss the ansatz for f_1, (9.13),

```
f₁₁[t_,x_]:=Exp[k₁(x-v₁t)]
f₁₂[t_,x_]:=Exp[k₂(x-v₂t)]
f₁₃[t_,x_]:=Exp[k₃(x-v₃t)]
f₁[t_,x_]:=f₁₁[t,x]+f₁₂[t,x]+f₁₃[t,x]
b₁[f₁_]:=hxt[1,f₁]+hxt[f₁,1]+hxm[f₁,1,4]hxm[1,f₁,4]
(* Bilinear form *)
```

We look for the relationships between k_i and v_i,

```
Reduce[b₁[f₁₁]==0&&b₁[f₂₁]==0&&b₁[f₁₃]==0,{v₁,v₂,v₃}]
```

Neglecting the trivial solutions, we find that

$$v_1 == k_1^2 \&\& v_2 == k_2^2 \&\& v_3 == k_3^2$$

We discuss the ansatz for f_2, (9.14),

```
f₂₁[t_,x_]:=a₁₂Exp[k₁(x-v₁t)+k₂(x-v₂t)]
f₂₂[t_,x_]:=a₁₃Exp[k₁(x-v₁t)+k₃(x-v₃t)]
f₂₃[t_,x_]:=a₂₃Exp[k₂(x-v₂t)+k₃(x-v₃t)]
f₂[t_,x_]:=f₂₁[t,x]+f₂₂[t,x]+f₂₃[t,x]
b₂[f₁_,f₂_]:=hxt[1,f₂]+hxt[f₂,1]+hxm[f₁,1,4]+hxm[1,f₂,4]
+hxm[f₂,1,4]+hxt[f₁,f₁]+hxm[f₁,f₁,4]  (* The bilinear
form *)
```

We look for the relationships between a_{ij} and k_n,

```
Reduce[(-1+a₁₂)k₁²+2(1+a₁₂)k₁k₂+(-1+a₁₂)k₂²==0&&
(-1+a₁₃)k₁²+2(1+a₁₃)k₁k₃+(-1+a₁₃)k₃²==0&&
(-1+a₂₃)k₂²+2(1+a₂₃)k₂k₃+(-1+a₂₃)k₃²==0,{a₁₂,a₁₃,a₂₃}]
```

One obtains that

$$a_{12}=\frac{(k_1-k_2)^2}{(k_1+k_2)^2}\,;\,a_{13}=\frac{(k_1-k_3)^2}{(k_1+k_3)^2}\,;\,a_{23}=\frac{(k_2-k_3)^2}{(k_2+k_3)^2}\,;$$

Finally, we discuss the ansatz for f_3, (9.15),

```
f₃[t_,x_]:=b₁₂₃Exp[k₁(x-v₁t)+k₂(x-v₂₁t)+k₃(x-v₃t)]]
f[t_,x_]:=1+f₁[t,x]+f₂[t,x]+f₃[t,x]
```

By (9.15),

```
b₃[f₁_,f₂_,f₃_]:=hxt[1,f₃]+hxt[f₃,1]+hxm[1,f₃,4]
+hxm[f₃,1,4]+hxt[f₁,f₂]+hxm[f₁,f₂,4]+hxt[f₂,f₁]
+hxm[f₂,f₁,4]  (* Bilinear form *)
```

one has that

$$b_3[f_1,f_2,f_3]=0 \implies b_{123}=a_{12}a_{13}a_{23}$$

The three-soliton solution is of the form

```
u[t_,x_]:=2D[Log[f[t,x]],x,x]//Simplify
```

For certain particular values of the parameters, we plot the corresponding three-soliton solution

```
Block[{k₁=1,k₂=Sqrt[2],k₃=1.3},
{Plot3D[##,ImageSize→190],
Plot3D[##,ImageSize→175,ViewPoint→{.03,2,20}]}
&[Evaluate[u[t,x]],{t,0,1},{x,-7,20},PlotPoints→50,
ColorFunction→Function[{x,y,z},Hue[z]],Mesh→None,
Ticks→{{0,1},{-7,20},{0,.8}},PlotStyle→Opacity[.4]]]
```

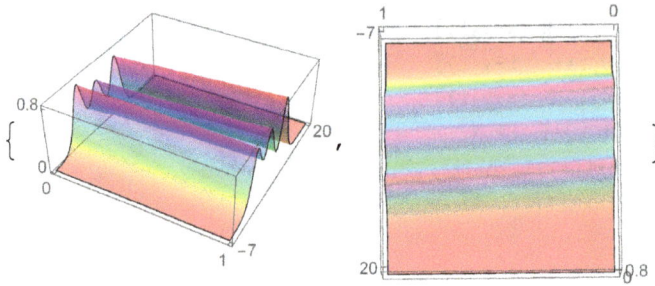

Figure 10.16: Plot of the three-soliton solution for problem 10.12.

See Figure 10.16.

We also show a dynamical plot of the three-soliton solution.

```
k₁=1;k₂=Sqrt[2];k₃=1.3;
```

```
Manipulate[
Plot3D[Evaluate[u[t,x]],{t,0,1},{x,-7,ξ},
ImageSize→175,Mesh→None, PlotPoints→25,
Ticks→{{0,1},{-7,N[ξ,2]},{0,.8}},
PlotStyle→Directive[Black,Opacity[.15]]],
{ξ,-6.8,20},SaveDefinitions→False,
SynchronousUpdating→False]
```

See Figure 10.17.

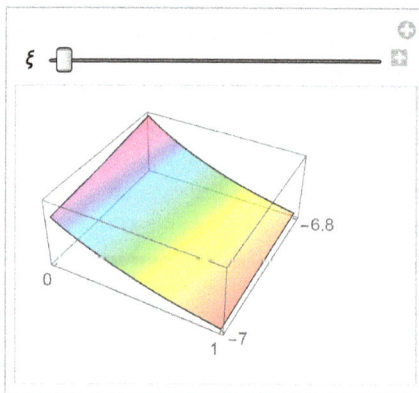

Figure 10.17: Dynamical plot of the three-soliton solution for problem 10.12.

10.1.3 Modified Korteweg-de Vries equation

The *modified Korteweg–de Vries equation* that we deal with is of the form

$$u_t(t,x) + u^2(t,x)u_x(t,x) + u_{x,x,x}(t,x) = 0, \quad t, x \in \mathbb{R}. \qquad (10.3)$$

Equation (10.3) is studied by the exp function method in paper [24].

10.13. We look for the soliton solution of equation (10.3) by the exp function method in Section 9.6.2.

10.13 The exp function method supposes that the soliton solution of equation (10.3) can be written under the form

$$w(\xi) = u(t,x) = \frac{\sum_{n=-d}^{c} a_n \exp(n\,\xi)}{\sum_{m=-q}^{p} b_m \exp(m\,\xi)}, \quad c, d, p, q \geq 0, \quad a_n, b_m, t, x, \xi \in \mathbb{R}, \qquad (10.4)$$

applying the transformation $\xi = k\,x + v\,t$.

```
<< Notation'
Symbolize[a_]
Symbolize[b_]

Clear[a,b,t,x,k,v]
```

By *Mathematica*, the left-hand side of the modified Korteweg–de Vries equation is written under the form

```
eqn=∂ₜu[t,x]+u[t,x]²∂ₓu[t,x]+∂ₓ,ₓ,ₓu[t,x];
```

and by the transformation $\xi=k\,x+v\,t$ it looks like

```
eqnv=v w'[ξ]+k w[ξ]²w'[ξ]+k³w'''[ξ];
```

If we substitute (10.4) in *eqnv* $= 0$ and balance the highest order of w''' and w^2w', we find that $p = c$. By balancing the lowest order of w''' and w^2w', we find that $q = d$. We set $p = c = 1$ and $q = d = 1$.
    We conclude that the solution is of the form

$$z[\xi\_] := \frac{a_1 e^{\xi} + a_0 + a_{-1} e^{-\xi}}{e^{\xi} + b_0 + b_{-1} e^{-\xi}}$$

We substitute it in the left-hand side of the equation

```
eqnz=v z'[ξ]+k z[ξ]²z'[ξ]+k³z'''[ξ];
eqnz1=(e^ξ+b₀+b₋₁e^{-ξ})⁴ eqnz/.Table[e^{nξ} →q^n,{n,-1,8}]
//Expand//Simplity
eqnz2=CoefficientList[Normal[Series[q³eqnz1,
{q,0,6}]],q] Simplify;
numberofcoeff=Length[eqnz2]
```

7

```
eqncom=Table[eqnz2[[k]]==0,{k,numberofcoeff}];
```

We transform the previous list into a system of algebraic equations and solve it.

```
Reduce[v+k a₁²+k³==0&&
k a₋₁²a₀ b₋₁+k³a₀b₋₁³+v a₀b₋₁³-k a₋₁³b₀-k³a₋₁b₋₁²b₀-v a₋₁b₋₁²b₀==0&&
k a₋₁³-k a₋₁a₀²b₋₁-k a₋₁²a₁b₋₁+4k³a₋₁b₋₁²+v a₋₁b₋₁²-4k³a₁b₋₁³
-v a₁b₋₁³+k a₋₁²a₀b₀+2k³a₀b₋₁²b₀-v a₀b₋₁²b₀-2k³a₋₁b₋₁b₀²
+v a₋₁b₋₁b₀²==0&&
-5k a₋₁²a₀+k a₀³b₋₁+6k a₋₁a₀a₁b₋₁-23k³a₀b₋₁²+v a₀b₋₁²-k a₋₁a₀²b₀
-k a₋₁²a₁b₀+18k³a₋₁b₋₁b₀-6v a₋₁b₋₁b₀+5k³a₁b₋₁²b₀+5v a₁b₋₁²b₀
+k³a₀b₋₁b₀²+v a₀b₋₁b₀²-k³a₋₁b₀³-v a₋₁b₀³==0&&
-k a₋₁a₀²-k a₋₁²a₁+8k³a₋₁b₋₁-v a₋₁b₋₁+k a₀²a₁b₋₁+k a₋₁a₁²b₋₁
-8k³a₁b₋₁²+v a₁b₋₁²-k³a₋₁b₀²-v a₋₁b₀²+k³a₁b₋₁b₀²
+v a₁b₋₁b₀²==0&&
-k a₀³-6k a₋₁a₀a₁+23k³a₀b₋₁-v a₀b₋₁+5k a₀a₁²b₋₁-5k³a₋₁b₀
-5v a₋₁b₀+k,a₀²a₁b₀+k a₋₁a₁²b₀-18k³a₁b₋₁b₀+6v a₁b₋₁b₀
-k³a₀b₀²-v a₀b₀²+k³a₁b₀³+v a₁b₀³==0&&
4k³a₋₁+v a₋₁+k a₀²a₁+k a₋₁a₁²-4k³a₁b₋₁-v a₁b₋₁-k a₁³b₋₁+v a₀b₀
-2k³a₀b₀-k a₀a₁²b₀+2k³a₁b₀²-v a₁b₀²==0&&
-k³a₀-v a₀-k a₀a₁²+k³a₁b₀+v a₁b₀+k a₁³b₀==0,
{a₀,a₋₁,b₋₁,v},Reals];
```

Then we get a feasible solution of the form (the parameters are a_1, b_0, k):

$$a_0=\frac{a_1^2b_0+3b_0k^2}{a_1}\ ;\ a_{-1}=\frac{b_0^2(2a_1^2+3k^2)}{8a_1}\ ;\ b_{-1}=\frac{b_0^2(2a_1^2+3k^2)}{8a_1^2}\ ;\ v=-a_1^2k-k^3\ ;$$

The solution that we look for is

$$u[t\_,x\_]:=\frac{a_1e^{kx+vt}+a_0+a_{-1}e^{-(kx+vt)}}{e^{kx+vt}+b_0+b_{-1}e^{-(kx+vt)}}$$

We check this solution.

$$\partial_t u[t,x]+u[t,x]^2\partial_x u[t,x]+\partial_{x,x,x}u[t,x]//\text{Simplify}$$

0

Next, we introduce the figure of the solution for certain values of the parameters.

```
Block[{a₁=1.5,b₀=1,k=.5},
Plot3D[u[t,x],{t,-5,5},{x,-5,5},ImageSize→175,
PlotStyle→Directive[Black,Opacity[.15]],PlotPoints→
50,Mesh→2,Ticks→{{-5,0,5},{-5,0,5},{1.5,1.7}}]]
```

See Figure 10.18.

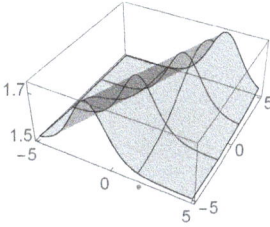

Figure 10.18: Plot of the one-soliton solution for problem 10.13.

10.1.4 Potential Korteweg–de Vries equation

The *potential Korteweg–de Vries equation* is of the form [3, (14)] and [73, (2)],

$$\partial_t u(t,x) + a\left(\partial_x u(t,x)\right)^2 + b\,\partial_{x,x,x} u(t,x) = 0, \quad a,b \in \mathbb{R}, \tag{10.5}$$

10.14. We show that the function introduced in [3, (14)], namely

$$u(t,x) = c\,\tanh(k(x - v\,t)),$$

for certain values of the parameters is a solution of the potential Korteweg–de Vries equation (10.5).

10.14

```
Clear[a,b,c,k,v,t,x]
```

We substitute function u into equation (10.5) to find out the suitable values of the parameters

```
∂ₜu[t,x]+a∂ₓu[t,x]²+b∂ₓ,ₓ,ₓu[t,x]//Simplify
1/2 c k(2a c k-8b k²-v+(4b k²-v)Cosh[2k(-t v+x)])
Sech[k(-t v+x)]⁴
```

The last expression is zero if and only if the next two equalities are valid

```
Reduce[-8b k²-v+2a k c==0&&4b k²-v==0,{c,k},Reals]
```

Table 10.1: Table of coefficients for problem 10.14.

n	a	b	C	V	k
1	<0	<0	$\frac{8bk^2+v}{2ak}$	<0	$\pm\frac{1}{2}\sqrt{\frac{v}{b}}$
2	<0	>0	$\frac{8bk^2+v}{2ak}$	>0	$\pm\frac{1}{2}\sqrt{\frac{v}{b}}$
3	>0	<0	$\frac{8bk^2+v}{2ak}$	<0	$\pm\frac{1}{2}\sqrt{\frac{v}{b}}$
4	>0	>0	$\frac{8bk^2+v}{2ak}$	>0	$\pm\frac{1}{2}\sqrt{\frac{v}{b}}$

The feasible solutions of the previous algebraic system are given by Table 10.1. Now we plot the solution for certain values of the parameters.

```
Manipulate[
Block[{a=1,b=1,k=1/2 Sqrt[v/b],c=(8bk²+v)/(2a k),v=w},
Plot3D[u[t,x],{t,0,5},{x,-20,20},ImageSize→175,
Mesh→2,Ticks→{{0,5},20{-1,1},Automatic},
PlotStyle→{Gray,Opacity[.15]}]],{{w,.1,"v"},.1,3}]
```

See Figure 10.19.

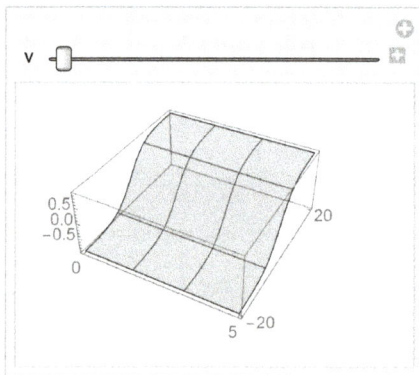

Figure 10.19: Plot of a solution to the potential Korteweg–de Vries equation in problem 10.14.

10.1.5 The generalized Kudryashov method

10.15. Below we show that the functions,

$$u_1(t,x) = \frac{1}{a\beta}\left(a\,a - 3v\,\beta\left(1 - \tanh\left(\frac{x-vt}{2}\right)\right)\right),$$

$$u_2(t,x) = \frac{1}{a\beta}\left(a\,a - 3v\,\beta\left(1 - \coth\left(\frac{x-vt}{2}\right)\right)\right)$$

for certain values of the parameters, are solutions of the potential Korteweg–de Vries equation (10.5).

10.15

```
Clear[a,b,v,t,x,α,β]
```

We introduce functions u_1, [73, (10)] and [73, (18)], and u_2, [73, (11)] and [73, (20)].

```
u₁[t_,x_]:=(1/aβ (aα-3vβ(1-Tanh[x-vt/2]))),
u₂[t_,x_]:=(1/aβ (aα-3vβ(1-Coth[x-vt/2])))
```

We successively substitute functions u_1 and u_2 in equation (10.5).

```
∂ₜu₁[t,x]+a(∂ₓu₁[t,x])²+b∂ₓ,ₓ,ₓu₁[t,x]//Simplify
∂ₜu₂[t,x]+a(∂ₓu₂[t,x])²+b∂ₓ,ₓ,ₓu₂[t,x]//Simplify
```

Both functions satisfy equation (10.5) for $v = b$.
    We plot the surfaces.

```
Block[{a=-1,b=1,α=-1,β=2,v=b},
{Plot3D[u₁[t,x],##,Ticks→{{0,10},{-10,20},{0,5}},
PlotLabel→Style["u₁",12,Black]],
Plot3D[u₂[t,x],##,Ticks→{{0,10},{-10,20},{-10,10}},
PlotLabel→Style["u₂",12,Black]]}
&[{t,0,10},{x,-10,20},ImageSize→175,Mesh→None,
PlotStyle→Directive[Black,Opacity[.15]],
PlotPoints→35]]
```

See Figure 10.20.

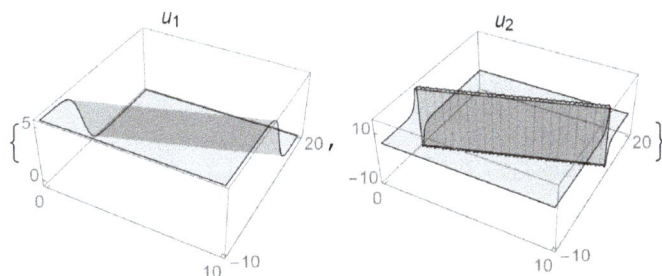

Figure 10.20: Plot of two solutions for the potential Korteweg–de Vries equation in problem 10.15.

10.16. We show that the functions [73, (35)], [73, (36)], and [73, (37)],

$$u_1(t,x) = \frac{1}{a\gamma}\left(a\,\alpha + 6b\,\beta - 3b\,\gamma\left(1 - \tanh\left(\frac{x-bt}{2}\right)\right)\right),$$

$$u_2(t,x) = \frac{1}{a\gamma}\left(a\,\alpha + 6b\,\beta - 3b\,\gamma\left(1 - \coth\left(\frac{x-bt}{2}\right)\right)\right),$$

$$u_3(t,x) = -\frac{\alpha}{2\beta} - \frac{3b}{a}\left(2 - \tanh\left(\frac{x-bt}{2}\right)\right)$$

are solutions of the potential Korteweg–de Vries equation (10.5).

10.16

```
Clear[a,b,α,β,γ,t,x]
```

We introduce functions u_1, u_2, and u_3,

```
u₁[t_,x_]:=1/(aγ)(a α+6b β-3b γ(1-Tanh[(x-bt)/2]));
u₂[t_,x_]:=1/(aγ)(a α+6b β-3b γ(1-Coth[(x-bt)/2]));
u₃[t_,x_]:=-α/(2β)-3b/a(2-Tanh[(x-bt)/2])
```

check out that they are indeed solutions of the potential Korteweg–de Vries equation (10.5),

```
{∂t#+a(∂x#)²+b∂x,x,x#}&/{u₁[t,x],u₂[t,x],u₃[t,x]}
//Simplify//Flatten
{0,0,0}
```

and plot them for some particular values of the parameters

```
Block[{a=1,b=1/2,c=-1,α=1,β=2,γ=-2},
{Plot3D[Evaluate[u₁[t,x]],##,
Ticks→{{0,10},{-10,10},{-6,-4}},
PlotLabel→Style["u₁",12,Black]],
Plot3D[Evaluate[u₂[t,x]],##,
Ticks→{{0,10},{-10,15},{-10,0}},
PlotLabel→Style["u₂",12,Black]],
Plot3D[Evaluate[u₃[t,x]],##,
Ticks→{{0,10},{-10,15},{-4,-2}},
PlotLabel→Style["u₃",12,Black]]}&[{t,0,10},{x,-10,10},
ImageSize→175,Mesh→None,PlotPoints→45,
PlotStyle→Directive[Black,Opacity[.15]]]]
```

See Figure 10.21.

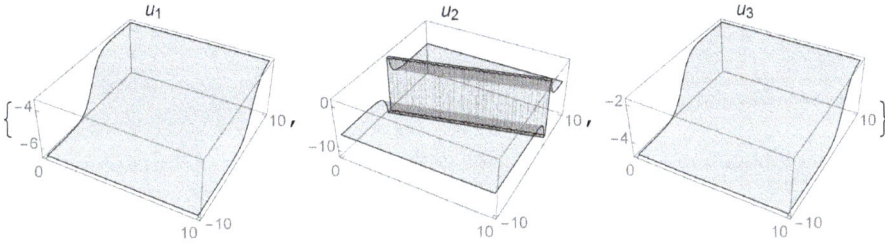

Figure 10.21: Plot of three solutions for the potential Korteweg–de Vries equation in problem 10.16.

10.17. Below we use the tan-cot method to study the potential Korteweg–de Vries equation (10.5), [73, (58)] and [73, (39)]. For it, the next functions are considered

$$u_1(t,x) = c \, \tan(k(x - v\,t)),$$
$$u_2(t,x) = c \, \cot(k(x - v\,t)).$$

10.17

```
Clear[a,b,c,t,x]
```

We introduce functions u_1 and u_2,

```
u₁[t_,x_]:=c Tan[k(x-v t)]
u₂[t_,x_]:=c Cot[k(x-v t)]
```

We substitute function u_2 in the potential Korteweg–de Vries equation (10.5).

```
∂ₜu₁[t,x]+a(∂ₓu₁[t,x])²+b∂ₓ,ₓ,ₓu₁[t,x]//Simplify
```

Then

```
Reduce[4b k²+v==0&&-2a c k-8b k²+v==0,{c,k}]
```

supplies

$$u_1[t\_,x\_]:= \frac{3\text{Sqrt}[b]\text{Sqrt}[-v]}{a} \; \text{Tan}\!\left[-\frac{\text{Sqrt}[-v]}{2\text{Sqrt}[b]}(x-v\,t)\right], \quad v<0.$$

Now we substitute function u_2 in the potential Korteweg–de Vries equation (10.5).

```
∂ₜu₂[t,x]+a(∂ₓu₂[t,x])²+b∂ₓ,ₓ,ₓu₂[t,x]//Simplify
```

Then

```
Reduce[4b k²+v==0&&-2a c k-8b k²+v==0,{c,k}]
```

supplies

$$u_2[t\_,x\_]:=\frac{3\text{Sqrt}[b]\text{Sqrt}[-v]}{a}\,\text{Cot}[-\frac{\text{Sqrt}[-v]}{2\text{Sqrt}[b]}\,(x-v\,t)],\quad v<0.$$

The plot of functions u_1 and u_2 follows for some particular values of the parameters,

```
Block[{a=1,b=2,v=-1},
{Plot3D[Evaluate[u₁[t,x]],##,PlotLabel→Style["u₁",12,
Black]],
Plot3D[Evaluate[u₂[t,x]],##,PlotLabel→Style["u₂",12,
Black]]}&[{t,-10,10},{x,-10,10},PlotPoints→50,
PlotStyle→Directive[Black,Opacity[.15]],
Ticks→Outer[Times,10{1,1,2},{-1,0,1}],
ImageSize→175,Mesh→None]]
```

See Figure 10.22.

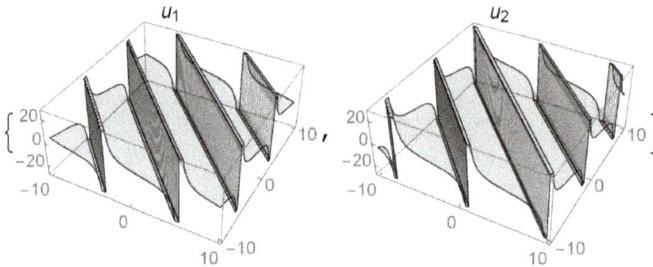

Figure 10.22: Plot of the solutions for the potential Korteweg–de Vries equation for problem 10.17.

10.2 The Dodd–Bullough–Mikhailov equation

The *Dodd–Bullough–Mikhailov equation* that we deal with is of the form [80, p. 640] and [24, (26)],

$$u_{t,x}(t,x) + e^{u(t,x)} + e^{-2u(t,x)} = 0, \quad t, x \in \mathbb{R}. \tag{10.6}$$

10.18. We look for the soliton solution of equation (10.6) by the exp function method in Section 9.6.3.

10.18

```
<< Notation`
Symbolize[b_]
```

The left-hand side of the Dodd–Bullough–Mikhailov equation is written by *Mathematica* under the form

eqn=∂ₜ,ₓu[t,x]+eᵘ⁽ᵗ,ˣ⁾+e⁻²ᵘ⁽ᵗ,ˣ⁾

If we try to solve it by *Mathematica*, then

```
DSolve[eqn==0,u[t,x],{t,x}];  (* No closed-form
solution is returned *)
```

We transform equation (10.6) by

```
z[t_,x_]:=Exp[u[t,x]]  (* z[t,x]>0 *)
```

and get

```
eqnz=z[t,x]∂t,xz[t,x]-∂tz[t,x]∂xz[t,x]+z[t,x]³+1;
```

New independent and dependent variables are introduced

```
ξ=k(x-v t);w[ξ]=z[t,x];
```

and the new equation is of the form

```
eqnw=-k²v w[ξ]w''[ξ]+k²v w'[ξ]²+w[ξ]³+1;
```

The exp function method supposes that the soliton solution of equation (10.6) can be written under the form

$$w(\xi) = \frac{\sum_{n=-d}^{c} a_n \exp(n\,\xi)}{\sum_{m=-q}^{p} b_m \exp(m\,\xi)}, \quad c,d,p,q \geq 0, \quad a_n, b_m \in \mathbb{R}. \tag{10.7}$$

We balance $v\,v''$ and v^3 to get $p = c$. Similarly, we find that $d = q$.

Along this problem, we take into account the case $p = c = d = q = 1$. Then the solution is of the form

w[ξ_]:=$\frac{a_1 e^{\xi}+a_0+a_{-1}e^{-\xi}}{e^{\xi}+b_0+b_{-1}e^{-\xi}}$

and is substituted in the left-hand side of the equation

```
eqnw==-k²v w[ξ]w''[ξ]+k²v w'[ξ]²+w[ξ]³+1;
```

then

```
eqnw1=((e^ξ+e^-ξ b_-1+b_0)^4 eqnw)/.
Table[e^nξ →q^n,{n,-1,5}]//Expand//Simplify;

eqnw2=Table[CoefficientList[Normal[
Series[q^4 eqnw1,{q,0,8}]],q][[i]]==0,{i,Length[eqnw2]}]
//Simplify;
Reduce[eqnw2,{a_1,a_0,a_-1,b_-1,v}]; (* System of algebraic
equations *)
```

From the list of returned solutions, we select the next values of parameters

$$a_1=-1; a_0=2b_0; a_0+b_0 \neq 0; a_{-1}=-\frac{b_0^2}{4}; b_{-1}=-a_{-1}; k\neq 0: v=-\frac{3}{k};$$

b_0 and k are free parameters.
Then we have

$$w[\xi\_]:=\frac{-e^\xi+2b_0-\frac{b_0^2}{4}e^{-\xi}}{e^\xi+b_0+b_{-1}e^{-\xi}}$$

and

$$z[t\_,x\_]:=\frac{-e^{k(x+\frac{3}{k}t)}+2b_0-\frac{b_0^2}{4}e^{-k(x+\frac{3}{k}t)}}{e^{k(x+\frac{3}{k}t)}+b_0+b_{-1}e^{-k(x+\frac{3}{k}t)}}$$

We remind that

```
u[t,x]=Log[z[t,x]],
```

and thus the figures of functions z and u are given below

```
Block[{b_0=.5,k=2},
{Plot3D[z[t,x],##,Ticks→Outer[Times,{5,5,1},{-1,0,1}],
PlotLabel→Style["z[t,x]",12,Black]],
Plot3D[u[t,x],##,Ticks→{{-5,0,5},{-5,3},-{10,1}},
PlotLabel→Style["u[t,x]",12,Black]]}
&[{t,-5,5},{x,-5,5},ImageSize→175,Mesh→2,
PlotStyle→Directive[Black,Opacity[.15]],
PlotRange→All,PlotPoints→50]]
```

See Figure 10.23.

Remark. We note that function z is defined on the entire real plane, while function u is defined only if $z > 0$. □

Remark. Other cases are possible as well, e. g., $p = c = 2$ and $d = q = 2$ [24]. □

10.19. We look for the soliton solution of equation (10.6) by the tanh method [80].

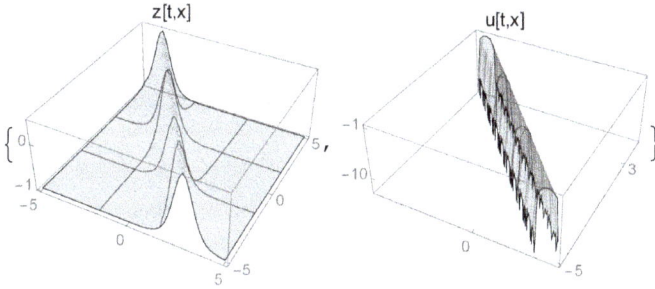

Figure 10.23: One-soliton solution for problem 10.18.

10.19 First, we transform the equation as we did before.

```
Clear[a,t,x,u,k,v,m,ζ]
z[t_,x_]:=Exp[u[t,x]] (* (>0) *)  ⟹
u[t,x]=Log[z[t,x]]
∂ₜu[t,x]=∂ₜz[t,x]/z[t,x]  ⟹
∂ₜ,ₓu[t,x]=-∂ₜz[t,x]∂ₓz[t,x]/z[t,x]²+∂ₜ,ₓz[t,x]/z[t,x]  ⟹
-∂ₜz[t,x]∂ₓz[t,x]/z[t,x]²+∂ₜ,ₓz[t,x]/z[t,x]+z[t,x]+1/z[t,x]²=0
```

Thus, we get the following nonlinear partial differential equation:

$$z[t,x]\partial_{t,x}z[t,x]-\partial_t z[t,x]\partial_x z[t,x]+z[t,x]^3+1=0$$

Next, we change the independent variable $\xi = k(x-vt)$, the dependent variable $z(t,x) = w(\xi)$, and the previous equation is transformed in an ordinary differential equation

```
∂ₜz[t,x]=-k v w'[ξ];∂ₓz[t,x]=k w'[ξ];
∂ₜ,ₓz[t,x]=-k²v w''[ξ]; (* and the equation *)
-k²v w[ξ]w''[ξ]+k²v w'[ξ]²+w[ξ]³+1=0;
```

We introduce a new independent variable and its corresponding function

```
ζ=Tanh[ξ];w[ξ]=s[ζ];w'[ξ]=(1-ζ²)s'[ζ]
w''[ξ]=-2ζ(1-ζ²)s'[ζ]+(1-ζ²)²s''[ζ] (* and the
new equation *)
2k²vζ(1-ζ²)s[ζ]s'[ζ]+k²v(1-ζ²)²s'[ζ]²
-k²v(1-ζ²)²s[ζ]s''[ζ]+s[ζ]³+1=0;
```

We apply the tanh method, Section 9.6.1 or [50], and consider the polynomial

$$f[\zeta\_]=a_0+\sum_{n=1}^{m}a_n\zeta^n,$$

where m is a natural number to be found. We substitute this polynomial in the equation to get

$$2k^2v\zeta(1-\zeta^2)f[\zeta]f'[\zeta]+k^2v(1-\zeta^2)^2f'[\zeta]^2$$
$$-k^2v(1-\zeta^2)^2f[\zeta]f''[\zeta]+f[\zeta]^3+1=0\,;$$

Balancing the highest degree term with the highest-order term, it follows that $3m = 2m + 2$, implying $m = 2$, so function f is a second degree polynomial,

$$f[\zeta\_]=a_0+a_1\zeta+a_2\zeta^2$$

We substitute this function in the equation to find the feasible coefficients

```
eq=2k²vζ(1-ζ²)f[ζ]f'[ζ]+k²v(1-ζ²)²f'[ζ]²
-k²v(1-ζ²)²f[ζ]f''[ζ]+f[ζ]³+1;
Series[eq,{ζ,0,7}]//Simplify;
Reduce[1+a₀³+k²v a₁²-2k²v a₀a₂==0&&
a₁(2k²v a₀+3a₀²+2k²v a₂)==0&&
3a₀²a2+2k²v a₂²+a₀(3a₁²+8k²v a₂)==0&&
a₁(a₁²+2k₂v a₂+a₀(-2k² v+6a₂))==0&&
3a₀a₂(-2k²v+a₂)+a₁²(-k²v+3a₂)==0&&
a₁a₂(-4k²v+3a₂)==0&&
a₂²(-2k₂v+a₂)==0,{a₀,a₁,a₂,k}]
```

From here, we get the following table of feasible coefficients, Table 10.2,

Table 10.2: Table of coefficients for problem 10.19.

n	a_0	a_1	a_2	k
1	-1	0	0	0
2	$\frac{1}{2}$	0	$-\frac{3}{2}$	$\frac{\sqrt{3}}{2\sqrt{-v}}$

In Table 10.2, we have two functions z and two functions u.

```
z₁[t_,x_]:=-1
u₁[t_,x_]:=Log[z₁[t,x]]
z₂[t_,x_]:=½-³⁄₂Tanh[√3/2√-v (x-v t)]²
u₂[t_,x_]:=Log[z₂[t,x]]
```

We check that both functions $u_1(t, x)$ and $u_2(t, x)$ are solutions of equation (10.3). Indeed,

```
D[#[t,x],t,x]+Exp[#[t,x]]+Exp[-2#[t,x]]&/@{u₁,u₂}
//Simplify
{0,0}
```

But function $u_1(t, x)$ has complex values, so we neglect it. Thus, we keep function $u_2(t, x)$ that has real values on a certain planar domain.

Now we introduce the figure containing the graphs of functions $z_2(t, x)$ and $u_2(t, x)$ for $v = -1$.

```
Block[{v=-1},
{Plot3D[z₂[t,x],##,Ticks→{5{-1,0,1},5{-1,0,1},{-1,0}},
PlotLabel→Style["z₂[t,x]",12,Black]],
Plot3D[u₂[t,x],##,Ticks→{5{-1,0,1},5{-1,0,1},{-3,-1}},
PlotLabel→Style["u₂[t,x]",12,Black]]}
&[{t,-5,5},{x,-5,5},ImageSize→175,Mesh→None,
PlotStyle→Directive[Black,Opacity[.15]],
PlotPoints→50]]
```

See Figure 10.24.

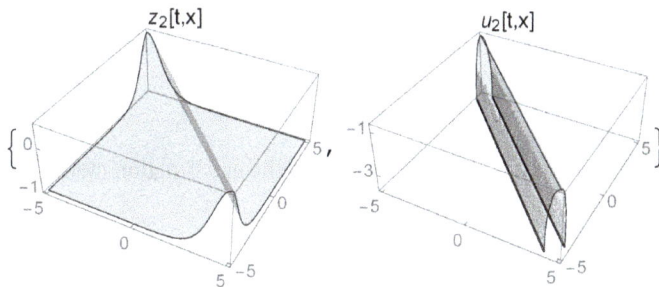

Figure 10.24: One-soliton solution for problem 10.19 by tanh function method.

The domain of definition of function u_2 is given by the set of pairs $\{t, x\} \subset \mathbb{R}^2$ satisfying the inequality

```
z₂[t,x]>0
```

For $v = -1$, the previous inequality is equivalent to

$$-\frac{1}{\sqrt{3}} < \text{Tanh}\left[\frac{\sqrt{3}}{2}(x+t)\right] < \frac{1}{\sqrt{3}}$$

that numerically is equivalent to

```
-.76<x+t<.76
```

Thus the domain of definition of function u_2 for $v = -1$ is represented below.

```
RegionPlot[ImplicitRegion[-.76<x+t<.76,{t,x}],
PlotStyle→Directive[Black,Opacity[.15]],ImageSize→
135,BoundaryStyle→Black]
```

See Figure 10.25.

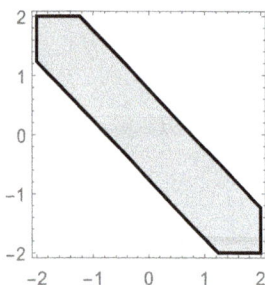

Figure 10.25: Domain of definition for function u_2.

10.3 The Tzitzeica–Dodd–Bullough equation

The *Tzitzeica–Dodd–Bullough equation* that we deal with is of the form, [80, p. 659],

$$u_{t,x}(t,x) = e^{-u(t,x)} + e^{-2u(t,x)}, \quad t, x \in \mathbb{R}. \tag{10.8}$$

10.20. We look for the soliton solution of equation (10.8) by the exp function method in Section 9.6.3 [24].

10.20

```
<< Notation`
Symbolize[a_]
Symbolize[b_]
```

The Tzitzeica–Dodd–Bullough equation is written by *Mathematica* under the form

eqn=$\partial_{t,x}$u[t,x]==e$^{-u[t,x]}$+e$^{-2u[t,x]}$

If we try to solve it by *Mathematica*,

```
DSolve[eqn,u[t,x],{t,x}];
```

then no closed-form solution is returned.
   We transform equation (10.8) by

```
z[t,x]=Exp[-u[t,x]];  (* (z[t,x]>0) *)
u[t,x]=-Log[z[t,x]];
```
$\partial_{t,x}$u[t,x]=$\frac{z_t[t,x]z_x[t,x]-z_{t,x}[t,x]z[t,x]}{z[t,x]^2}$;

and get

$$eqnz=z[t,x]\partial_{t,x}z[t,x]-\partial_t z[t,x]\partial_x z[t,x]+z[t,x]^3+z[t,x]^4;$$

New independent and dependent variables are introduced

$$\xi=k(x-v\,t);w[\xi]=z[t,x];$$

and the equation is of the form

$$eqnw=-k^2v\,w[\xi]w''[\xi]+k^2v\,w'[\xi]^2+w[\xi]^3+w[\xi]^4;$$

The exp function method supposes that the soliton solution of equation (10.8) can be written under the form

$$w[\xi]=\frac{\sum_{n=-d}^{c}a_n\,e^{n\xi}}{\sum_{m=-q}^{p}b_m\,e^{m\xi}}$$

Then because we are only interested in the relations between exponents, the exact values of the coefficients are neglected for a while.

$$w[\xi]^4=\frac{a\,e^{4c\xi}+\ldots}{b\,e^{4p\xi}+\ldots}$$

$$w'[\xi]=\frac{a\,e^{(c+p)\xi}+\ldots}{b\,e^{2p\xi}+\ldots}\implies w'[\xi]^2=\frac{a\,e^{2(c+p)\xi}+\ldots}{b\,e^{4p\xi}+\ldots}\implies w''[\xi]=\frac{a\,e^{(c+3p)\xi}+\ldots}{b\,e^{4p\xi}+\ldots}$$

$$\implies w[\xi]\,w''[\xi]=\frac{a\,e^{(2c+3p)\xi}+\ldots}{b\,e^{5p\xi}+\ldots}\implies$$

$$w[\xi]^4=\frac{a\,e^{(4c+5p)\xi}+\ldots}{b\,e^{9p\xi}+\ldots}\quad(*\text{ and }*)\quad w[\xi]\,w''[\xi]=\frac{a\,e^{(2c+7p)\xi}+\ldots}{b\,e^{9p\xi}+\ldots}$$

$$\implies 4c+5p=2c+7p\implies c=p$$

In a similar way, we found that $d = q$. In this case, we may suppose that $c = p = d = q = 1$ and conclude that

$$w[\xi\_]:=\frac{a_1e^{\xi}+a_0+a_{-1}e^{-\xi}}{e^{\xi}+b_0+b_{-1}e^{-\xi}}$$

This is substituted into the left-hand side of the earlier introduced equation

$$eqnw=-k^2v\,w[\xi]w''[\xi]+k^2v\,w'[\xi]^2+w[\xi]^3+w[\xi]^4;$$

Then

```
eqnw1=((e^ξ+e^−ξ b_−1+b_0)^4 eqnw)/.
Table[e^nξ →q^n,{n,−1,5}]//Expand//Simplify;
```

and

```
eqnw2=Table[CoefficientList[Normal[
Series[q⁴ eqnw1,{q,0,8}]],q][[i]]==0,{i,Length[eqnw2]}]
//Simplify;
Reduce[eqnw2,{a₁,a₀,a₋₁,b₋₁,v}]; (* System of algebraic
equations *)
```

From the list of solutions returned, we select the next one

$$a_1=-1\,;a_{-1}=0\,;b_0\neq0\,;b_{-1}=-b_0a_0-a_0^2\,;k(b_0+a_0)\neq0\,;v=\frac{1}{k^2}\,;$$

Then functions w, z, and u are as follows:

```
w[ξ_]:=  -eˆξ+a₀
         ─────────────────────
         eˆξ+b₀-(b₀a₀+a₀²)e⁻ˆξ
z[t_,x_]:=w[ξ]/.ξ →k(x-─₁t)
                        k²
u[t_,x_]:=-Log[z[t,x]]
```

Function u satisfies the equation (10.8). Indeed,

```
D[u[t,x],t,x]==eˆ-u[t,x]+eˆ-2u[t,x]//Simplify
True
```

Now we plot functions u and z for certain values of parameters.

```
Block[{a₀=.2,b₀=-2,k=.5},
{Plot3D[u[t,x],##,PlotLabel→Style["z[t,x]",12,
Black]],
Plot3D[z[t,x],##,PlotLabel→Style["u[t,x]",12,Black]]}
&[{t,-5,5},{x,-10,10},ImageSize→175,Mesh→None,
PlotStyle→Directive[Black,Opacity[.15]],
Ticks→Outer[Times,5{1,2,2},{-1,0,1}],
PlotPoints→50]]
```

See Figure 10.26.

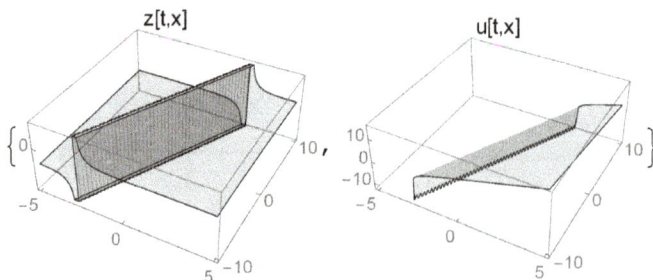

Figure 10.26: Plot of a solution for problem 10.20.

Now we want to find the domain of definition for function u. This domain is given by the set $\{t, x\} \subset \mathbb{R}^2$ satisfying the inequality

```
0<z[t,x]=w[ξ]/.ξ →k(x-1/k² t)
```

For $a_0 = .2$, $b_0 = -2$, and $k = .5$, $w(\xi) > 0$ if and only if $\xi < .587787$. For the same value of parameters, the previous inequality is equivalent to

```
(k(x-1/k²t)/.k→.5)<.587787
0.5(-4. t+x)<0.587787
```

This planar domain is suggested below:

```
RegionPlot[ImplicitRegion[.5(-4. t+x)<.587787,{t,x}],
PlotStyle→Directive[Black,Opacity[.15]],ImageSize→
135,BoundaryStyle→Black]
```

See Figure 10.27.

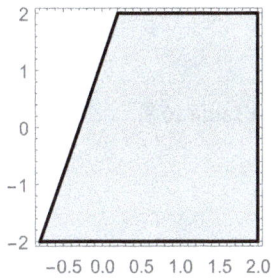

Figure 10.27: Domain of definition for function u in problem 10.20.

10.21. We look for the soliton solution of equation (10.8) by the tanh function method.

10.21

```
Clear[a,k,v,t,x,y]
```

```
∂t,x u=e⁻ᵘ+e⁻²ᵘ  (* The equation *)
(* If *)    0<z=e⁻ᵘ,  (* then *)    u = -Log[z] (* and *)
z z t,x -z t z x +z³+z⁴=0 (* the new equation *)
```

We introduce the traveling wave variable ξ and the corresponding function $w(\xi) :=$ $z(t, x)$.

```
ξ=k(x-v t)  (* and *)   w[ξ]=z[t,x]  (* then *)
k² v w''-k² v w'²-w³-w⁴=0  (* The equation *)
```

We introduce the tanh variable $y = \text{Tanh}(\xi)$ and the corresponding function $w(\xi) = f(y)$, thus resulting the equation under the form

```
k²v(1-y²)²f[y] f''[y]-2k²v y(1-y²)f[y] f'[y]
-k²v(1-y²)²f'[y]²-f[y]³-f[y]⁴=0
```

Supposing that f is an m degree polynomial in y and that we balance the linear terms of highest derivative order with the highest degree nonlinear terms, then $m = 1$. Thus, we have the form of function f,

```
f[y_]=a₀+a₁y
```

and look for the coefficients of it

```
eqny[y_]:=k²v(1-y²)²f[y] f''[y]-2k²v y(1-y²)f[y] f'[y]
-k²v(1-y²)²f'[y]²-f[y]³-f[y]⁴
```

Going on, we have

```
Table[CoefficientList[Normal[Series[eqnt[y],
{y,0,4}]],y][[i]]==0,{i,5}];
Reduce[%,{a₀,a₁,v}]
```

and the real values of the coefficients and the parameter v (see Table 10.3).

	a_0	a_1	v
1	$-\frac{1}{2}$	$-\frac{1}{2}$	$\frac{1}{4k^2}$
2	$-\frac{1}{2}$	$\frac{1}{2}$	$\frac{1}{4k^2}$

By Table 10.3, we write two functions z,

$$z_1[t_-,x_-]:=-\tfrac{1}{2}(1+\text{Tanh}[k(x-\tfrac{1}{4k^2}t)]);$$
$$z_2[t_-,x_-]:=-\tfrac{1}{2}(1-\text{Tanh}[k(x-\tfrac{1}{4k^2}t)]);$$

We remark that both functions are negative, so the corresponding functions u_{1-2} do not exist. The fact that functions z_{1-2} are negative for a value of parameter k follows from the next figure.

```
Block[{k=.5},
{Plot3D[z₁[t,x],##,PlotLabel→Style[z₁[t,x]",12,
Black]],
Plot3D[z₂[t,x],##,PlotLabel→Style[z₂[t,x]",12,Black]]}
&[{t,-5,5},{x,-10,10},Mesh→None,ImageSize→175,
PlotStyle→Directive[Gray,Opacity[.15]],
Ticks→Outer[Times,{5,10,1},{-1,0,1}]]]
```

See Figure 10.28.

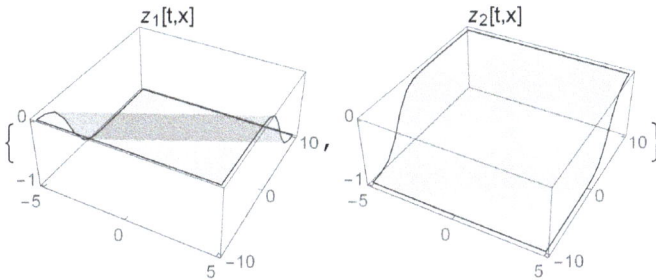

Figure 10.28: Negative functions z_{1-2} in problem 10.21.

Remark. We conclude that this approach does not supply any solution. □

10.4 The modified Kawahara equation

The standard *Kawahara equation* is of the form [37],

$$u_t + a\,u\,u_x + b\,u_{3x} - c\,u_{5x} = 0, \quad a,b,c \in \mathbb{R}, \text{ and } c \neq 0$$

whereas the *modified Kawahara equation* is of the form [81],

$$u_t + a\,u^2\,u_x + b\,u_{3x} - c\,u_{5x} = 0, \quad a,b,c \in \mathbb{R}, \text{ and } c \neq 0. \tag{10.9}$$

Remark. If function u is a solution of equation (10.9), then $-u$ is a solution as well. □

10.22. We study the existence of periodic solutions to equation (10.9) by the sine-cosine method, Section 9.6.4, [81], and [97].

10.22

```
Clear[a,b,c,k,β,λ,t,x]

eqn=∂ₜu[t,x]+a u[t,x]²∂ₓu[t,x]+b ∂{x,3}u[t,x]-c ∂{x,5}u[t,x]
(* Equation in t and x *)
```

By the transformations $\xi = k(x - vt)$ and $w(\xi) = u(t, x)$, the equation in t and x looks like

```
eqn1=-k v w'[ξ]+a k w[ξ]²w'[ξ]+b k³w'''[ξ]-c k⁵w'''''[ξ]
(* Equation in ξ *)
```

We can integrate this equation, and thus reduce by one its order of derivatives

```
eqn2=-v w[ξ]+ᵃ⁄₃ w[ξ]³+b k²w''[ξ]-c k⁴w''''[ξ]
(* Reduced equation in ξ *)
```

Now we use the sine-cosine method in Section 9.6.4, i. e., introduce the function

```
w[ξ_]:=λ Sin[ξ]ᵝ
```

and substitute it into eqn2.

```
eqn3=eqn2//Simplify
```
$\frac{1}{24}$ λ Sin[k(-t v+x)]ᵝ(-24v+12 b k²β(-2+β+βCos[2 k(-tv+x)])
Csc[k (-t v+x)]²-3 c k⁴β(-32+56β-24β²+3β³+
4 (-4+8β-6β²+β³)Cos[2 k(-t v+x)]+β³
Cos[4 k(-t v+x)])Csc[k(-t v+x)]⁴+8 aλ²Sin[k(-t v+x)]²ᵝ)

We transform eqn3

```
eqn4=eqn3/.{Cos[2ξ]→1-2 Sin[ξ]²,
Cos[4ξ]→1-8 Sin[ξ]²+8 Sin[ξ]⁴}
```
$\frac{1}{24}$ λ Sin[ξ]ᵝ(-24 v+8 aλ²Sin[ξ]²ᵝ+12 b k²βCsc[ξ]²
(-2+β+β(1-2Sin[ξ]²))-3 c k⁴βCsc[ξ]⁴(-32+56β-24β²+3β³
+4(-4+8β-6β²+β³)(1-2 Sin[ξ]²)+β³(1-8 Sin[ξ]²+8 Sin[ξ]⁴)
))

and successively

```
eqn5=eqn4/.{Sin[ξ]→y,Csc[ξ]→1/y}//Simplify
```
$\frac{1}{3}$λ yᵝ⁻⁴ (y²(aλ²y²ᵝ⁺²-3 b βk²(β(y²-1)+1))
-3βc k⁴(β³-6β²+11β
+β³y⁴-2(β³-3β²+4β-2)y²-6) -3 v y⁴)

```
eqn6=Normal[Series[eqn5,{y,0,5}]]
```
yᵝ⁻⁴ ($\frac{1}{3}$ aλ³y²ᵝ⁺⁴ − λy⁴(bβ²k² + β⁴c k⁴+v)
+$\frac{1}{3}$ λy²(3 bβ²k²-3 b βk²+6 β(β³-3β²+4β-2)c k⁴)
-β(β³-6β²+11β-6)ck⁴λ)

We balance the terms and find that $\beta = -2$. Then we have eqn7=eqn6/.β →-2

$$\frac{\frac{a u^3}{3} - \lambda y^4 (4bk^2 + 16ck^4 + v) + \frac{1}{3}\lambda y^2 (18bk^2 + 360ck^4) - 120c\lambda k^4}{y^6}$$

Equating the coefficients of like powers of y function eqn7 leads to the next system of algebraic equations

Reduce[-360c k⁴+aλ²==0&&18b k²+360c k⁴==0&&
-4b k²-16c k⁴-v==0,{k,v,λ}]

The next results follow:

$$\lambda = \pm\frac{3b}{\sqrt{10}\sqrt{a}\sqrt{c}}\,;k=\pm\frac{i\sqrt{b}}{2\sqrt{5}\sqrt{c}}\,;v=\frac{4b^2}{25c}\,;$$

Remark. If $b < 0$, k is real for positive c. In this case, we take $a > 0$. If $b > 0$, we need considering that $c < 0$ and $a < 0$. ☐

We select the case when $b < 0$.

$$\lambda = \frac{3b}{\sqrt{10}\sqrt{a}\sqrt{c}}\,;k=\frac{\sqrt{-b}}{2\sqrt{5}\sqrt{c}}\,;v=\frac{4b^2}{25c}\,;$$

The plot of two solutions is given below; see also [81, (22) and (21)].

```
u₁[t_,x_]:=λCsc[k(x-v t)]²
u₂[t_,x_]:=λSec[k(x-v t)]²
Block[{a=3,b=-.5,c=1},
{Plot3D[u₁[t,x],##,PlotLabel→Style["u₁",12,Black],
Ticks→{10{-1,1},10{-1,1},{0,{2*10³⁰,"2*10³⁰"}}}],
Plot3D[u₂[t,x],##,PlotLabel→Style["u₂",12,Black],
Ticks→{10{-1,1},10{-1,1},{0,{3*10⁵,"3*10⁵"}}}]}
&[{t,-10,10},{x,-10,10},ImageSize→175,Mesh→None,
PlotStyle→Directive[Black,Opacity[.15]],
PlotRange→All]
```

See Figure 10.29.

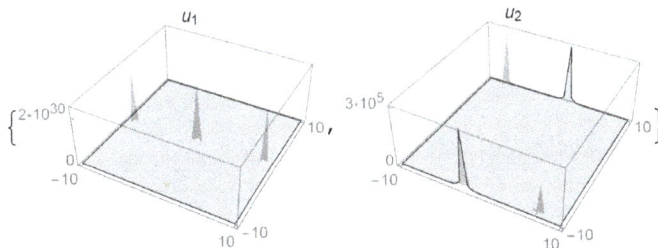

Figure 10.29: Peak solutions for problem 10.22.

10.5 The Benjamin equation

The *Benjamin equation* is of the form [3],

$$\partial_{t,t}u(t,x) + a\,\partial_x\big(u(t,x)\partial_x u(t,x)\big) + b\,\partial_{x,x,x,x}u(t,x) = 0, \quad a, b \in \mathbb{R}, \quad b \neq 0. \qquad (10.10)$$

10.23. Below we show that the functions

$$u(t,x) = \lambda\,\sec^2(k(x - v\,t)) \quad \text{and} \quad u(t,x) = \lambda\,\mathrm{sech}^2(k(x - v\,t))$$

are solutions of the Benjamin equation (10.10) for certain values of the parameters.

10.23 We approach the problem using two methods.
1.
    First, we apply the sine-cosine method, Section 9.6.4.

```
Clear[a,b,k,v,t,x]
```

To apply the sine-cosine method, suppose that equation (10.10) has a solution of the form

```
u[t_,x_]:=λCos[k(x-vt)]^β
```

and substitute it into the equation

```
eqn=D[u[t,x],t,t]+a D[u[t,x]D[u[t,x],x],x]
+b D[u[t,x],{x,4}]
```

By a substitution, we try to simplify the expression of the equation

```
eqn1=eqn/.{Cos[k(x-tv)]→q}
eqn2=Expand[eqn1,q]
```

Balancing the exponentials, one has that

```
2β=β-2 ⟹ β=-2 (* and *)
eqn3=eqn2/.{β →-2}
eqn4=q^6*Series[eqn3,{q,0,4}]
```

We have the system of algebraic equations

```
Reduce[120b k²+10a λ==0&&-120b k²+6v²-8a λa==0&&
16b k²-4v²==0,{λ,k}]
```

Neglecting the trivial solutions, we have

$$\lambda = -\frac{3v^2}{a}\,; k=\pm\frac{v}{2\sqrt{b}}\,; \quad (* \ b>0 \ *)$$

Then the solution is of the form

```
u[t_,x_]:=λ Sec[k(x-v t)]²
```

and satisfies the equation

```
∂t,tu[t,x]+a∂x(∂xu[t,x])+b∂x,x,x,xu[t,x]//Simplify
0
```

We plot the solution for two sets of the parameters.

```
Block[{a=#[[1]],b=2,v=#[[2]]},
Plot3D[u[t,x],{t,-20,20},{x,-25,25},ImageSize→175,
Mesh→None,Ticks→{{-20,20},{-25,25},#[[3]]},
PlotStyle→Directive[Black,Opacity[.15]],PlotPoints→
45]]&/{{1,-.5,{-22,0}},{-1,.5,{0,22}}}
```

See Figure 10.30.

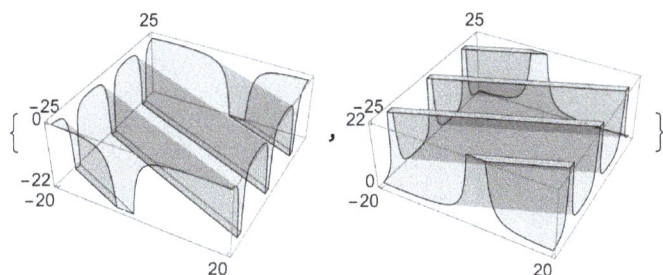

Figure 10.30: Two solutions of the Benjamin equation.

2.

Below we study the same equation but using the Cosh function.

```
Clear[a,b,v,β,λ,t,x]
```

We consider the form of the solution

```
u[t_,x_]:=λ Cosh[k(x-v t)]ᵝ
```

and substitute it into the equation.

```
eqn=D[u[t,x],t,t]+a D[u[t,x] D[u[t,x],x],x]
+b D[u[t,x],{x,4}]
```

We transform it to find β

```
eqn1=eqn/.{Sinh[k(x-tv)]² →Cosh[k(x-tv)]²-1,
Sinh[k(x-tv)]⁴ →(Cosh[k(x-tv)]²-1)²};
eqn2=Expand[eqn1/.Cosh[k(x-tv)]→q,q];
eqn3=Series[eqn2,{q,0,6}]
2β-2=β-4 ⟹ β=-2
```

The parameters of the solution follows by

```
Reduce[120b k⁴λ-10a k²λ²==0&&
-120b k⁴λ+8a k²λ²-6k²v²λ==0&&
16b k⁴λ+4k²v²λ==0,{λ,k}]  (* It results *)
λ = - 3v²/a ;k= v/2Sqrt[-b] ;  (* b<0 *)
```

By these values of the parameters, we write the solution, check its validity, and plot it for two sets of parameters.

```
u[t_,x_]:=λSech[k(x-v t)]²
D[u[t,x],t,t]+a D[u[t,x] D[u[t,x],x],x]
+b D[u[t,x],{x,4}]//Simplify
0

Block[{a=#[[1]],b=-3,v=.25},
Plot3D[u[t,x],{t,0,10},{x,60,60},ImageSize→175,
Mesh→None,Ticks→{{0,10},{-60,0,60},#[[2]]},
PlotStyle→Directive[Black,Opacity[.15]]]]
&/{{-1,{0,.17}},{1,{0,-.17}}}
```

See Figure 10.31.

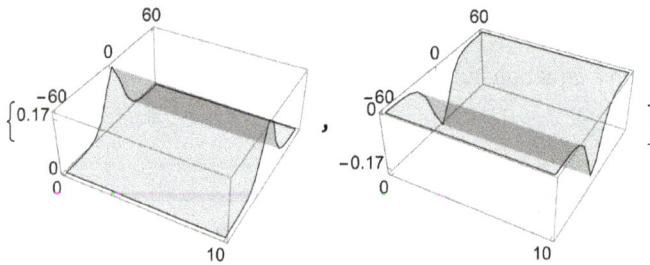

Figure 10.31: Another solution of the Benjamin equation.

10.24. Below we show that the Benjamin equation (10.10) has another solution as well. This approach is based on the tanh function method.

10.24

```
Clear[a,b,k,v,t,x]
```

By the transformations,

```
ξ=k(x-v t);u[t,x]=w[ξ]; (* One has that *)
```
$$v^2 \frac{d^2 w}{d\xi^2} + a\left(\frac{dw}{d\xi}\right)^2 + a\, u[t,x]\frac{d^2 w}{d\xi^2} + b\, k^2 \frac{d^4 w}{d\xi^4} = 0 \quad (* \text{ Equation in}$$
```
variable ξ *)
```

and

```
ζ=Tanh[ξ];s[ζ]=w[ξ];
```

one obtains function $s(\zeta)$.

We apply the tanh function method, Section 9.6.1, consider polynomial $f(\zeta\_) := c_0 + \sum_{n=1}^{m} c_n \zeta^n$, m being a positive integer number, and substitute this finite sum into the function $s(\zeta)$. Balancing the highest degree term $2m+2$ with the highest-order derivative $m + 4$, it follows that $m = 2$. Thus,

```
f[ζ_]:=c₀+∑²ₙ₌₁ cₙζⁿ
```

We substitute it into the equation and get the following system of algebraic equations:

```
Reduce[a c₁²-16b k²+2v²c₂+2a c₀ c₂==0&&
16b k²c₁-2v²c₁-2a c₀c₁+6a c₁c₂==0&&
-4a c₁²+136b k²c₂-8v²c₂-8a c₀c₂+6a c₂²==0&&
-40b k²c₁+2v²c₁+2a c₀c₁-18a c₁c₂==0&&
3a c₁²-240b k²c₂+6v²c₂+6a c₀c₂-16a c₂²==0&&
24b k²c₁+12a c₁c₂==0&&120b k²c₂+10a c₂²==0,
{k,v,c₀,c₁,c₂}]
```

Two solutions result. One solution is the constant function and we neglect it. The other solution of the algebraic system results by using the parameters

$$a\, b\, k \neq 0; c_0 = \frac{8b\, k^2 - v^2}{a}; c_1 = 0; c_2 = -\frac{12b\, k^2}{a};$$

This solution is written as

$$u[t\_,x\_] := \frac{8b\, k^2 - v^2}{a} - \frac{12b\, k^2}{a}\, \text{Tanh}[k(x-v\, t)]^2$$

and satisfies the equation

```
∂_{t,t}u[t,x]+a∂ₓ(u[t,x]∂ₓu[t,x])+∂_{x,4}u[t,x]//Simplify
0
```

Now we plot the solution under two assumptions

```
Block[{a=#,b=-3,k=.5,v=.25},
Plot3D[u[t,x],{t,-15,15},{x,-10,10},ImageSize→175,
Mesh→None,Ticks→Outer[Times,5{3,2,10},{-1,0,1}],
PlotStyle→Directive[Black,Opacity[.15]],
PlotRange→All,PlotPoints→35]]&/@{1,-1}
```

See Figure 10.32.

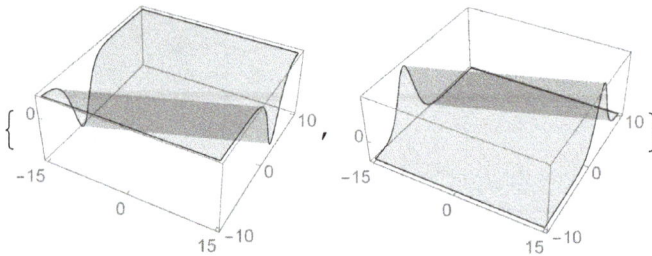

Figure 10.32: Another solutions of the Benjamin equation.

Remark. Since

```
(8bk²-v²)/a-(12bk²)/a Tanh[k(x-v t)]²
==(-4bk²-v²)/a+(12bk²)/a Sech[k(x-v t)]²//Simplify
True
```

instead of the Tanh solution one may prefer the Sech solution. □

10.6 The Kadomtsev–Petviashvili equation

Kadomtsev and Petviashvili introduced their equation in paper [34]. Their result has been largely extended; we mention here only a few titles: [46, 52, 98, 19, 55], and [96].

The *Kadomtsev–Petviashvili equation* or *KP equation* is a nonlinear partial differential equation. We already saw that the Korteweg–de Vries equation assumes that there are two independent variables: a temporal variable and a spatial variable. In the case of the KP equation, there are one temporal variable and two spatial variables. Its form is

$$\begin{cases} (u_t(t,x,y) + 6u(t,x,y)u_x(t,x,y) + u_{x,x,x})_x + \lambda\, u_{y,y}(t,x,y) = 0, \\ t,x,y,\lambda \in \mathbb{R}. \end{cases} \tag{10.11}$$

The KP equation is classified as the *KPI equation* when $\lambda = 1$ and the *KPII equation* when $\lambda = -1$. The equation can be applied to model water waves of long wavelengths with weakly nonlinear restoring forces and frequency dispersion. It is a two-dimensional generalization of the one-dimensional Korteweg–de–de Vries equation.

Remark. Equation (10.11) is also called as the *2D Korteweg–de Vries equation*, [20]. □

10.6.1 Single soliton solution

10.25. Using the power of *Mathematica*, we study the existence of solutions to equation (10.11).

10.25

```
Clear[t,x,y,u,λ]
```

The Kadomtsev–Petviashvili equation (10.11) is written by *Mathematica* as

```
eqn=∂ₓ(∂ₜu[t,x,y]+6u[t,x,y]∂ₓu[t,x,y]+∂₍3ₓ₎u[t,x,y])
+λ∂ᵧ,ᵧu[t,x,y]==0;
```

We solve the equation,

```
sol=DSolve[eqn,u,{t,x,y}]//Simplify//Flatten
{u→Function[{t,x,y},
```
$$-\frac{C[1]C[2]-8C[2]^4+\lambda C[3]^2+12C[2]^4\text{Tanh}[tC[1]+xC[2]+yC[3]+C[4]]^2}{6C[2]^2}]}$$

select a particular solution,

```
uparticular[t_,x_,y_]:=u[t,x,y]/.sol/.{C[1]→1,C[2]→1,
C[3]→1,C[4]→3,λ→1}
```

and plot it for *t* = 0

```
Block[{t=0},
Plot3D[uparticular[t,x,y],{x,-10,10},{y,-10,10},
Mesh→None,PlotStyle→Directive[Black,Opacity[.15]],
ImageSize→175,Ticks→Outer[Times,{10,10,1},{-1,0,1}]]]
```

See Figure 10.33.
    Now one contour plot of the same particular solution is introduced.

```
Block[{a=1,b=-1,x₀=0},
ContourPlot3D[u[t,x,y],{t,0,2},{x,-2,2},{y,-8,10},
PlotRange→All,Mesh→None,Contours→2,
ContourStyle→Directive[Black,Opacity[.15]],
Ticks→None,ImageSize→175]]
```

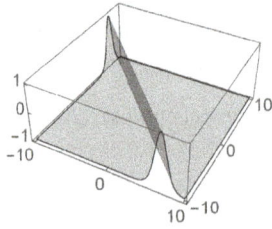

Figure 10.33: Plot of a particular solution for problem 10.25.

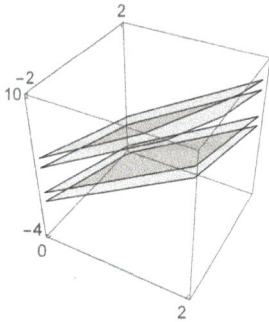

Figure 10.34: Contour plot of a particular solution for problem 10.25.

See Figure 10.34.

10.26. Below we verify by *Mathematica* that function

$$u(t,x,y) = \frac{1}{2}a^2 \operatorname{sech}^2\left(\frac{1}{2}a\left(x - by - \frac{(a^3 + 3ab^2)t}{a} - x_0\right)\right) \tag{10.12}$$

introduced in [7] and [8] is a solution to the Kadomtsev–Petviashvili equation (10.11) with $\lambda = 3$.

10.26

```
<< Notation`
Symbolize[x_]

Clear[a,b,t,x,y,u]

u[t_,x_,y_]:=1/2 a²Sech[1/2 a(x-by-(a³+3a b²)t/a-x₀)]²
(* Define the function *)

∂ₓ(∂ₜu[t,x,y]+6u[t,x,y]∂ₓ+∂ₓ,ₓ,ₓu[t,x,y])
+3∂ᵧ,ᵧu[t,x,y]//Simplify (* Check the equation *)
0
```

We plot function *u* for some particular values of parameters

```
Block[{a=1,b=-1,x₀=0},
Plot3D[u[1,x,y],{x,-10,10},{y,-8,10},Mesh→None,
PlotRange→All,PlotStyle→Directive[Gray,Opacity[.15]],
ImageSize→175,Ticks→{10{-1,0,1},2{-4,0,5},{0,.4}}]]
```

See Figure 10.35.

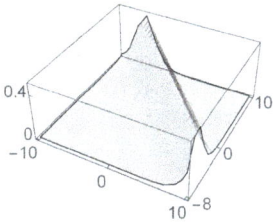

Figure 10.35: One-soliton solution to KP equation for $\lambda = 3$ in problem 10.26.

We represent function u by ContourPlot3D command.

```
Block[{a=1,b=-1,x₀=0},
ContourPlot3D[u[t,x,y],{t,0,2},{x,-2,2},{y,-4,10},
PlotRange→All,Mesh→None,Contours→2,
Ticks→{{0,2},{-2,2},{-4,10}},ImageSize→175,
ContourStyle→Directive[Gray,Opacity[.15]]]]
```

See Figure 10.36.

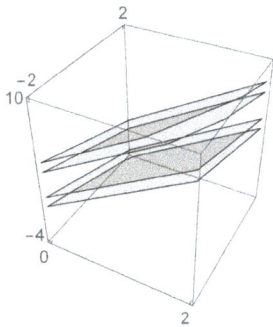

Figure 10.36: Representation of the one-soliton solution in problem 10.26 by ContourPlot3D.

10.6.1.1 Algebraic solutions to the KP equation
Here, we introduce some algebraic solutions to the Kadomtsev–Petviashvili equation.

10.27. Below we introduce the first algebraic solution to the KPII equation [46, (2.13)],

$$u(t,x,y) = \frac{4(-(t(a^2 - b^2) + ay + c + x)^2 + b^2(2at + d + y)^2 + 3/b^2)}{((t(a^2 - b^2) + ay + c + x)^2 + b^2(2at + d + y)^2 + 3/b^2)^2}$$

10.27

```
Clear[a,b,c,d,t,x,y,u]
```

$$u[t\_,x\_,y\_]:=\frac{4(-(t(a^2-b^2)+ay+c+x)^2+b^2(2at+d+y)^2+3/b^2)}{((t(a^2-b^2)+ay+c+x)^2+b^2(2at+d+y)^2+3/b^2)^2}$$

```
(* Define the function *)
∂ₓ(∂ₜu[t,x,y]+6u[t,x,y]∂ₓ+∂ₓ,ₓ,ₓu[t,x,y])-∂ᵧ,ᵧu[t,x,y]
//Simplify (* Check the equation *)
0
```

We plot this solution for some particular values of the parameters.

```
Block[{t=0,a=1,b=-1,c=1,d=0},
Plot3D[u[t,x,y],{x,-15,15},{y,-10,10},Mesh→None,
ColorFunction→"Temperature",PlotRange→All,
Ticks→Outer[Times,{12,10,1},{-1,0,1}],ImageSize→175]]
```

See Figure 10.37.

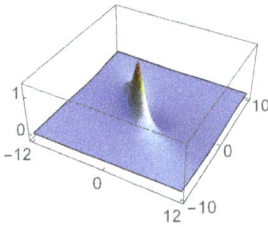

Figure 10.37: Algebraic solution to the KPII equation at a given instant.

10.28. Below we introduce the second algebraic solution to the KPII equation,

$$\partial_x(\partial_t u(t,x,y) + 6u(t,x,y)\partial_x u(t,x,y) + \partial_{x,x,x}u(t,x,y)) - \partial_{y,y}u(t,x,y) = 0, \qquad (10.13)$$

[46, (2.17)].

10.28

```
<< Notation`
Symbolize[a_]

Clear[a,t,x,y,u]
```

We define some parameters

$$a_3 = \frac{a_1 a_2^2 - a_1 a_6^2 + 2a_2 a_5 a_6}{a_1^2 + a_5^2} \; ; a_7 = \frac{a_5 a_6^2 - a_5 a_2^2 + a_1 a_2 a_6}{a_1^2 + a_5^2} \; ;$$

$$a_9 = \frac{3(a_1^2 + a_5^2)^3}{(a_1 a_6 - a_2 a_5)^2}$$

and the ancillary function

```
g[t_,x_,y_]:=a₁x+a₂y+a₃t+a₄
h[t_,x_,y_]:=a₅x+a₆y+a₇t+a₈
f[t_,x_,y_]:=g[t,x,y]²+h[t,x,y]²+a₉
```

We check that the function

```
u[t_,x_,y_]:=2 D[Log[f[t,x,y]],x,x]
```

satisfies equation (10.13).

```
D[(D[u[t,x,y],t]+6u[t,x,y]D[u[t,x,y],x]+
D[u[t,x,y],x,x,x]),x]-D[u[t,x,y],y,y]//Simplify
0
```

Solid and planar representations of a particular solution of this family are presented for $t = 1$.

```
Block[{a₁=1,a₂=2,a₄=0,a₅=1,a₆=-1,a₈=0},
{Plot3D[Evaluate[u[1,x,y]],{x,-15,15},{y,-5,5},
ImageSize→175,Mesh→2,ColorFunction→"Temperature",
Ticks→Outer[Times,{15,5,2},{-1,0,1}],PlotRange→All],
Plot[Evaluate[u[1,x,1]],{x,-15,15},ImageSize→175,
Ticks→{15{-1,1},{.4}},PlotRange→All]}]
```

See Figure 10.38.

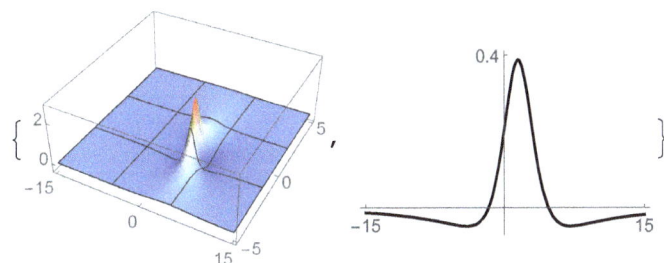

Figure 10.38: Algebraic solution for the KPII equation (10.13).

We introduce solid and planar representations of the solution by contour commands.

```
Block[{a₁=1,a₂=2,a₄=0,a₅=1,a₆=-1,a₈=0},
{ContourPlot3D[Evaluate[u[t,x,y]],{t,0,2},{x,0,5},
{y,-4,2},ImageSize→160,Mesh→2,PlotRange→All,
ColorFunction→"Temperature",
Ticks→{{0,1,2},{0,5},{-3,-1,1}}],
ContourPlot[Evaluate[u[1,x,y]],{x,0,5},{y,-4,2},
ColorFunction→"Temperature",PlotRange→All,
ImageSize→160]}]
```

See Figure 10.39.

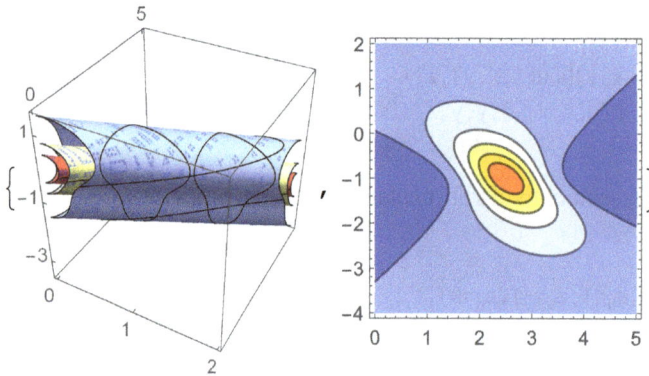

Figure 10.39: Contour representations of the algebraic solution for the KPII equation (10.13).

10.6.2 Hirota bilinear method to the KPI equation

10.29. By Hirota bilinear method, we study the two-soliton solution to Kadomtsev–Petviashvili equation (10.11) with $\lambda = 1$.

10.29

```
<< Notation`
Symbolize[k_]
Symbolize[v_]

Clear[t,x,y,a,k,v]
kp[t_,x_,y_]:=D[D[u[t,x,y],t]+6u[t,x,y]D[u[t,x,y],x]+
D[u[t,x,y],x,x,x],x]+D[u[t,x,y],y,y] (* KPI equation
*)
```

```
hxm[f_,g_,m_]:=∑ᵢ₌₀ᵐ(-1)ⁱBinomial[m,i]D[f,{x,m-i}]*
D[g,{x,i}] (* Define Hirota derivatives *)
hym[f_,g_,m_]:=∑ᵢ₌₀ᵐ(-1)ⁱBinomial[m,i]D[f,{y,m-i}]*
D[g,{y,i}]
htm[f_,g_,m_]:=∑ᵢ₌₀ᵐ(-1)ⁱBinomial[m,i]D[f,{t,m-i}]*
D[g,{t,i}]
hxt[f_,g_]:=f D[g,x,t]-D[f,x]D[g,t]-D[f,t]D[g,x]
+D[f,x,t]g
```

One checks whether the bilinear transform is correct,

```
kpcheck[t_,x_,y_]:=D[(hxt[f[t,x,y],f[t,x,y]]+
hxm[f[t,x,y],f[t,x,y],4]+
hym[f[t,x,y],f[t,x,y],2])/f[t,x,y]²,x,x]
kp[t,x,y]-kpcheck[t,x,y]//Simplify (* Same equation *)
0
```

and introduce the ansatz for f_1,

```
f₁₁[t_,x_,y_]:=Exp[k₁x+v₁y-w₁t]
f₁₂[t_,x_,y_]:=Exp[k₂x+v₂y-w₂t]
f₁[t_,x_,y_]:=f₁₁[t,x,y]+f₁₂[t,x,y]
b₁[f1_]:=hxt[1,f1]+hxt[f1,1]+hxm[1,f1,4]+hxm[f1,1,4]+
3(hym[1,f1,2]+hym[f1,1,2]) (* Bilinear form *)
b₁[f₁][t,x,y]//Simplify;
```

We look for w_1 and w_2,

```
Reduce[b₁[f₁₁[t,x,y]]==0&&b₁[f₁₂[t,x,y]]==0,{w₁,w₂}]
```

It follows that

$$w_1 = \frac{k_1^4 + 3v_1^2}{k_1}; w_2 = \frac{k_2^4 + 3v_2^2}{k_2}.$$

We introduce the ansatz for f_2,

```
f₂[t_,x_,y_]:=a₁₂Exp[k₁x+v₁y-w₁t+k₂x+v₂y-w₂t]
```

and define the two-soliton ancillary function.

```
f[t_,x_,y_]:=1+f₁[t,x,y]+f₂[t,x,y]
b₂[f1_,f2_]:=hxt[1,f2]+hxt[f2,1]+hxm[1,f2,4]
+hxm[f2,1,4]+3(hym[1,f2,2]+hym[f2,1,2])
Reduce[b₂[f₁[t,x,y],f₂[t,x,y]]==0,a₁₂]//Simplify
```

$$a_{12} = \frac{(-k_1 v_2 + k_2 v_1 - k_2 k_1^2 - k_2^2 k_1)(k_1 v_2 - k_2 v_1 + k_2 k_1^2 - k_2^2 k_1)}{(k_1 v_2 - k_2 v_1 + k_2 k_1^2 + k_2^2 k_1)(-k_1 v_2 + k_2 v_1 + k_2 k_1^2 + k_2^2 k_1)}$$

The equation of the two-soliton solution is given below

```
Clear[x,y]
u[t_,x_,y_]:=2D[Log[f[t,x,y]],x,x]
```

and its plot follows for three instances:

```
Block[{k₁=1,k₂=Sqrt[2],v₁=1/5,v₂=5/6},
{Plot3D[Evaluate[u[5,x,y]],{x,-10,25},{y,30,70},##,
PlotLabel→Style["t=5",12,Black]],
Plot3D[Evaluate[u[10,x,y]],{x,-10,25},{y,30,70},##,
PlotLabel→Style["t=10",12,Black]],
Plot3D[Evaluate[u[15,x,y]],{x,-10,25},{y,30,70},##,
PlotLabel→Style["t=15",12,Black]]}&[Mesh→None,
ImageSize→175,PlotPoints→50,PlotPoints→50,
PlotRange→{{-10,25},{30,70},{0,1.1}},
Ticks→{5{-2,5},10{3,7},{0,1}},
ColorFunction→"Temperature",PlotStyle→Opacity[.8]]]
```

See Figure 10.40.

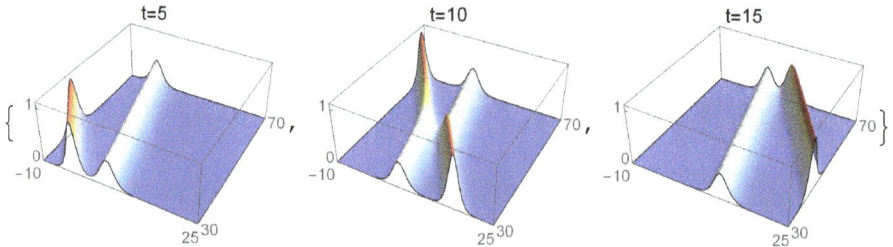

Figure 10.40: A two-soliton solution for KPI equation at instances $t = 5, 10, 15$.

10.7 The Sawada–Kotera equation

The *Sawada–Kotera equation* is a fifth-order nonlinear partial differential equation of the form [9, (5)] and [10, (4)],

$$u_t + u_{x,x,x,x,x} + a\,u_{x,x,x}u + a\,u_{x,x}u_x + \frac{a^2}{5}u^2 u_x = 0, \quad a,t,x \in \mathbb{R}, \ a \neq 0. \tag{10.14}$$

10.30. We study a one-soliton solution of the Sawada–Kotera equation (10.14); see [10, (12)].

10.30

```
Clear[a,t,x,k,ϕ] (* Clean values and definitions *)
u[t_,x_]:=30k²/a Sech[-16k⁵t+k x-ϕ]² (* Define a
solution *)
```

Now we check that the above defined function u satisfies the equation (10.14) for all real parameters a, k, and ϕ,

```
∂ₜu[t,x]+∂₍ₓ,₅₎u[t,x]+a ∂₍ₓ,₃₎u[t,x] u[t,x]
+a ∂ₓ,ₓu[t,x] u[t,x]+a²/5 u[t,x]²∂ₓu[t,x]//Simplify
0
```

We plot solution u for some particular values of the parameters.

```
a=15;k=.5;ϕ=-5;
{Plot3D[u[t,x],{t,0,20},{x,-20,20},Mesh→None,
PlotStyle→Directive[Gray,Opacity[.15]],ImageSize→175,
Ticks→{10{0,1,2},20{-1,0,1},{0,.5}},PlotRange→All,
PlotPoints→50],
Manipulate[
Plot[Evaluate[u[τ,x]],{x,-15,15},Mesh→None,
ImageSize→160,PlotStyle→Black,PlotRange→All,
Ticks→{15{-1,0,1},{0,.5}}],{τ,5,15},
SaveDefinitions→True]}
```

See Figure 10.41.

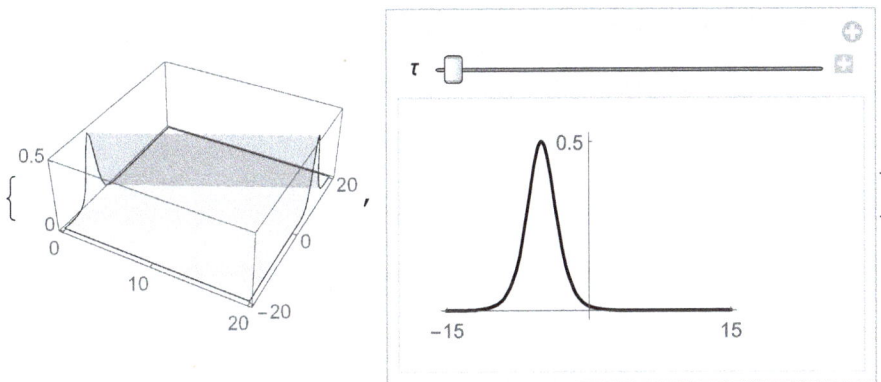

Figure 10.41: A particular solution of the Sawada–Kotera equation and a cross-section.

10.31. We study the existence of a solution of the Sawada–Kotera equation (10.14) with the sine-cosine method in Section 9.6.4 supposing that $a = 15$, [10, (4)] and [99, p. 179].

10.31

```
Clear[k,q,t,v,x,β,λ,ϕ]
```

Let us introduce the initial assumption of the problem that the Sawada–Kotera equation admits a traveling solution of the form

```
ξ=k(x-v t)+ϕ;
u[t_,x_]:=λCos[ξ]^β
a=15;
```

Then

```
eqn=∂ₜu[t,x]+∂_{x,5}u[t,x]+a∂_{x,3}u[t,x]u[t,x]
+a∂_{x,x}u[t,x]u[t,x]+a²/5 u[t,x]²∂ₓu[t,x]//Simplify;
eqn1=eqn/.{Cos[2(k(-t v+x)+ϕ)]→(2 Cos[k(-t v+x)+ϕ]²-1),
Cos[4(k(-t v+x)+ϕ)]→(2(2 Cos[k(-t v+x)+ϕ]²-1)²-1)}//
Simplify;
```

　　Successively, we have

```
eqn2=eqn1/.{Cos[k(-t v+x)+ϕ]→q,Cos[k (-t v+x)+ϕ]² →q²,
Cos[2(k(-t v+x)+ϕ)]→2 q²-1}//Simplify;
eqn3=Expand[k⁴(24-50β+35β²-10β³+β⁴)
-2 k⁴q²(4-10β+10β²-5β³+β⁴)+q⁴(-v+k⁴β⁴)+45q^{2(2+β)}λ²
-15 k²q^{2+β}λ(-2+4 β-β²+β²(2 q²-1)),q]
(β⁴-10β³+35β²-50β+24)k⁴+q⁴(β⁴k⁴-v)
-2(β⁴-5β³+10β²-10β+4)k⁴q²+30β²k²λq^{β+2}-30β²k²λq^{β+4}
+30 k²λq^{β+2}-60βk²λq^{β+2}+45λ²q^{2(β+2)}
```

　　Balancing the exponents, we find that $\beta = -2$. To simplify the solving process, we expand the equation as follows:

```
eqn4=eqn3/.β →-2;
Series[eqn4,{q,0,4}];
```

Then we solve the system of algebraic equations of parameters,

```
Reduce[360k⁴+270k²λ+45λ²==0&&2k²+λ==0&&16k⁴-v==0,
{λ,k}];
```

that returns the solutions

$$\lambda = \pm\frac{\sqrt{v}}{2}; k = \pm\frac{1}{2}i\sqrt{\pm\sqrt{v}}$$

We consider the following parameters:

```
λ=-Sqrt[v]/2;k=Sqrt[Sqrt[v]]/2;(* v>0 *)
```

and the corresponding solution of the Sawada–Kotera equation (10.14) with $a = 15$,

```
u[t_,x_]:=λSec[k(x-v t)+ϕ]^2
```

The validity of this solution follows immediately:

```
D[u[t,x],t]+D[u[t,x],{x,5}]+15D[u[t,x],x,x,x]u[t,x]
+15D[u[t,x],x,x]D[u[t,x],x]+45u[t,x]^2D[u[t,x],x]
//Simplify
0
```

and we plot the solution.

```
Block[{v=.25,ϕ=.55},
Plot3D[u[t,x],{t,-10,10},{x,-20,20},ImageSize→175,
Mesh→None,PlotPoints→35,PlotStyle→Opacity[.4],
Ticks→Outer[Times,10{1,2,.7},{-1,0,1}],
PlotStyle→Directive[Black,Opacity[.15]]]]
```

See Figure 10.42.

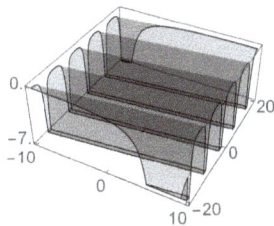

Figure 10.42: Particular solution for problem 10.31.

10.32. Here, we study a steady state solution of the Sawada–Kotera equation (10.14), [9, (32)],

$$u(t,x) = \frac{60\,k}{a}\,\frac{1 - k\,x^2}{(1 + k\,x^2)^2}, \quad a, k, t, x \in \mathbb{R}, \quad a \neq 0.$$

10.32

```
Clear[a,t,x,k]
u[t_,x_]:=60k/a (1-kx^2)/(1+kx^2)^2  (* Define the solution *)
```

Remark. We note that this solution does not depend on t. □

Now we check that the above defined function u satisfies equation (10.14) for all real parameters a and k,

```
∂ₜu[t,x]+∂{x,5}u[t,x]+a∂{x,3}u[t,x] u[t,x]
+a∂x,xu[t,x] u[t,x]+a²/5 u[t,x]²∂ₓu[t,x]//Simplify
0
```

We plot the solution u for some particular values of the parameters.

```
Block[{a=15,k=.5},
{Plot3D[u[t,x],{t,-10,10},{x,-10,10},Mesh→None,
PlotRange→All,ImageSize→200,PlotPoints→50,
PlotStyle→Directive[Black,Opacity[.15]],
Ticks→Outer[Times,{10,10,2},{-1,0,1}],
Manipulate[
Plot[u[1,x],{x,-10,10},PlotStyle→Black,ImageSize→200,
PlotRange→All,Ticks→{10{-1,1},{-.2,2}}],{t,-10,10},
SaveDefinitions→True]}]
```

See Figure 10.43.

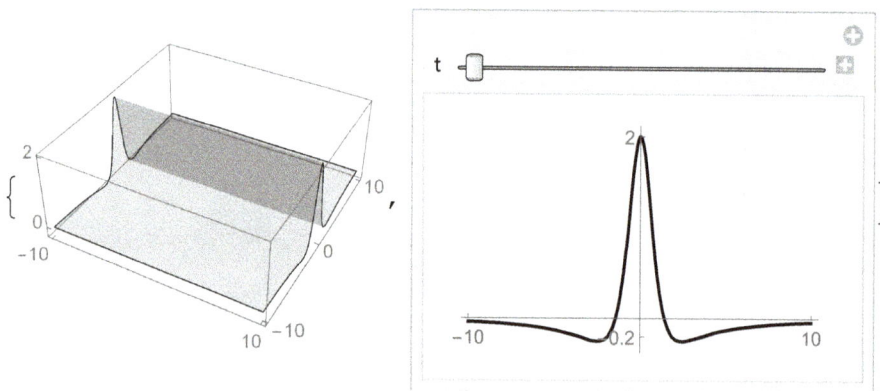

Figure 10.43: Steady-state solution and cross-section through Sawada–Kotera equation for $a = 15$.

10.33. Now we study a static soliton of the Sawada–Kotera equation (10.14), [10, (14)],

$$\begin{cases} u(t,x) = \dfrac{30\,k^2}{a}\,\dfrac{-3+k^2x^2+k^4x^4\,\mathrm{sech}^2(\xi)+6\,k\,x\,\tanh(\xi)-3k^2x^2\,\tanh^2(\xi)}{(3+k^2x^2-3\,k\,x\,\tanh(\xi))^2}, \\ \xi = -16k^5t + k\,x - \phi, \quad a,k,\phi \in \mathbb{R}. \end{cases}$$

10.33

```
Clear[a,k,t,x,ϕ]
u[t_,x_]:=30 k² -3+k²x²+k⁴x⁴Sech[ξ]²+6 k x Tanh[ξ]-3k²x²Tanh[ξ]²
             a              (3+k²x²-3 k x Tanh[ξ])²
(* Define the solution *)
```

We check that the above defined function u satisfies the equation (10.14) for all real parameters a, k, t, x, and ϕ,

```
∂ₜu[t,x]+∂{x,5}u[t,x]+a ∂{x,3}u[t,x] u[t,x]
+a ∂ₓ,ₓu[t,x] u[t,x]+a²/5 u[t,x]²∂ₓu[t,x]//Simplify
0
```

We plot the solution u for some particular values of the parameters.

```
Block[{a=15,k=.5,ϕ=-5},
Plot3D[u[t,x],{t,-5,25},{x,-25,25},Mesh→None,
PlotRange→All,ImageSize→200,PlotPoints→50,
PlotStyle→Directive[Gray,Opacity[.15]],
Ticks→{10{0,1,2},25{-1,0,1},{0,1}}]]
```

See Figure 10.44.

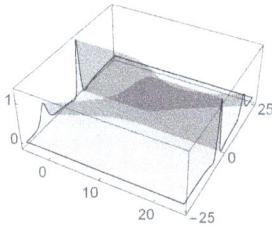

Figure 10.44: Static soliton solution of the Sawada–Kotera equation.

10.8 The Kaup–Kupershmidt equation

The *Kaup–Kupershmidt equation* (named after David J. Kaup and Boris A. Kupershmidt) is a fifth-order nonlinear partial differential equation of the form [36, 42], [9, (7)], [10, (5)], [25, (21)], and [38, (2.7)],

$$u_t + u_{x,x,x,x,x} + a\, u_{x,x,x} u + b\, u_{x,x} u_x + c\, u^2\, u_x = 0, \quad a, b, c \in \mathbb{R}.$$

This equation can be written as

$$u_t + \partial_x \left(u_{x,x,x,x} + a\, u\, u_{x,x} + \frac{b-a}{2}\, u_x^2 + \frac{c}{3}\, u^3 \right) = 0.$$

The Kaup–Kupershmidt equation is also introduced as [9, (5)],

$$u_t = u_{x,x,x,x,x} + \beta\, u_{x,x,x} u + \frac{5}{2}\beta\, u_{x,x} u_x + \frac{\beta^2}{5}\, u^2\, u_x.$$ (10.15)

We apply the previous equation for $\beta = 10$ as follows [25, (22)]:

$$u_t = u_{x,x,x,x,x} + 10\, u_{x,x,x} u + 25\, u_{x,x} u_x + 20\, u^2\, u_x.$$ (10.16)

For further discussions, we pass equation (10.16) in *Mathematica* language.

$$\partial_t u[t,x]+10\partial_{x,x,x}u[t,x]\,u[t,x]+25\,\partial_{x,x}u[t,x]\partial_x u[t,x]$$
$$+20\,u[t,x]^2\,\partial_x u[t,x] + \partial_{\{x,5\}}u[t,x] = 0.$$ (10.17)

10.34. We study the existence of a single soliton solution of equation (10.17) by the Hirota bilinear method in 9.6.2 [25, (37)], and [9].

10.34

```
Clear[t,x,k,δ]
ξ[t_,x_]:=-k⁵t+k x+δ
f[t_,x_]:=1+Exp[ξ[t,x]]+1/16 Exp[2ξ[t,x]] (* [25, (37)] *)
u[t_,x_]:=3/2∂x,xLog[f[t,x]] (* Single soliton
solution of equation (10.17) *)
```

Function u just defined satisfies equation (10.17). Indeed

```
∂tu[t,x]+10∂{x,3}u[t,x] u[t,x]+25∂x,xu[t,x]∂xu[t,x]
+20 u[t,x]²∂xu[t,x]+∂{x,5}u[t,x]//Simplify
0
```

We introduce an one-soliton solution generated by this equation.

```
Block[{k=1,δ=1},
Plot3D[Evaluate[u[t,x]],{t,-5,10},{x,-15,25},
PlotStyle→Directive[Black,Opacity[.15]],Mesh→None,
Ticks→{5 Range[-1,2],5{-3,0,3,5},{0,.5}},
PlotRange→All,ImageSize→200,PlotPoints→50]]
```

See Figure 10.45.

10.35. We study the two-soliton solution of the Kaup–Kuperschmidt equation (10.17) by the exp function method [25].

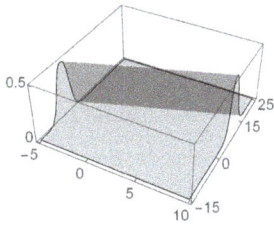

Figure 10.45: Single soliton solution for the Kaup–Kuperschmidt equation (10.17).

10.35

```
<< Notation`
Symbolize[k_]
Symbolize[δ_]
```

Define

```
Clear[t,x,k,δ]
ξ₁[t_,x_]:=-k₁⁵t+k₁x+δ₁;ξ₂[t_,x_]:=-k₂⁵t+k₂x+δ₂;
```

$$a=\frac{2k_1^4-k_1^2k_2^2+2k_2^2}{2(k_1+k_2)^2(k_1^2+k_1k_2+k_2^2)}\;;b=\frac{2(k_1-k_2)(k_1^2-k_1k_2+k_2^2)}{16(k_1+k_2)^2(k_1^2+k_1k_2+k_2^2)}$$

Now comes the ancillary function.

```
f[t_,x_]:=1+Exp[ξ₁[t,x]]+Exp[ξ₂[t,x]]
+1/16 Exp[2ξ₁[t,x]]+1/16 Exp[2ξ₂[t,x]]+a Exp[ξ₁[t,x]+ξ₂[t,x]]
+b (Exp[2ξ₁[t,x]+ξ₂[t,x]]+Exp[ξ₁[t,x]+2ξ₂[t,x]])
+b² Exp[2ξ₁[t,x]+2ξ₂[t,x]]  (* [25, (49)] *)
```

```
u[t_,x_]:=3/2 ∂x,x Log[f[t,x]]  (* Two-soliton solution of
equation (10.17) *)
```

We check the solution.

```
∂t u[t,x]+10 ∂{x,3}u[t,x] u[t,x]+25 ∂x,x u[t,x] ∂x u[t,x]
+20 u[t,x]² ∂x u[t,x]+∂{x,5}u[t,x]//Simplify
0
```

For some particular values of the parameters, we plot the surface illustrating the solution.

```
Block[{k₁=1,k₂=5/4,δ₁=δ₂=0},
Plot3D[Evaluate[u[t,x]],{t,-4,4},{x,-20,35},Mesh→None,
PlotStyle→Directive[Black,Opacity[.15]],
Ticks→{4{-1,0,1},10{-2,0,2,3},{0,.6}},
ImageSize→175,PlotRange→All,PlotPoints→50]]
```

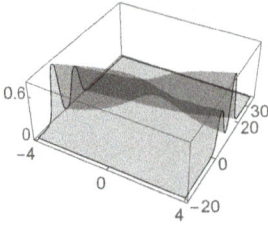

See Figure 10.46.

10.36. We study the three-soliton solution of the Kaup–Kuperschmidt equation (10.17), [25].

10.36 We need some calculations to write down the three-soliton solution.

```
<< Notation`
Symbolize[k_]
Symbolize[δ_]

Clear[t,x,k,δ]
```

We introduce

$$\xi_1[t_-,x_-]:=k_1x-k_1^5t+\delta_1\,;\xi_2[t_-,x_-]:=k_2x-k_2^5t+\delta_2\,;$$
$$\xi_3[t_-,x_-]:=k_3x-k_3^5t+\delta_3\,;$$

$$a_{12}=\frac{2k_1^4-k_1^2k_2^2+2k_2^4}{2(k_1+k_2)^2(k_1^2+k_1k_2+k_2^2)}\,;a_{13}=\frac{2k_1^4-k_1^2k_3^2+2k_3^4}{2(k_1+k_3)^2(k_1^2+k_1k_3+k_3^2)}\,;$$
$$a_{23}=\frac{2k_2^4-k_2^2k_3^2+2k_3^4}{2(k_2+k_3)^2(k_2^2+k_2k_3+k_3^2)}\,;$$

$$b_{12}=\frac{(k_1-k_2)^2(k_1^2-k_1k_2+k_2^2)}{16(k_1+k_2)^2(k_1^2+k_1k_2+k_2^2)}\,;b_{13}=\frac{(k_1-k_3)^2(k_1^2-k_1k_3+k_3^2)}{16(k_1+k_3)^2(k_1^2+k_1k_3+k_3^2)}\,;$$
$$b_{23}=\frac{(k_2-k_3)^2(k_2^2-k_2k_3+k_3^2)}{16(k_2+k_3)^2(k_1^2+k_2k_3+k_3^2)}\,;$$

Other constants are introduced below:

$$d=4(k_1+k_2)^2(k_1^2+k_1k_2+k_2^2)(k_1+k_3)^2(k_1^2+k_1k_3+k_3^2)$$
$$\times(k_2+k_3)^2(k_2^2+k_2k_3+k_3^2)\,;$$
$$c=\frac{1}{d}((2k_1^4-k_1^2k_2^2+2k_2^4)(k_3^8+k_1^4k_2^4)+(2k_1^4-k_1^2k_3^2+2k_3^4)(k_2^8+k_1^4k_3^4)$$
$$+(2k_2^4-k_2^2k_3^2+2k_3^4)(k_1^8+k_2^4k_3^4))$$
$$-\frac{1}{2d}((k_1^2+k_2^2)(k_1^4+k_2^4)(k_3^6+k_1^2k_2^2k_3^2)+(k_1^2+k_3^2)(k_1^4+k_3^4)(k_2^6+k_1^2k_2^2k_3^2)$$
$$+(k_2^2+k_3^2)(k_2^4+k_3^4)(k_1^6+k_1^2k_2^2k_3^2)+12k_1^4k_2^4k_3^4)\,;$$

We introduce the ancillary function.

```
f[t_,x_]:=1+Exp[ξ₁[t,x]]+Exp[ξ₂[t,x]]+Exp[ξ₃[t,x]]
+1/16(Exp[2ξ₁[t,x]]+Exp[2ξ₂[t,x]]+Exp[2ξ₃[t,x]])
+a₁₂ Exp[ξ₁[t,x]+ξ₂[t,x]]+a₁₃ Exp[ξ₁[t,x]+ξ₃[t,x]]
+a₂₃ Exp[ξ₂[t,x]+ξ₃[t,x]]
+b₁₂ (Exp[2ξ₁[t,x]+ξ₂[t,x]]+Exp[ξ₁[t,x]+2ξ₂[t,x]])
+b₁₃ (Exp[2ξ₁[t,x]+ξ₃[t,x]]+Exp[ξ₁[t,x]+2ξ₃[t,x]])
+b₂₃ (Exp[2ξ₂[t,x]+ξ₃[t,x]]+Exp[ξ₂[t,x]+2ξ₃[t,x]])
+c Exp[ξ₁[t,x]+ξ₂[t,x]+ξ₃[t,x]]
+b₁₂²Exp[2ξ₁[t,x]+2ξ₂[t,x]]+b₁₃²Exp[2ξ₁[t,x]+2ξ₃[t,x]]
+b₂₃²Exp[2ξ₂[t,x]+2ξ₃[t,x]]
+16(a₂₃ b₁₂ b₁₃ Exp[2ξ₁[t,x]+ξ₂[t,x]+ξ₃[t,x]]
+a₁₃b₁₂b₂₃ Exp[ξ₁[t,x]+2ξ₂[t,x]+ξ₃[t,x]]
+a₁₂b₁₃b₂₃ Exp[ξ₁[t,x]+ξ₂[t,x]+2ξ₃[t,x]])
+16²b₁₂b₁₃b₂₃ (b₁₂Exp[2ξ₁[t,x]+2ξ₂[t,x]+ξ₃[t,x]]
+b₁₃Exp[2ξ₁[t,x]+ξ₂[t,x]+2ξ₃[t,x]]
+b₂₃Exp[ξ₁[t,x]+2ξ₂[t,x]+2ξ₃[t,x]])
+16(16b₁₂ b₁₃ b₂₃)²Exp[2ξ₁[t,x]+2ξ₂[t,x]+2ξ₃[t,x]];
```

The solution is defined as

```
u[t_,x_]:=3/2∂_{x,x}Log[f[t,x]]
```

and we check it

```
∂_t u[t,x]+10∂_{x,3}u[t,x] u[t,x]+25 ∂_{x,x}u[t,x] ∂_x u[t,x]
+20 u[t,x]² ∂_x u[t,x]+∂_{x,5}u[t,x]//Simplify
```
0

For some particular values of the parameters, we plot the surface illustrating the solution.

```
Block[{k₁=1,k₂=5/4,k₃=3/2,δ₁=δ₂=δ₃=0},
{{Plot3D[Evaluate[u[t,x]],##,ImageSize→175],
Plot3D[Evaluate[u[t,x]],##,ViewPoint→{0,7.5,55},
ImageSize→150]}
&[{t,-5,5},{x,-30,45},PlotStyle→Opacity[.4],
ColorFunction→Function[{x,y,z},Hue[Abs[y+30]/75+z]],
PlotRange→All,PlotPoints→45,Mesh→None,
Ticks→{5{-1,0,1},20Range[-1,2],{0,1}},
PlotPoints→45],
Plot[Evaluate[u[4,x]],{x,-10,40},PlotStyle→Black,
ImageSize→160,Ticks→{{-10,40},{.6,1.1}}]}]
// Flatten
```

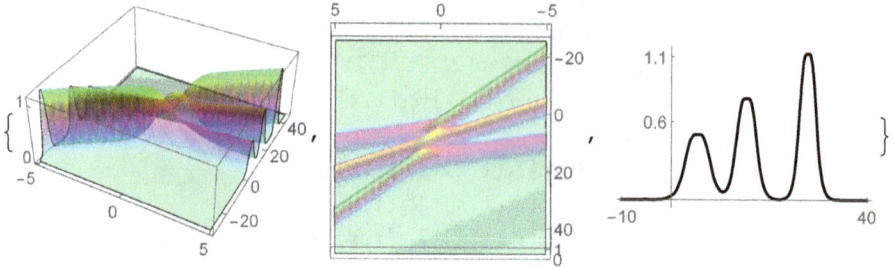

Figure 10.47: Three-soliton solution for the Kaup–Kuperschmidt equation.

See Figure 10.47.

Now we introduce some "anomalous" solitons to the Kaup–Kupershmidt equation. Solitary wave solutions of the Kaup–Kupershmidt equation are frequently called "anomalous" soliton solutions, [38], to emphasize the difference from the common "regular" sech^2 solitons. These "anomalous" solitary wave solutions have the form [9, (8)] and [10, (5)]:

$$u(t,x) = \frac{60\,k^2}{\beta}\,\frac{2\cosh(2\,\xi)+1}{(\cosh(2\,\xi)+2)^2}, \quad \text{where } \xi = k(x - 16\,k^4\,t). \qquad (10.18)$$

10.37. Here, we study the "anomalous" solitons solution defined by (10.18).

10.37

```
Clear[k,t,x,β]
ξ=k(x-16k⁴t);
u[t_,x_]:=60k²/β 2Cosh[2ξ]+1/(Cosh[2ξ]+2)²
```

Now we check that the above defined function u satisfies the equation (10.15),

```
∂ₜu[t,x]+∂{x,5}u[t,x]+β ∂{x,3}u[t,x] u[t,x]
+5/2 β∂x,xu[t,x] u[t,x]+β²/5 u[t,x]²∂xu[t,x]//Simplify
0
```

We plot the solution for some particular values of the parameters.

```
Block[{β=15,k=.5},
Plot3D[u[t,x],{t,-5,25},{x,-15,40},PlotRange→All,
PlotStyle→Directive[Black,Opacity[.15]],
Ticks→{10{0,1,2},20{-1,0,1,2},{0,.3}},Mesh→None,
ImageSize→175,PlotPoints→50]]
```

See Figures 10.48.

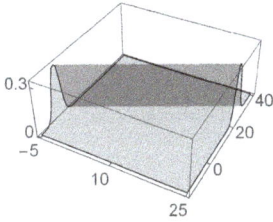

y

Figure 10.48: "Anomalous" soliton in the Kaup–Kupershmidt equation.

Further, we show a cross-section through the surface in Figure 10.48 as well as a moving section.

```
Clear[k,t,x,β]
ζ[t_,x_]:=k(x-16 k⁴t);
w[t_,x_]:=60k²/β · (2 Cosh[ζ]+1)/(Cosh[ζ]+2)²

β=15;k=.5;

{Plot[Evaluate[Table[w[τ,x],{τ,-5,5,5}]],{x,-15,15},
Mesh→None,ImageSize→175,PlotRange→All,
PlotStyle→Black,Ticks→{15{-1,0,1},{0,.35}}],
(* Discrete cross section *)
Manipulate[
Plot[w[τ,x],{x,-15,15},Mesh→None,ImageSize→160,
PlotStyle→Black,PlotRange→All,
Ticks→{15{-1,0,1},{0,.35}}],{τ,-5,5},
SaveDefinitions→True] (* Continuous cross section *)}
```

See Figure 10.49.

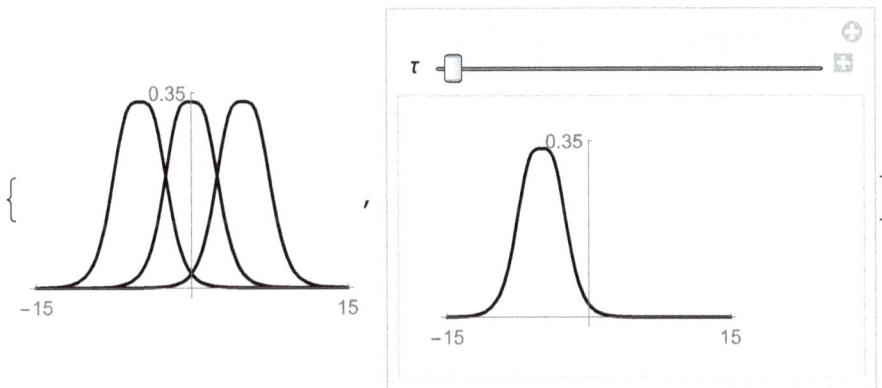

Figure 10.49: Cross-sections through the "anomalous" soliton.

Bibliography

[1] Ablowitz, M. J., Clarkson, P. A.: Solitons, Nonlinear Evolution Equations and Inverse Scattering. No. 149 in London Math. Soc. Lec. Note Ser. Cambridge University Press, Cambridge, UK (1991).

[2] Abramowitz, M., Stegun, I. A. (eds.): Handbook of Mathematical Functions with Functions, Graphs, and Mathematical Tables. Dover, New York, NY (1972).

[3] Ak, T., Dhawan, S.: A practical and powerful approach to potential KdV and Benjamin equation. Beni-Suef University J. Basic Appl. Sci. **6**(4), 383–390 (2017). https://doi.org/10.1016/j.bjbas.2017.07.008.

[4] Arfken, G., Weber, H., Harris, F. E.: Mathematical Methods for Physicists. A Comprehensive Guide, 7 edn. Academic Press, Waltham, MA (2012).

[5] Bailey, D. H., Borwein, J. M., Borwein, P. B., Plouffe, S.: The quest for Pi. Math. Intell. **19**(1), 50–57 (1997). Crd.lbl.gov/~dhbailey/dhbpapers/pi-quest.pdf.

[6] Bell, W. W.: Special Functions for Scientists and Engineers. Dover, Mineola, NY (2016).

[7] Biondini, G.: Line soliton interactions of the Kadomtsev-Petviashvili equation. Phys. Rev. Lett. **99**, 064,103–1–064,103–4 (2007). https://doi.org/10.1103/PhysRevLett.99.064103.

[8] Biondini, G., Pelinovsky, D.: Kadomtsev–Petviashvili equation. Scholarpedia **3**(10:6539), 1–10 (2008). Revision #50387.

[9] Burde, G. I.: Generalized Kaup–Kupershmidt solitons and other solitary wave solutions of the higher order KdV equations. J. Phys. A **43**(8), 13 (2010). https://doi.org/10.1088/1751-8113/43/8/085208.

[10] Burde, G. I.: Static algebraic solitons in Korteweg-de Vries type systems and the Hirota transfornation. Phys. Rev. E **84**(2), 12 (2011). https://doi.org/10.1103/PhysRevE.84.026615.

[11] Calogero, F.: The evolution partial differential equation $u_t = u_{xxx} + 3(u_{xx}u^2 + 3u_x^2 u) + 3u_x u^4$. J. Math. Phys. **28**(3), 538–555 (1987). https://doi.org/10.1063/1.527639.

[12] Calogero, F., Degasperis, A.: Spectral Transform and Solitons: Tools to Solve and Investigate Nonlinear Evolution Equations, *Studies in mathematics and its applications, 13*, vol. 1. North-Holland Publishing Company, Amsterdam, New York, Oxford (1982).

[13] Craig, W.: A Course on Partial Differential Equations. No. 197 in GSM. Amer. Math. Soc., Providence, RI (2018).

[14] Dai, H., Fan, E., Geng, X.: Constructing periodic wave solutions of nonlinear equations by Hirota bilinear method. Tech. rep., Cornell University (2006). arXiv:nlin/0602015.pdf.

[15] Dobrushkin, V.: *Mathematica* tutorial for the Second Course. Part II: Phase portrait. Brown University, Providence, RI. www.cfm.brown.edu/people/dobrush/am34/Mathematica/ch2/portrait.html.

[16] Drazin, P. G.: Solitons. No. 85 in London Math. Soc. Lec. Note Ser. Cambridge University Press, Cambridge, UK (1983).

[17] Elsgolts, L.: Differential Equations and the Calculus of Variations. Mir, Moscow (1977).

[18] Fan, E., Hon, Y. C.: Generalized tanh method extended to special types of nonlinear equations. Z. Naturforsch. A **57a**, 692–700 (2002). https://doi.org/10.1515/zna-2002-0809.

[19] Fard, N. Y., Foroutan, M. R., Eslami, M., Mirzazadeh, M., Biswas, A.: Solitary waves and other solutions to Kadomtsev-Petviashvili equation with spatio-temporal dispersion. Rom. J. Phys. **60**(9–10), 1337–1360 (2015).

[20] Freeman, N. C.: A two dimensional distributed soliton solution of the Korteweg-de Vries equation. Proc. R. Soc. Lond. A. **366**(1725), 185–204 (1979).

[21] Graham, R. L., Knuth, D. E., Patashnik, O.: Concrete Mathematics. A Foundation for Computer Science, 2 edn. Addison-Wesley, Reading, MA (1994).

[22] Griffiths, G. W., Schiesser, W. E.: Traveling Wave Analysis of Partial Differential Equations Numerical and Analytical Methods with MATLAB®and Maple™. Elsevier, Academic Press, Amsterdam, Boston, Heidelberg, London (2012).

[23] Hartman, P.: Ordinary Differential Equations. Wiley, Hoboken, NJ (1964).

https://doi.org/10.1515/9783111411392-013

[24] He, J. H., Wu, X. H.: Exp-function method for nonlinear wave equations. Chaos Solitons Fractals **30**(3), 700–708 (2006). https://doi.org/10.1016/j.chaos.2006.03.020.

[25] Hereman, W., Nuseir, A.: Symbolic methods to construct exact solutions of nonlinear partial differential equations. Math. Comput. Simul. **43**(1), 13–27 (1997).

[26] Hereman, W., Zhuang, W.: Symbolic software for soliton theory. Acta Appl. Math. **39**(1-3), 361–378 (1995).

[27] Hietarinta, J.: Introduction to the hirota bilinear method. In: Y. Kosmann-Schwarzbach, B. Grammaticos, K. Tamizhmani (eds.) Integrability of Nonlinear Systems, Lect. Notes Phys. 495, pp. 95–105. Springer, Berlin Heilderberg New York (2004). arXiv:solv-int/9708006.

[28] Hirota, R.: Exact solution of the Korteweg—de Vries equation for multiple collisions of solitons. Phys. Rev. Lett. **27**(18), 1192–1193 (1971). https://doi.org/10.1103/PhysRevLett.27.1192.

[29] Hirota, R.: A new form of Bäcklund transformations and its relations to the inverse scattering problem. Prog. Theor. Phys. **52**(5), 1498–1512 (1974).

[30] Hirota, R., Ramani, A.: The Miura transformations of Kaup's equation and of Mikhailov's equation. Phys. Lett. A **76**(2), 95–96 (1980). https://doi.org/10.1016/0375-9601(80)90578-2.

[31] Hirota, R., Satsuma, J.: N-soliton solutions of model equations for shallow water waves. J. Phys. Soc. Jpn. **40**(2), 611 (1976). https://doi.org/10.1143/JPSJ.40.611.

[32] Hubbard, J. H., West, B. H.: Differential Equation. A Dynamical Systems Approach. Part I: Ordinary Differential Equations. No. 5 in TAM. Springer, New York, NY (1991).

[33] Ince, E. L.: Ordinary Differential Equations. Dover, New York, NY (1956).

[34] Kadomtsev, B. B., Petviashvili, V. I.: On the stability of solitary waves in weakly dispersive media. Sov. Phys. Dokl. **15**(6), 539–541 (1970).

[35] Kasman, A.: Glimpses of Soliton Theory. The Algebra and Geometry of Nonlinear PDEs. No. 54 in SML. Amer. Math. Soc., Providence, RI (2010).

[36] Kaup, D. J.: On the inverse scattering problem for cubic eigenvalue problems of the class $\psi_{x,x,x}+6q\psi_x+6r\psi = \lambda\psi$. Stud. Appl. Math. **62**(3), 189–216 (1980).

[37] Kawahara, T.: Oscillatory solitary waves in dispersive media. J. Phys. Soc. Jpn. **33**(1), 260–264 (1972). https://doi.org/10.1143/JPSJ.33.260.

[38] Kichenassamy, S., Olver, P. J.: Existence and nonexistence of solitary wave solutions to higher-order model evolution equations. SIAM J. Math. Anal. **23**(5), 1141–1166 (1992). https://doi.org/10.1137/0523064.

[39] Knobel, R.: An Introduction to the Mathematical Theory of Waves. No. 3 in SML. Amer. Math. Soc., Providence, RI (2000).

[40] Korteweg, D. J., de Vries, G.: On the change of form of long waves advancing in a rectangular canal, and on a new type of long stationary waves. Philos. Mag. **39**(5), 422–443 (1895).

[41] Krasnov, M. L., Kiselyov, A. I., Makarenko, G. I.: A Book of Problems in Ordinary Differential Equations. Mir, Moscow (1981).

[42] Kuperschmidt, B. A.: A super Korteweg-de Vries equation: An integrable system. Phys. Lett. A **102**(5-6), 213–215 (1984).

[43] Lawden, D. F.: Elliptic Functions and Applications. No. 80 in Applied mathematical sciences. Springer-Verlag, New York Berlin Heidelberg (1989).

[44] Lebedev, N. N.: Special Functions & Their Applications. Dover, New York, NY (2018).

[45] Lungu, N., Mureşan, M., Ciupa, A.: Higher Mathematics. Problem Book. Politechnical Instit., Cluj-Napoca (1984).

[46] Ma, W. X.: Lump solutions to the Kadomtsev-Petviashvili equation. Phys. Lett. A **379**(36), 1975–1978 (2015). https://doi.org/10.1016/j.physleta.2015.06.061.

[47] Malfliet, W.: Solitary wave solutions of nonlinear wave equations. Am. J. Phys. **60**, 650–654 (1992). https://doi.org/10.1119/1.17120.

[48] Malfliet, W.: The tanh method: a tool for solving certain classes of nonlinear evolution and wave equations. J. Comput. Appl. Math. **164-165**, 529–541 (2004). https://doi.org/10.1016/S0377-0427(03)00645-9.

[49] Malfliet, W.: The tanh method: a tool for solving certain classes of non-linear PDEs. Math. Methods Appl. Sci. **28**(17), 2031–2035 (2005). https://doi.org/10.1002/mma.650.

[50] Malfliet, W., Hereman, W.: The tanh method: I. Exact solutions of nonlinear evolution and wave equations. Phys. Scr. **54**(6), 563–568 (1996). https://doi.org/10.1088/0031-8949/54/6/003.

[51] Malfliet, W., Hereman, W.: The tanh method: II. Perturbation technique for conservative systems. Phys. Scr. **54**(6), 569–575 (1996). https://doi.org/10.1088/0031-8949/54/6/004.

[52] Manukure, S., Zhou, Y., Ma, W. X.: Lump solution to a (2 + 1)-dimensional extended KP. Comput. Math. Appl. **75**(7), 2414–2419 (2018).

[53] Mathieu, E. L.: Mémoire sur le mouvement vibratoire d'une membrane de forme elliptique. J. Math. Pures Appl. (9) pp. 137–203 (1868).

[54] McLachlan, N. W.: Theory and Applications of Mathieu Functions. Oxford Press, London, UK (1947).

[55] Minzoni, A. A., Smyth, N. F.: Evolution of lump solutions for the KP equation. Wave Motion **24**(3), 291–305 (1996). https://doi.org/10.1016/S0165-2125(96)00023-6.

[56] Mureşan, M.: A Concrete Approach to Classical Analysis. CMS Books in Mathematics. Springer, New York, NY (2009). https://doi.org/10.1007/978-0-387-78933-0.

[57] Mureşan, M.: Some remarks on the brachistochrone problem. Schriftenreihe der Fakultät für Mathematik, SM-DU-735, Universität Duisburg–Essen, Duisburg, Germany (2011).

[58] Mureşan, M.: Soft landing on the Moon with *Mathematica*. The Mathematica® Journal **14**, 1–12 (2012). https://doi.org/10.3888/tmj.14-16. Published in Japanese: https://www.hulinks.co.jp/software/math_develop/mj/mj_1208.

[59] Mureşan, M.: Some remarks on the brachistochrone problem with Coulomb friction. Filomat **26**(4), 697–711 (2012).

[60] Mureşan, M.: On Zermelo's navigation problem with *Mathematica*. J. Appl. Funct. Anal. **9**(3–4), 359–365 (2014).

[61] Mureşan, M.: On the maximal orbit transfer problem. The Mathematica Journal® **17**, 1–11 (2015). https://doi.org/10.3888/tmj.17-4. Published in Japanese: https://www.hulinks.co.jp/mj/mj-1607.html.

[62] Mureşan, M.: Introduction to *Mathematica* with Applications. Springer, London, UK (2017). https://doi.org/10.1007/978-3-319-52003-2.

[63] Mureşan, M.: *Mathematica* with Differential Equations. Amazon (2021).

[64] Olariu, V., Stănăşilă, T.: Differential Equations and with Partial Derivative. Ed. Technică, Bucharest (1982).

[65] Olver, P. J.: Introduction to Partial Differential Equations. UTM. Springer, Cham, Heidelberg (2016). https://doi.org/10.1007/978-3-319-02099-0.

[66] Palais, R. S.: The visualization of mathematics: towards a mathematical exploratorium. Not. Am. Math. Soc. **46**(6), 647–658 (1999).

[67] Palais, R. S.: An Introduction to Wave Equations and Solitons. Tech. rep., The Morningside Center of Mathematics, Chinese Academy of Sciences, Beijing, China (2000).

[68] Palais, R. S.: Linear and Nonlinear Waves and Solitons. Tech. rep., Dept. of Math., Univ. of Ca., Irvine, CA 92697 (2010).

[69] Polyanin, A. D.: http://eqworld.ipmnet.ru/en/solutions/ (2004).

[70] Polyanin, A. D., Zaitsev, V. F.: Handbook of Exact Solutions for Ordinary Differential Equations, 2 edn. Chapman and Hall/CRC, Boca Raton, FL (2003).

[71] Polyanin, A. D., Zaitsev, V. F.: Handbook of Nonlinear Partial Differential Equations, 2 edn. Chapman & Hall, Boca Raton, FL (2011).

[72] Schaback, R.: On COVID-19 modelling. Jahresber. Dtsch. Math.-Ver. **122**(3), 167–205 (2020).

[73] Triki, H., Ak, T., Ekici, M., et al.: Some new exact wave solutions and conservation laws of potential Korteweg–de Vries equation. Nonlinear Dyn. **89**(1), 501–508 (2017). https://doi.org/10.1007/s11071-017-3467-4.

[74] Trott, M.: The *Mathematica* GuideBook for Graphics. Springer, New York, NY (2004).

[75] Trott, M.: The *Mathematica* GuideBook for Programming. Springer, New York, NY (2004).

[76] Trott, M.: The *Mathematica* GuideBook for Numerics. Springer, New York, NY (2006).

[77] Trott, M.: The *Mathematica* GuideBook for Symbolics. Springer, New York, NY (2006).

[78] Wang, M., Li, X.: Exact solutions to the double Sine-Gordon equation. Chaos Solitons Fractals **27**(2), 477–486 (2006). https://doi.org/doi.org/10.1016/j.chaos.2005.04.027.

[79] Wazwaz, A. M.: The tanh method for traveling wave solutions of nonlinear equations. Appl. Math. Comput. **154**(3), 713–723 (2004). https://doi.org/10.1016/S0096-3003(03)00745-8.

[80] Wazwaz, A. M.: The tanh method:solitons and periodic solutions for the Dodd-Bullough-Mikhailov and the Tzitzeica-Dodd-Bullough equations. Chaos Solitons Fractals **25**, 55–63 (2005). https://doi.org/10.1016/j.chaos.2004.09.122.

[81] Wazwaz, A. M.: New solitary wave solutions to the modified Kawahara equation. Phys. Lett. A **360**(4-5), 588–592 (2007). https://doi.org/10.1016/j.physleta.2006.08.068.

[82] Wazwaz, A. M.: Partial Differential Equations and Solitary Waves Theory. Nonlinear Physical Science. Springer, Dordrecht Heidelberg London New York (2009).

[83] Weisstein, E. W.: Associated Legendre Polynomial. From MathWorld–A Wolfram Web Resource. http://mathworld.wolfram.com/AssociatedLegendrePolynomial.html.

[84] Weisstein, E. W.: Bessel Function of the First Kind. From MathWorld–A Wolfram Web Resource. http://mathworld.wolfram.com/BesselFunctionoftheFirstKind.html.

[85] Weisstein, E. W.: Bessel Function of the Second Kind. From MathWorld–A Wolfram Web Resource. http://mathworld.wolfram.com/BesselFunctionoftheSecondKind.html.

[86] Weisstein, E. W.: Beta Function. From MathWorld–A Wolfram Web Resource. http://mathworld.wolfram.com/BetaFunction.html.

[87] Weisstein, E. W.: Gamma Function. From MathWorld–A Wolfram Web Resource. http://mathworld.wolfram.com/GammaFunction.html.

[88] Weisstein, E. W.: Pendulum. From MathWorld–A Wolfram Web Resource. http://scienceworld.wolfram.com/physics/Pendulum.html.

[89] Weisstein, E. W.: Tangent. From MathWorld–A Wolfram Web Resource. http://mathworld.wolfram.com/Tangent.html.

[90] Whitham, G. B.: Linear and Nonlinear Waves. Pure & Applied Mathematics. John Wiley & Sons, New York, NY (1999).

[91] Wolfram, S.: An Elementary Introduction to the Wolfram Language. Wolfram Media, Champaign, IL (2015).

[92] Wolfram Research, Champaign, IL: Advanced Numerical Differential Equation Solving in *Mathematica* (2008). Wolfram Mathematica® Tutorial Collection.

[93] Wolfram Research, Champaign, IL: Differential Equation Solving with DSolve (2008). Wolfram Mathematica® Tutorial Collection.

[94] Wolfram Research, Champaign, IL: Differential Equation (2018). Wolfram Mathematica® Tutorial Collection.

[95] Wolfram Research, Champaign, IL: Numerical Solution of Differential Equation (2018). Wolfram Mathematica® Tutorial Collection.

[96] Yu, J., Ma, W. X., Chen, S. T.: Lump solutions of a new generalized Kadomtsev-Petviasvili equation. Mod. Phys. Lett. B **33**(10), 1–9 (2019). https://doi.org/10.1142/S0217984919501264.

[97] Yusufoğlu, E., Bekir, A., Alp, M.: Periodic and solitary wave solutions of Kawahara and modified Kawahara equations by using Sine–Cosine method. Chaos Solitons Fractals **37**(4), 1193–1197 (2008). https://doi.org/10.1016/j.chaos.2006.10.012.

[98] Zhou, Y., Manukure, S., Ma, W. X.: Lump and lump-soliton solutions to the Hirota-Satsuma-Ito equation. Commun. Nonlinear Sci. Numer. Simul. **68**, 56–62 (2019).

[99] Zwillinger, D.: Handbook of Differential Equations, 3 edn. Academic Press, San Diego, CA (1997).

Index

https://doi.org/10.1515/9783111411392-014